Hydrogen Fuel Cell Technology for Mobile Applications

Raluca Andreea Felseghi
Department of Mechanical Engineering, Faculty of Automotive, Mechatronics, and Mechanical Engineering, Technical University of Cluj-Napoca, Romania

A volume in the Practice, Progress, and Proficiency in Sustainability (PPPS) Book Series

IGI Global
PUBLISHER of TIMELY KNOWLEDGE

Published in the United States of America by
IGI Global
Engineering Science Reference (an imprint of IGI Global)
701 E. Chocolate Avenue
Hershey PA, USA 17033
Tel: 717-533-8845
Fax: 717-533-8661
E-mail: cust@igi-global.com
Web site: http://www.igi-global.com

Library of Congress Cataloging-in-Publication Data

Names: Felseghi, Raluca Andreea, 1978- editor.
Title: Hydrogen fuel cell technology for mobile applications / edited by Raluca Andreea Felseghi.
Description: Hershey, PA : Engineering Science Reference, [2023] | Includes bibliographical references and index. | Summary: "This book is intended to collect and disseminate the relevant knowledge regarding HYDROGEN ECONOMY dedicated to the MOBILE APPLICATIONS (clean green mobility). This book aims to provide a platform for researchers and practitioners from both academia and industry to present their latest state-of-art research results and methodologies addressing top concerns in the areas of Hydrogen Fuel Cell Technology for Mobile Applications and to establish worldwide cooperative research and development"-- Provided by publisher.
Identifiers: LCCN 2023002170 (print) | LCCN 2023002171 (ebook) | ISBN 9781668467213 (h/c) | ISBN 9781668467220 (s/c) | ISBN 9781668467237 (eISBN)
Subjects: LCSH: Fuel cell vehicles. | Hydrogen cars.
Classification: LCC TL221.13 .H93 2023 (print) | LCC TL221.13 (ebook) | DDC 629.22/974--dc23/eng/20230202
LC record available at https://lccn.loc.gov/2023002170
LC ebook record available at https://lccn.loc.gov/2023002171

This book is published in the IGI Global book series Practice, Progress, and Proficiency in Sustainability (PPPS) (ISSN: 2330-3271; eISSN: 2330-328X)

British Cataloguing in Publication Data
A Cataloguing in Publication record for this book is available from the British Library.

All work contributed to this book is new, previously-unpublished material.
The views expressed in this book are those of the authors, but not necessarily of the publisher.

For electronic access to this publication, please contact: eresources@igi-global.com.

Practice, Progress, and Proficiency in Sustainability (PPPS) Book Series

Ayman Batisha
International Sustainability Institute,
Egypt

ISSN:2330-3271
EISSN:2330-328X

MISSION

In a world where traditional business practices are reconsidered and economic activity is performed in a global context, new areas of economic developments are recognized as the key enablers of wealth and income production. This knowledge of information technologies provides infrastructures, systems, and services towards sustainable development.

The **Practices, Progress, and Proficiency in Sustainability (PPPS) Book Series** focuses on the local and global challenges, business opportunities, and societal needs surrounding international collaboration and sustainable development of technology. This series brings together academics, researchers, entrepreneurs, policy makers and government officers aiming to contribute to the progress and proficiency in sustainability.

COVERAGE

- E-Development
- Strategic Management of IT
- Knowledge clusters
- Intellectual Capital
- Eco-Innovation
- Green Technology
- Sustainable Development
- ICT and knowledge for development
- Innovation Networks
- Socio-Economic

IGI Global is currently accepting manuscripts for publication within this series. To submit a proposal for a volume in this series, please contact our Acquisition Editors at Acquisitions@igi-global.com or visit: http://www.igi-global.com/publish/.

The Practice, Progress, and Proficiency in Sustainability (PPPS) Book Series (ISSN 2330-3271) is published by IGI Global, 701 E. Chocolate Avenue, Hershey, PA 17033-1240, USA, www.igi-global.com. This series is composed of titles available for purchase individually; each title is edited to be contextually exclusive from any other title within the series. For pricing and ordering information please visit http://www.igi-global.com/book-series/practice-progress-proficiency-sustainability/73810. Postmaster: Send all address changes to above address.

Titles in this Series

For a list of additional titles in this series, please visit: http://www.igi-global.com/book-series/

Handbook of Research on Safe Disposal Methods of Municipal Solid Wastes for a Sustainable Environment
Prateek Srivastava (Chitkara University Institute of Engineering and Technology, Chitkara University, India) Durgaprasad Ramteke (Fibre and Particle Engineering, Faculty of Technology, University of Oulu, Finland) Ankush Kumar Bedyal (Department of Physics and Astronomical Sciences, Central University of Jammu, India) Monika Gupta (Chitkara Business School, Chitkara University, India) and Jasminder Kaur Sandhu (Chitkara University Institute of Engineering and Technology, Chitkara University, India)
Engineering Science Reference • © 2023 • 400pp • H/C (ISBN: 9781668481172) • US $295.00

Financing Regions Toward Sustainability in the Midst of Climate Change Risks and Uncertainty
Beata Zofia Filipiak (University of Szczecin, Poland) Dominika Kordela (University of Szczecin, Poland) and Izabela Nawrolska (University of Szczecin, Poland)
Information Science Reference • © 2023 • 330pp • H/C (ISBN: 9781668476208) • US $225.00

Pragmatic Legal and Policy Implications of Environmental Lawmaking
Hussein Movahedian (Islamic Azad University, UAE & Department of Private Law, Islamic Studies and Law Faculty, Imam Sadiq University, Tehran, Iran) and Nima Norouzi (Islamic Azad University, UAE & Law and Political Science Department, University of Tehran, Tehran, Iran)
Information Science Reference • © 2023 • 269pp • H/C (ISBN: 9781668441589) • US $215.00

Leadership and Governance for Sustainability
Neeta Baporikar (Namibia University of Science and Technology, Namibia & SP Pune University, India)
Information Science Reference • © 2023 • 350pp • H/C (ISBN: 9781668497111) • US $250.00

IGI Global
PUBLISHER of TIMELY KNOWLEDGE

701 East Chocolate Avenue, Hershey, PA 17033, USA
Tel: 717-533-8845 x100 • Fax: 717-533-8661
E-Mail: cust@igi-global.com • www.igi-global.com

Editorial Advisory Board

Table of Contents

Detailed Table of Contents

Today hydrogen is recognized as a non-polluting energy carrier because it does not contribute to global warming if it is produced from renewable energy resources. Hydrogen is the only secondary energy carrier that is suitable for wide applications. At the center of attention is the fact that hydrogen can be obtained from a wide range of primary energies. It can be used advantageously for a wide range of applications. Hydrogen can be used in decentralized systems without emitting CO2. Hydrogen is already a part of today's chemical industry, but as an energy resource, its rare benefits can only be achieved through fuel cell technology. The next generations of energy systems for mobile applications based on hydrogen fuel cells have the potential of using and implementing clean energy in the mobility domain, as well as in the tertiary and industrial sector, thus having a significant impact on greenhouse gas emissions decreasing, specific characteristics of hydrogen technology having an important role in the decarbonization of transport sector.

The number of disasters caused by global warming has increased in recent years, making it important to accelerate the studies on alternative and clean energy conversion systems to decrease these detrimental effects. Conventional transportation systems mainly based on fossil fuels seem to be the most critical issue for the environment. Electric vehicles are promising clean candidates that can be used for transportation. Electric vehicle technology can be classified into battery electric vehicles (BEVs), hybrid electric vehicles (HEVs), plug-in hybrid electric vehicles (PHEVs), and fuel cell electric vehicles (FCEVs). In recent years, due to the development in nanotechnology and material science engineering, newly designed battery and fuel cell-based technologies draw attention. This chapter aims to give brief information about the electric vehicle types, a comparison of the existing systems, hydrogen fuel cell electric vehicles and drawbacks of these systems, and the current status of the hydrogen economy.

A high-quality energy carrier like hydrogen can be created by thermochemical processes involving the conversion of hydrocarbons, such as coal or natural gas. It can also be obtained by water electrolysis using renewable energy sources, such as wind and solar power. Due to climate change, air pollution, and energy deficiency, the exploration of new transportation fuels is very important. Despite the various challenges that hydrogen faces, it is still regarded as a clean energy source that can help solve the environmental and energy problems of the future. It can be used in fuel cells to provide high-quality electricity and heat. This chapter aims to discuss the various techniques involved in the production of hydrogen and their applications in fuel cells. It also explores the different types of fuel cells, such as PEMFCs, MCFCs, and SOFCs as well as to understand the effect towards sustainable development goals.

The improvement in the standard of living due to urbanization and industrialization can without a doubt be accredited to the discovery and use of fossil fuels over the centuries. Although this energy source has had a huge positive impact on the human population, the negatives are just as huge. The continuous use of fossil has been detrimental to the environment and has pushed the world into a paradigm shift where green technology has to be implemented. Hydrogen as an alternative source has intrigued many researchers due to its zero emission characteristic and higher energy content compared to traditional energy sources. Fuel cells which use hydrogen as a fuel source, are the fastest growing clean technologies. Fuel cells are used in both

Industrial revolution and sustainable development have intensified the need for clean resources of energy. Among all the possible alternative sources of green energy, hydrogen energy is the most promising one because of its abundancy, maximum energy density per unit mass, and clean combustion. The energy content per unit mass of hydrogen is estimated to be the highest among all known chemical fuels. Therefore, it has enough potential to meet the rising energy demand if some major barriers in its production, storage, and effective commercial or vehicular use are resolved efficiently. One of the most significant challenges in the development of a global hydrogen-based economy is the hydrogen storage problem, which essentially means minimizing the enormous volume of hydrogen gas to achieve optimum gravimetric and volumetric density synergistically. This chapter briefly discusses the significant challenges, especially the hydrogen storage problem in the path of the development of a hydrogen-based economy and the possible approaches to meet the challenges.

The energy demands of the future are ever increasing, and hydrogen as an ideal energy carrier can fulfil these demands. The production, purification, delivery, storage, and application are the significant measures of the hydrogen-based economy. The utmost challenge to utilize hydrogen as a fuel lies in the improvement of storage techniques. Hydrogen storage technologies comprise of high-pressure compression, cryogenic liquefaction, and absorption in solid state such as metal hydrides and complex hydrides. As compared with other techniques, hydrogen storage in solid form seems to be one of the utmost likely solutions. However, it involves extremely coupled transport processes such as chemical kinetics, heat, and mass transfer. Complex hydrides are capable substitute aspirants for solid state hydrogen storage because of many advantages, but many of such hydrides suffer from poor kinetics as well as great thermodynamic stability. Significant heat transfer techniques and issues associated with hydrogen storage methods are discussed, with emphasis on metal hydride.

Hydrogen energy is a promising source to meet the global energy demand. The adoption of hydrogen energy by society is possible with the installation and operation of reliable hydrogen infrastructures. Hydrogen has some risk factors inherently. The wide flammability range in air, high flame speed, and low minimum ignition energy cause fire and explosion accidents. In addition, hydrogen embrittlement in metals leads to ruptures in equipment and the formation of leakages. Hydrogen refueling stations, commercial hydrogen facilities, nuclear power plants, and laboratories are places that are open to accidents caused by hydrogen equipment failure or wrong human interventions. Determination of standards for safe operation and making quantitative risk analysis assessments help eliminate accidents or mitigate their damaging consequences. This chapter aims to raise awareness about the safety of hydrogen energy by addressing the risks of hydrogen, past accidents, the lessons learned, and the applicable precautions and recommendations against hazards.

Today, hydrogen is recognized as a non-polluting energy carrier because it does not contribute to global warming if it is produced from renewable energy resources. This chapter presents one of the possibilities of using hydrogen fuel cells in the energy support of electric vehicle charging stations. A comparative study was presented for five situations in which fuel cells have different nominal powers, a study whose objective is to investigate the equipment capability, assuming their use as the first and only source of electricity in a future economy based on hydrogen, in which case the hydrogen fuel used by the fuel cell is available to the final consumer through a centralized distribution infrastructure. The main decision criteria taken into account include electricity generated by fuel cell, degree of autonomy from the classic electricity grid, total hydrogen consumption, energy purchased from the grid, excess energy, total CO2 emissions, initial investment, electricity unit prices, total cost calculated for 25 years of system life.

Foreword

The use of hydrogen fuel cell technology for mobile application presents a challenge for our days in which humanity is facing energy crisis and environmental problems. The present of the mobility seem to be dominated by the electric vehicles powered by electricity form batteries, but fuel cell still presents potential on this huge market, since several prototypes are running based on this technology. Automobiles, but also trucks, buses, trains, ships or even airplanes are planned or even successfully tested all over the world. Since the fuel cell produces electricity from hydrogen and oxygen, the single residual product is water, with no negative environmental impact. The challenges associated with the fuel cell technology are related to different aspects, such as the source of hydrogen that should be produced in so called green conditions, the need of hydrogen transportation from the production plants to the usage locations, the use of materials capable to increase the fuel cell efficiency, the economic criteria, and the cost constraints. In this context *Hydrogen Fuel Cell Technology for Mobile Applications* is a modern and useful monography that provides answers and solutions to many problems, but also raises new ideas to be solved in the future.

Mugur Balan
Department of Mechanical Engineering, Faculty of Automotive, Mechatronics, and Mechanical Engineering, Technical University of Cluj-Napoca, Romania

Preface

Worldwide, increasing energy consumption—regardless of its type—makes it harder for the energy sector to secure the primary resources it needs. As a result of this fact, as well as the environmental pollution factor, respectively global warming, this problem has the potential to become a considerable concern. There has been a significant increase in the emissions of carbon dioxide into the atmosphere in the current context because of the construction of more and more residential and non-residential buildings, offices, and factories, as well as the intensification usage of cars and trucks around the world. This has caused undesirable effects on human health and the environment. Scientists and researchers must find solutions for the entire planet to be green, as protecting the environment for future generations is a major challenge.

Hydrogen is an important raw material for the chemical, petrochemical, or agrochemical industry, with the best chance of becoming an energy carrier alongside electricity in the near future. The story of energy is marked by the transition to fuels characterized by a higher hydrogen content: from wood to coal, oil and finally natural gas, all characterized by a higher calorific value for the presence of the majority of carbon and hydrogen. Among the chemical fuels, the highest calorific value is that of hydrogen. Hydrogen has a very high energy density so 1 kg of hydrogen contains the same amount of energy as 2.6 kg of natural gas or 3.1 kg of gasoline. This phenomenon makes hydrogen an ideal fuel for various applications where weight and not volume is an important factor. With the transition to a sustainable energy system, the potential of hydrogen to become an energy vector or an alternative fuel will lead to an obvious development of the associated technologies in the next time period.

Hydrogen production is not something recent, to obtain it there are already well-known technologies, but the latest trends in hydrogen extraction technology are based on producing hydrogen from renewable sources, however being the subject of various demonstrative studies that will validate this concept as an alternative to the classic ones, having a real potential to become a solution in providing access for every citizen of this planet to pure, clean, non-polluted energy or fuel. The electrolysis of water is one of the most basic technologies to generate almost pure hydrogen. Water electrolysis can potentially be a very important process for producing hydrogen as an energy carrier for our energy needs. At present, many technologies and equipment have been developed that offer the possibility of obtaining hydrogen through mechanisms of water dissociation under the action of renewable energy.

Fuel cells have enabled hydrogen-based energy generation, making hydrogen one of the world's energy alternatives. With a wide range of primary energy sources available for production as well as an increasing number of applications, hydrogen will become a major energy source, just like electricity. One of the advantages of hydrogen over electricity is that it can be stored for a long time. As a result, we are entering an era of advanced technologies and new fuels. We will discover new and completely different manners of generating and using energy, and hydrogen can be considered as a secondary carrier, synthetic fuel in a future era, following the fossil fuel economy, within a sustainable energy system.

The hydrogen economy represents all the energy infrastructures at the national and international levels which are based on hydrogen as a fuel or energy resource. Building an infrastructure that allows, easily and efficiently, the transport and delivery of hydrogen fuel or energy is a critical step toward a future hydrogen-based economy. Hydrogen has great potential as a non-polluting fuel and is also a means of supporting the reduction of energy resource imports in certain cases. Before hydrogen plays an important energetic role and becomes an alternative to the use of oil, an infrastructure must be built that contains both centralized or decentralized production facilities, as well as distribution systems and networks. Some facilities will be needed to produce hydrogen, store and transport it, but also to build economical fuel cells, and consumers will need the necessary education to use it safely. To emphasize the complexity of a hydrogen-based economy, it is necessary to analyze the main aspects related to the production, transport, distribution, storage, and use of hydrogen.

Taking into account the research and development programs supported in this domain, the technical problems regarding the production, storage and distribution of hydrogen, simultaneously with the decrease in costs and the increase in the lifespan of the equipment used in the generation of energy based on hydrogen, will be solved in a short time, and hydrogen will become a possible solution for providing fuels, at the same time being an alternative energy resource to the traditional one. Although it is not yet widely used for this scope, given the decarbonisation targets in the transport sector, hydrogen will play a direct fuel role in public transport, road rail and freight transport, as well as as a raw material for synthetic fuels for the maritime and aviation sectors.

The book *Hydrogen Fuel Cell Technology for Mobile Applications* was developed based on the editor's motivation to provide an overview of the topic in the context of global interest increasing for the use of hydrogen energy as a vector-carrier of clean energy, to give effect to sustainable mobile applications, since the role of hydrogen in the decarbonization of our society and in the independence of mobile applications from fossil fuels can be determinative.

TARGET AUDIENCE

This book targets advanced-level students, researchers, specialists and practitioners in areas of interest: green alternative energy resources, clean alternative fuels, electromobility, sustainable transport, chemistry, physics, mechanics, machine building and plant engineering, automotive engineering, electrical engineering, environmental engineering, process and materials engineering, and anyone interested in new technologies of clean energy or green power generation systems. Other stakeholders who are interested in associated technologies specific to sustainable automotive and mobility will also find this book very useful in their research.

OBJECTIVE OF THE BOOK

This book's key objective is to inventory and disseminate the relevant knowledge regarding hydrogen economy dedicated to the mobile applications, through reviews or overviews on the main components of the hydrogen economy and the fuel cell typology, through theoretically and empirically based studies or through research approaches that highlight the practical applicability of hydrogen fuel cell in mobile applications. The significance of the book resides in its capacity to enlarge our knowledge by emphasizing the complexity of a hydrogen-based economy by presenting in detail and analysing the main aspects related to the production, transport, distribution, storage and use of hydrogen as an energy vector or as an alternative fuel for practical mobil applications. Therefore, the book's main target is to provide a broad and comprehensive understanding that will deepen the ontology of the topic of 'Hydrogen Fuel Cell Technology for Mobile Applications'.

ORGANIZATION OF THE BOOK

This book consists of 10 chapters, each addressing an issue related to the main objective of the proposed topic, specifically the usage of hydrogen as alternative fuel and fuel cell technology in mobile applications. Chapters are organized so that they outline a logical structure and provide an overview of hydrogen economy and the possibilities for using fuel cell technology in mobile applications. Each chapter presents a broad perspective on a particular topic, advancing a comprehensive perspective on it. Each chapter is briefly described as follows:

Chapter 1 serves as an introduction to the book, providing an overview of the main properties of hydrogen which recommends it as a clean sustainable energy carrier compared to conventional fossil fuels. The main arguments in favor of using hydrogen as a synthetic fuel, obtained from renewable sources, are the following: it has the highest energy/unit of mass of all types of fuel; it is ecological, from its combustion resulting water vapors, observing that for hydrogen the amount of carbon emissions is zero; it has the highest energy reserve factor, respectively the highest conversion factor into electricity, being considered for this reason the best of the actual fuels, and the energy efficiency is very high. Hydrogen is expected to play an important role in future global energy scenarios. Also, the chapter briefly presents the main components of hydrogen economy, namely: hydrogen production methods (thermal methods, electrochemical methods and biological methods), insisting on renewable energy-based hydrogen production technologies; hydrogen transport and distribution and hydrogen storage. The authors conclude that the future hydrogen-based economy presents hydrogen as an feasible energy carrier for a clean and sustainable automotive sector.

To decrease the negative effects of global warming, the energy vicinities accelerate the studies on alternative and clean energy conversion systems in light of the increased number of disasters caused by this phenomenon. Conventional transportation systems mainly based on fossil fuels seem to be the most critical issue for the environment. Electric vehicles are promising clean candidates that can be used for transportation. Electric vehicle technology can be classified into battery electric vehicles (BEVs), hybrid electric vehicles (HEVs), plug-in hybrid electric vehicles (PHEVs), and fuel cell electric vehicles (FCEVs). In recent years, due to the development in nanotechnology and material science engineering newly designed battery and fuel cell-based technologies draw attention. Chapter 2 aims to give brief information about the electric vehicle types, a comparison of the existing systems, hydrogen fuel cell electric vehicles and drawbacks of these systems, and the current status of the hydrogen economy.

A high-quality energy carrier like hydrogen can be created by thermochemical processes involving the conversion of hydrocarbons, such as coal or natural gas. It can also be obtained by water electrolysis using renewable energy sources, such as wind and solar power. Due to climate change, air pollution, and energy deficiency, the exploration of new transportation fuels is very important. Despite the various challenges that hydrogen faces, it is still regarded as a clean energy source that can help solve the environmental and energy problems of the future. It can be used in fuel cells to provide high-quality electricity and heat. Chapter 3 aims to discuss the various techniques involved in the production of hydrogen and their applications in fuel cells. It also explores the different types of fuel cells, such as PEMFCs, MCFCs, and SOFCs as well as to understand effect towards sustainable development goals.

In Chapter 4, Athule Ngqalakwezi discusses the hydrogen economy and fuel cell technology in the South African context and whether or not the complete replacement of fossil fuel with hydrogen is feasible or not. The improvement in the standard of living due to urbanization and industrialization can without a doubt be accredited to the discovery and use of fossil fuels over the centuries. Although this energy source has had a huge positive impact on the human population, the negatives are just as huge. The continuous use of fossil has been detrimental to the environment and has pushed the world into a paradigm shift where green technology has to be implemented. Hydrogen as an alternative source, has intrigued many researchers due to its zero emission characteristic and higher energy content compared to traditional energy sources. Fuel cells which use hydrogen as a fuel source, are the fastest growing clean technologies. Fuel cells are used in both stationary and on-board applications.

Chapter 5 summarizes the recent development of the hydrogen production from waste, current state of art in hydrogen fuel cells, hydrogen fuel cell vehicles and their economical and sustainability benefits. It also includes the key challenges of these technologies for the commercial applications. The increasing rate of industrialization and urbanization increases the waste and wastewater generation along with global energy demand. That makes the necessity of the current waste treatment and process technologies to be upgraded into the sustainable or green energy generation such as hydrogen production from the waste materials and hydrogen fuel cells-based technologies. Fuel cells are electrochemical devices that use hydrogen as fuel cells for the electricity or heat generation. The fuel cells-based technologies are in the infant stage that need to be explored for the vital utilization of hydrogen for the various energy related applications. Stationary application of fuel cells or hydrogen fuel cell vehicles related technologies are hotspot research areas in the 21st century. The authors emphasizes that the hydrogen generation from waste or wastewater can unquestionably be regarded as the progenitor of the decisive clean, efficient and climate-neutral energy systems. That is one step forward to the sustainability and green process along with capability to ensure the increasing global energy demand. Hydrogen and fuel cells may serve as a breeding ground for new science and technology, innovations, companies, products, and employment.

In a case study, Chapter 6 presents scenarios developed with the aim of analyzing two types of energy management strategies dedicated to charging stations for electric vehicles based on photovoltaic energy, having the power reserve provided with or without hydrogen energy support and its associated technologies (fuel cell and electrolyser). The proposed solution, to solve the problem between the amount of available energy and the number of parking spaces for the electric vehicle charging station, works according to the following principle: the system management is designed to work with one to five electric vehicle charging stations. It can be installed from one to five electric vehicle charging stations with a power of 22 kW. In this research, the state of charge of electric vehicles and the management of distributed energy for electric vehicles were monitored.

Industrial revolution and sustainable development have intensified the need for clean resources of energy. Among all the possible alternative sources of green energy, hydrogen energy is the most promising one because of its abundancy, maximum energy density per unit mass and clean combustion. The energy content per unit mass of hydrogen is estimated to be the highest among all known chemical fuels. Therefore, it has enough potential to meet the rising energy demand if some major barriers in its production, storage, and effective commercial or vehicular use are resolved efficiently. One of the most significant challenges in the development of a global hydrogen-based economy is the hydrogen storage problem, which essentially means minimizing the enormous volume of hydrogen gas to achieve optimum gravimetric and volumetric density synergistically. Chapter 7 briefly discusses the significant challenges especially the hydrogen storage problem in the path of the development of a hydrogen-based economy and the possible approaches to meet the challenges. The authors establish the following conclusions: Despite having maximum energy density and clean combustion, hydrogen economy has not yet developed with full fledge. One of the major challenges behind this lag is the hydrogen storage problem. Various hydrogen storage methods have been employed till date, but none of them completely fulfils the US-DoE criterions. Hydrogen stored in high pressure tanks are associated with drawbacks such as low gravimetric density, poor efficiency and accidental risk. Although hydrogen stored under high pressure in cylinders does not meet the US-DOE requirements but still it is the current choice for industrial and vehicular applications. Further, storage at cryogenic temperature may improve the gravimetric and volumetric density up to some extent but still it is not optimum for the industrial applications and is limited by severe drawbacks like easy boil off of hydrogen even at little thermal fluctuations and flammability risk. In such scenario, hydrogen storage in material-based systems has proven to be quite efficient with minimum risk but are very hard to realize in practical applications. Thus, development of hydrogen based global economy solely depends on the efficient hydrogen storge mechanism.

The energy demands of future are ever increasing and hydrogen as an ideal energy carrier can fulfil these demands. The production, purification, delivery, storage and application are the significant measures of the hydrogen based economy. The utmost challenge to utilize hydrogen as a fuel lies in the improvement of storage techniques. Hydrogen storage technologies comprise of high pressure compression, cryogenic liquefaction and absorption in solid state such as metal hydrides and complex hydrides. As compared with other techniques, hydrogen storage in solid form seems to be one of the utmost likely solutions. However, it involves extremely coupled transport processes such as chemical kinetics, heat and mass transfer. Complex hydrides are capable substitute aspirants for solid state hydrogen storage because of many advantages. But, many of such hydrides suffer from poor kinetics as well as great thermodynamic stability. Rahul Uday Urunkar and Sharad D. Patil, in Chapter 8, discuss significant heat transfer techniques and issues associated with hydrogen storage methods, with emphasis on Metal Hydride.

Chapter 9 aims to raise awareness about the safety of hydrogen energy by addressing the risks of hydrogen, past accidents, the lessons learned, and the applicable precautions and recommendations against hazards. Hydrogen energy is a promising source to meet the global energy demand. The adoption of hydrogen energy by society is possible with the installation and operation of reliable hydrogen infrastructures. Hydrogen has some risk factors inherently. The wide flammability range in air, high flame speed, and low minimum ignition energy cause fire and explosion accidents. In addition, hydrogen embrittlement in metals leads to ruptures in equipment and the formation of leakages. Hydrogen refueling stations, commercial hydrogen facilities, nuclear power plants, and laboratories are places that are open to accidents caused by hydrogen equipment failure or wrong human interventions. Determination of standards for safe operation and making quantitative risk analysis assessments help eliminate accidents or mitigate their damaging consequences.

Today, hydrogen is recognized as a non-polluting energy carrier because it does not contribute to global warming if it is produced from renewable energy resources. Chapter 10 presents one of the possibilities of using hydrogen fuel cells in the power supply of electric vehicle charging stations. A multicriteria comparative study was presented for five situations in which fuel cells have different nominal powers, a study whose objective is to investigate the equipment capability, assuming their use as the first and only source of electricity in a future economy based on hydrogen, in which case the hydrogen fuel used by the fuel cell is available to the final consumer through a centralized distribution infrastructure. The technical-economic and ecological study carried out aims to determine and highlight a set of general criteria that allow the characterization of these types of systems, but also to approach this energy solution of using the fuel cell in the field of green mobility from the point of view of sustainability. The main decision criteria taken into account: electricity generated by fuel cell, degree of autonomy from the classic electricity grid, total hydrogen consumption, energy purchased from the grid, excess energy, total CO_2 emissions, initial investment, electricity unit prices, total cost calculated for 25 years of system life.

The topics addressed in the chapters cover the main aspects related to a hydrogen economy - production, storage, transport, distribution of hydrogen; reviewing the state-of-the-art fuel cell technologies developed; discussing the main problems related to safety in the use of hydrogen, but also the main integration paths of the fuel cell based on hydrogen into sustainable mobile practical applications. The chapters examined the main technical, environmental and economic characteristics, but also the key challenges in adopting energy systems based on clean hydrogen and evaluated various opportunities to implement hydrogen fuel cell technologies in practical mobile applications in order to promote the development of clean alternative fuel while protecting the environment.

The technical information, discussions, studies, analyzes and research results presented during this book, but also the problems identified, the technical limitations encountered allow the identification and establishment of new future approaches in the field of 'Hydrogen Fuel Cell Technology for Mobile Applications', each chapter representing a starting point in the elaboration of exhaustive studies specific to the addressed subject that to further explore ways to reduce the gaps in the studies and the specialized technical literature.

Raluca Andreea Felseghi
Department of Mechanical Engineering, Faculty of Automotive, Mechatronics, and Mechanical Engineering, Technical University of Cluj-Napoca, Romania

Chapter 1
Hydrogen-Energy Vector Within a Sustainable Energy System for Mobile Applications

Raluca Andreea Felseghi
https://orcid.org/0000-0001-6310-5932
Technical University of Cluj-Napoca, Romania

ABSTRACT

Today hydrogen is recognized as a non-polluting energy carrier because it does not contribute to global warming if it is produced from renewable energy resources. Hydrogen is the only secondary energy carrier that is suitable for wide applications. At the center of attention is the fact that hydrogen can be obtained from a wide range of primary energies. It can be used advantageously for a wide range of applications. Hydrogen can be used in decentralized systems without emitting CO_2. Hydrogen is already a part of today's chemical industry, but as an energy resource, its rare benefits can only be achieved through fuel cell technology. The next generations of energy systems for mobile applications based on hydrogen fuel cells have the potential of using and implementing clean energy in the mobility domain, as well as in the tertiary and industrial sector, thus having a significant impact on greenhouse gas emissions decreasing, specific characteristics of hydrogen technology having an important role in the decarbonization of transport sector.

DOI: 10.4018/978-1-6684-6721-3.ch001

INTRODUCTION

The context of environmental problems, the necessity to switch from conventional fuels intensely polluting currently used to the usage of alternative fuels and energies for power generation, and current energy policies create a suitable context for the implementation and development of new clean and green technologies. The hydrogen-based economy and, implicitly, hydrogen fuel cell technology are of particular interest in global energy strategies. Recently, interest has been expressed in hydrogen and fuel cells along with the potential to obtain electrolytic green hydrogen in relation to alternative energy storage in the context of the unprecedented development of the renewable resource sectors. (Azam, W., et al., 2023; Genovese, M., et al., 2023; Wolniak, R., and Skotnicka-Zasadzień, B., 2022).

The Hydrogen Economy Miami Energy Conference, USA March 18-20, 1974 was the first international discussion where hydrogen was introduced as a standing resource to conventional fuels depletion and global environmental issues, an event that laid the bases of the International Association for Hydrogen Energy (Midilli, A., et al., 2005; Momirlan, M., et al., 2022; Tiwari, A.,2021; Veziroğlu, T. N., 2000; Veziroğlu, T. N., 2004).

Despite the reality that conventional energy has been constantly subsidized by governments, even if it harms human health, the well-being and integrity of our world, and the fact that equivalent subsidies have not been extended to the hydrogen energy sector, in the quarter of a century since the hydrogen movement began, however, it has made significant progress in several directions in the energy sector, due to its unmatched, high-level properties and its characteristics as a carrier, a secondary energy vector. (Pingkuo, L., and Xue, H., 2022; Razi, F., and Dincer, I., 2022; Veziroğlu, T. N., 2000; Veziroğlu, T. N., 2004).

GENERAL ASPECTS

Unconventional energy sources have gained and will continue to gain an increasing share in energy systems around the world, both due to the research effort and political will involved in their development and due to the rising price of energy obtained through the traditional methods. Primary energy sources, generally called renewable, are those sources in the natural environment, available in virtually unlimited quantities or that are regenerated by natural processes, at a faster rate than the one in which they are consumed. (Badea, G., et al., 2015; Felseghi, R.A., et al., 2019, a; Noor, M., et al., 2023; Stennikov, V., 2023). Officially recognized renewable energies originate from the sun's rays, the earth's internal temperature, or the gravitational interactions of the sun and moon with the oceans (Felseghi, R.A., et al., 2019 a; Li, Y., et al.,2020; Filote, C., et al.,2020;Hagen, M., et al., 2022; Shadman, M., et al., 2023).

Processes and procedures for the production or capture of these types of alternative energies under development, high investment costs and low returns on conversion processes have made renewable energy sources a small part of the needs on a global scale. Optimistic forecasts estimate renewable energy production to account for 30-50% of the total energy market around 2050, but this depends on reducing production costs and finding opportunities for massive storage of electricity. In addition, all these forms of energy cannot provide fuels in satisfactory quantities for various industrial uses. (Atems, B., et al. 2023; Colbertaldo, P., et al., 2019; Felseghi, R.A., et al., 2019, a)

In this context, alternatives are currently being sought to obtain electricity by using technologies that provide maximum efficiency, high reliability and minimum pollution. Such a technology, currently considered the cleanest, through which sustainable energy can be obtained, is based on fuel cells. With the development of fuel cells, the production of hydrogen-based energy is becoming a reality, hydrogen being one of the safest energy alternatives for mankind. The future hydrogen-based economy presents hydrogen as an energy carrier in a safe and sustainable energy system. (Felseghi, R.A., et al., 2019, b; Ren, X., et al., 2020; Cloete, S., Ruhnau, O., & Hirth, L., 2021; Scheller, F., et al., 2023; Genovese, M., et al., 2023)

Humanity is on the threshold of a new era characterized by advanced technologies and new fuels. We will witness new and completely different ways of producing and using energy. Electricity could be generated by sources with virtually no pollution. Hydrogen can be considered as a synthetic fuel, a secondary energy carrier in a future era, after the fossil fuel economy. (Bellocchi, S., et al., 2023; Felseghi, R.A., et al., 2019, b; Hosseini, S. E., & Wahid, M. A., 2020; Pareek, A., 2020; Sazali, N., 2020).

HYDROGEN-ENERGY VECTOR IN A SUSTAINABLE ENERGY SYSTEM

Today hydrogen is recognized as a clean energy carrier because it does not contribute to global warming if it is produced from renewable sources. In addition, hydrogen is the only secondary energy carrier suitable for wide applications on the market. The focus is on the fact that hydrogen can be obtained from a wide range of primary energies. It can be used advantageously for a wide range of applications, from transport and portable to stationary. In addition, hydrogen can also be used in decentralized systems without emitting carbon dioxide. Hydrogen is already a component of today's chemical industry, but as an energy source, its rare advantages can only be achieved with the help of technologies such as fuel cells. *(Badea, G., et al., 2021; Felseghi, R.A., et al., 2019, a; Lagioia, G., et al., 2023; Weeda, M., & Segers, R., 2020).*

As hydrogen can be produced from a wide range of primary energies and can be consumed in an even greater number of applications, it will become an energy center, just as electricity is today. The advantage of hydrogen over electricity is that it can be stored for long periods of time and transported away from the place of production. Consequently, an energy carrier helps to increase the stabilization of energy security and price, giving rise to competition between different energy sources. *(Felseghi, R.A., et al., 2019, a; Zhiznin, S. Z., Vassilev, S., & Gusev, A. L., 2019; Midilli, A., Ay, M., Dincer, I., & Rosen, M. A., 2005).*

T. N. Veziroglu, editor of the journal specializing in hydrogen technology and energy, International Journal of Hydrogen Energy briefly presents some features that recommend the use of hydrogen as an energy secondary vector produced on the basis of unconventional technologies:

- hydrogen concentrates primary energy sources, which it makes available to the consumer in a convenient form;
- the production cost is relatively low as a result of technological developments and improvements;
- offers the possibility of transformation into various other forms of energy through conversion processes characterized by high efficiency;
- it is an inexhaustible source if it is obtained electrolytically from water; hydrogen production and consumption is a closed cycle, the source of production - water remains constant and is a classic cycle of recirculation of this type of raw material;
- is the lightest and cleanest fuel; hydrogen combustion is almost entirely free of pollutant emissions;
- has a much higher gravimetric energy density compared to other fuels;
- hydrogen can be stored in various ways, such as: gas at normal or high pressure, in the form of liquid hydrogen or in the form of solid hydrides;
- can be transported over long distances stored in the form or in one of the ways presented above;
- the equipment for converting hydrogen into electricity, of fuel cell type, has an efficiency of over 60%. (Midilli, A., et al., 2005; Veziroğlu, T. N., 2000; Veziroğlu, T. N., 2004).

Hydrogen is considered by more and more specialists to be a true fuel of the future. Hydrogen condenses at -252.77 °C, and the specific gravity of liquefied hydrogen is 71 g/L, which gives it the highest energy density per unit mass of all fuels and energy carriers: 1 kg of hydrogen it contains as much energy as 2.1 kg of natural gas or 2.8 kg of oil. This feature has made hydrogen the fuel used in the propulsion and power supply of spacecraft. Unlike other fuels, such as oil, natural gas and coal, hydrogen is renewable and non-toxic when used in fuel cells. Hydrogen has great potential as an environmentally friendly fuel and in reducing the import of energy resources. Even if the use of hydrogen now seems unprofitable, following the improvement of technologies, we could witness in the not too distant future the development of a hydrogen-based economy. (Badea, G., et al., 2021; Molnarne, M., & Schroeder, V.,2019)

Hydrogen-Carrier of Energy and Synthetic Fuel

Etymologically, the word hydrogen is a combination of two Greek words, meaning "to make water". *(Iordache, I., & Stefănescu, I., 2010)*. Produced from non-fossil sources and raw materials, by using various forms of alternative energy (solar, wind, hydroelectric, geothermal, biomass, etc.), hydrogen is considered to be a prime fuel in the supply of so-called "green energy". Thus, hydrogen-powered systems can be considered one of the best solutions for accelerating and ensuring global stability *(Dincer, I., & Zamfirescu, C., 2012)*. Hydrogen is expected to play an important role in the world's future energy scenarios, the most important factor that will determine the specific role of hydrogen will probably be energy demand. At the same time, hydrogen can, to some extent, replace fossil fuels and become the preferred clean, non-toxic energy carrier in the near future. (*McCarty, R. D., Hord, J., & Roder, H. M., 1981; Felseghi, R.A., et al., 2019, a)*

The main properties of hydrogen, presented in Table 1, recommend it as an alternative fuel to the classic ones.

In order to highlight the advantages of hydrogen compared to other fuels, Table 2 summarizes the main properties of the various fuels currently used.

Analyzing the tabular information, it can be concluded that the main arguments in favor of using hydrogen as a synthetic fuel, obtained from renewable sources, are the following: it has the highest energy/unit of mass of all types of fuel; it is ecological, from its combustion resulting water vapors, observing that for hydrogen the amount of carbon emissions is zero; it has the highest energy reserve factor, respectively the highest conversion factor into electricity, being considered for this reason the best of the presented fuels, and the energy efficiency is very high. Hydrogen is expected to play an important role in future global energy scenarios. (Notteboom, T., and Haralambides, H.,2023; Zhiznin, S. Z., et al., 2023)

Table 1. Properties of hydrogen

Characteristics	Measure Unit	Value
Density	kg/m^3	0,0838
HVV/ LH_2	MJ/kg (liquid)	141,90 - 119,90
HVV/ CGH_2	MJ/m^3 (volumetric)	11,89 - 10,05
Boiling point	K	20,41
Freezing point	K	13,97
Density (liquid)	kg/m^3	70,8
Air diffusion coefficient	cm^2/s	0,61
Specific heat	kJ/kg K	14,89
Ignition limits in air	% (volume)	4 - 75
Ignition energy in air	Millijoule	0,02
Ignition temperature	K	585,00
Flame temperature in air	K	2318,00
Energy exploding	kJ/g TNT	58,823
Flame emissivity	%	17 - 25
Stoichiometric mixture in air	%	29,53
Stoichiometric air/fuel	kg/kg	34,30/1
Burning speed	cm/s	2,75
Energy reserve factor	-	1,00

Source: (Badea, G., et al., 2021; Felseghi, R.A., et al., 2019, a)

Table 2. Comparison between the main properties of hydrogen and other fuels

Fuel type	Energy/ mass unit (J/kg)	Energy/ volume unit (J/m^3)	Energy reserve factor	Specific carbon emission (kgC/kg eq.)
Liquid hydrogen	141,90	10,10	1,00	0,00
Gaseous hydrogen	141,90	0,013	1,00	0,00
Fuel oil	45,50	38,65	0,78	0,84
Benzine	47,40	34,85	0,76	0,86
Jet fuel	46,50	35,30	0,75	-
GPL	48,80	24,40	0,62	-
GNL	50,00	23,00	0,61	-
Methanol	22,30	18,10	0,23	0,50
Ethanol	29,90	23,60	0,37	0,50
Biodiesel	37,00	33,00	-	0,50
Natural gases	50,00	0,04	0,75	0,46
Coal	30,00	-	-	0,50

Source: (Badea, G., et al., 2021; Felseghi, R.A., et al., 2019, a)

The advantages that promote it as an energy vector compared to other forms of energy are as follows:

- hydrogen can be transported remotely through pipelines, in safe conditions;
- hydrogen is a non-toxic energy transport operator, with a high specific energy per mass unit (for example, the energy obtained from 9.5 kg of hydrogen is equivalent to that of 25 kg of gasoline);

- hydrogen can be generated from various energy sources, including renewable ones;
- compared to electricity, hydrogen can be stored for relatively long periods of time;
- hydrogen can be used advantageously in all sectors of the economy (as a raw material in the industry, as a fuel for automobiles and as an energy carrier in sustainable energy systems to generate electricity through fuel cells). *(Badea, G., et al., 2021; Felseghi, R.A., et al., 2019, a)*

Barriers to be overcome concern issues related to:

- hydrogen burns in the presence of air, and this can cause safety problems in operation;
- storage of hydrogen in liquid form is difficult because very low temperatures are needed to liquefy hydrogen;
- high costs of hydrogen technologies and processes;
- high costs of technologies for converting hydrogen into electricity through fuel cells;
- the viability/cost ratio of hydrogen and fuel cell production technologies is relatively low;
- lack of logistics, hydrogen transmission and distribution infrastructure to final consumers involves costly investments. *(Badea, G., et al., 2021; Felseghi, R.A., et al., 2019, a)*

Given the research and development programs supported in this field, the deficient technical problems regarding the production, storage and distribution of hydrogen, at the same time with the decrease of costs and the increase of the life of the equipment used in hydrogen-based energy generation, will be solved shortly. and hydrogen will become a possible solution for providing fuel, while also being an alternative energy resource to the traditional one. (*Badea, G., et al., 2017; Badea, G., et al., 2021)*

HYDROGEN ECONOMY

The hydrogen economy represents the totality of energy infrastructures at the national and international level that have hydrogen as their energy resource *(Posso, F., et al., 2023; Ren, J., et al., 2023)*. Building an infrastructure that allows, easily and efficiently, the transport and delivery of hydrogen energy is a critical step towards a future hydrogen-based economy *(Cloete, S., Ruhnau, O., & Hirth, L., 2021*

Scheller, F., et al., 2023). Hydrogen has great potential as a clean fuel and is also a means of supporting the reduction of energy imports in some cases. Before hydrogen can play an important energy role and become an alternative to the use of oil, and infrastructure must be built that contains both centralized or decentralized production facilities and distribution systems and networks *(Linssen, J., & Hake, J. F., 2016; Pezhmani, Y., et,al. 2023).* Some facilities will be needed for the production of hydrogen, for its storage and transportation, but also for the construction of economical fuel cells, and consumers will need the education necessary to use it safely. (*Eiler, J. M., 2020; Nuttall, W. J., & Bakenne, A. T., 2020;* Dillman, K., and Heinonen, J., 2023; Mneimneh, F., et al., 2023).

In order to emphasize the complexity of a hydrogen-based economy, it is necessary to analyze the main aspects related to the production, transport, distribution, storage and use of hydrogen.

Hydrogen Production Processes

Hydrogen cannot be found in nature in its pure state, so it cannot be mined in the same way as oil or coal. Hydrogen must be extracted from chemical compounds. Regardless of the resource from which hydrogen is extracted, a production process is needed and it involves energy consumption. The big advantage, however, is that for the generation of hydrogen it is not strictly necessary to use energy from fossil fuels, you can use renewable, hydraulic, solar, wind, geothermal energy (Badea, G., et al., 2021; Liu, W., et al., 2020; Sarker, A. K., et al., 2023).

The main processes currently used for hydrogen manufacture are summarized in Table 3.

Hydrogen production is not new. Every year, approx. 500 billion m^3 of hydrogen is produced, stored, transported and used, mainly in the chemical and petrochemical industry (Acar, C., & Dincer, I., 2014). However, most of it is obtained as a by-product from the processing of fossil fuels (oil and natural gas).

There are already established technologies in the manufacture of hydrogen, used in the chemical industry, as well as new processes developed especially for fuel cells. Each of the existing technologies has its advantages and disadvantages, so research is needed to optimize them.

All processes for obtaining hydrogen are based on separating it from the products that contain it. The nature of the raw material dictates the choice of the separation method. Today, two main methods of hydrogen separation are used: thermo-chemical and electrochemical. The third method, the biological one, is in the research-development phase. *(Iordache, I., & Stefănescu, I., 2010; Dawood, F., et al., 2020, Sillero, L., et al., 2022; Xu, X., Zhou, Q., and Yu, D., 2022).*

Table 3. Processes used for hydrogen manufacture

Basic method	Process	Raw material	Energy type	Pollutant emissions
Thermal method	Steam reforming	Natural gases	High temperature steam	Some emissions, carbon sequestration may decrease their effect.
	Thermochemical splitting of water	Water	High temperature obtained from the cooling of nuclear reactors	No emissions.
	Gasification	Coals, Biomass	Steam and oxygen at high temperature and pressure	Some emissions, carbon sequestration may decrease their effect.
	Pyrolysis	Biomass	Moderate temperature steam	Some emissions, carbon sequestration may decrease their effect.
Electrochemical method	Electrolysis	Water	Wind, solar, hydro and nuclear electricity	No emissions.
	Electrolysis	Water	Electricity produced from coal and natural gas	Some emissions from electricity production.
	Photochemical	Water	Direct sunlight	No emissions.
Biological method	Photobiological	Water and algae species	Direct sunlight	No emissions.
	Anaerobic fermentation	Biomass	High temperature	No emissions.
	Fermentative microorganisms	Biomass	High temperature	No emissions.

Source: (Iordache, I., & Stefănescu, I., 2010)

Renewable Energy-Based Hydrogen Production Technologies

Numerous technologies and equipment have currently been developed that offer the possibility of obtaining hydrogen through mechanisms of water splitting under the action of renewable energy, especially solar and wind energy. In this part of the chapter are briefly presented only the technologies that use water as raw material, and the resulting products are hydrogen and oxygen, although in practice there are other types of raw materials used in the production of organic hydrogen, and as objectives of this presentation aims to analyze and compare the most important water-splitting technologies under the action of various components of renewable energy.

Thermal and Thermochemical Technologies for Hydrogen Production

In the case of using the thermal component related to renewable energy, the water molecules can be cleaved directly under the action of thermal energy or indirectly with the help of intermediate auxiliary chemicals. In principle, when the temperature reaches higher than 2000 °C, the direct decomposition of water begins without the use of any other auxiliary chemicals. However, the temperature of direct thermolysis is too high, requiring special construction of refractory equipment *(Cobîrzan, N., et al., 2018)*. In addition, the thermolysis product is a mixture of hydrogen and oxygen gases, which indicates a considerable risk of explosion. (Wang, Z. et al.,2012; Safari, F.& Dincer, I., 2020).

Therefore, in order to reduce the temperature required for the dissociation of the water molecule, some auxiliary chemicals can be introduced. This type of water decomposition is often called the thermochemical cycle because water enters the process along with the chemical compounds that will be recovered and reused, and the results of the process are hydrogen and oxygen. If a small part of the cycle's input energy is electricity, then we are talking about a type of hybrid thermochemical cycle, rather than a pure thermochemical cycle. Currently, more than 300 thermochemical cycles have been indicated, either purely thermal or hybrid. (Wang, Z. et al.,2012; Safari, F.& Dincer, I., 2020).

Currently, the most studied thermochemical cycle is the sulfur - iodine cycle, which consists of three chemical reactions in which sulfur and iodine compounds decompose water at high temperatures into oxygen and hydrogen. Sulfur and iodine compounds are recycled, the process requiring a heat source. In the first stage of the process, iodine, sulfur dioxide and water vapor react to produce a mixture of sulfuric and iodine acid. The iodine acid is then separated by distillation. The reaction that requires the highest temperature, 800-900 °C, is the decomposition of H_2SO_4. Water, sulfur dioxide and residual acid are separated from the oxygen produced by condensation. The third reaction is the breakdown of iodine acid into iodine and hydrogen, the reaction takes place at about 450 °C. Iodine, possible water residues or sulfur dioxide are separated by condensation, hydrogen being the only gaseous product of the separation process. (Ying, Z., et al.,2020; Mandal, S. & Jana, A. K., 2020).

The coldest cycle is copper - chlorine-based, where the maximum required temperature does not exceed 550 °C. The Cu-Cl thermochemical cycle is represented by a series of chemical reactions that lead to the cleavage of the water molecule into hydrogen and oxygen, using interim compounds based on copper and chlorine that will be recovered and regenerated. Basically, the process consumes thermal energy and water, and the copper and chlorine compounds undergo transformations during the cycle without being consumed. (Mehrpooya, M. & Habibi, R.,2020; Kiani, R., et al., 2020).

As mentioned there is a very wide variety of thermochemical cycles from a theoretical point of view, but practically only a limited number of such cycles have technological potential. In addition to those mentioned above, the thermochemical cycles that use metal oxides (cerium, iron, zinc, etc.) may be of interest from certain perspectives. (Carrillo, R. J., & Scheffe, J. R., 2017; Ishaq, H., & Dincer, I., 2019).

Different cycles require different inlet temperatures, which also determine the availability of the energy source. At present, a high temperature of up to 3500 °C can be obtained in solar concentration furnaces only in the pilot phase laboratory. By comparison, the maximum temperature of future generation planned IV reactors is up to 1000 °C. This suggests that solar thermal energy has a more feasible potential to power a thermochemical cycle to obtain hydrogen. (Wang, Z. et al.,2012; Oruc, O.& Dincer, I., 2021).

Water Electrolysis

Electrolysis comprises all the processes that take place when the direct electric current passes through the solution or melt of an electrolyte, in other words, it is the process of orienting and separating the ions of an electrolyte with the help of the direct electric current. In the electrolysis process, the positive ions are directed towards the negative pole, and the negative ions towards the positive pole, where they lose their charge and deposit or enter a chemical reaction, at the anode there is an oxidation process, while at the cathode a reduction one. (Yuzer, B., et al., 2020; Kakoulaki, G., et al., 2021)

In water electrolysis, it is separated by direct current to produce hydrogen. Direct current passes through two electrodes immersed in water, anode and cathode, and hydrogen is obtained on the surface of the cathode when the electric potential is high enough. In this regard, in order to avoid confusion with other terminologies, such as photoelectrolysis or other photoelectrochemical methods, it is suggested that the electrolysis terminology will be used only when electricity is fully received from an external power source (eg energy electricity generated by photovoltaic panels). Therefore, electrolyzers do not receive sunlight to dissociate water. (Wang, Z. et al.,2012; Pareek, A., et al.,2020)

Conventional electrolysis is a mature technology, and the efficiency of a commercially available electrolyzer is 50-80%. Regarding the electrolytic production of hydrogen, there are three types of technologies and implicitly three types of distinct electrolyzers: alkaline electrolyzers, electrolyzers with ion exchange membranes and electrolyzers with solid oxide electrolyte. Currently, the efficiency of electricity production, either with photovoltaic panels or other types of renewable energy conversion equipment, is about 10-20%, therefore, the efficiency of hydrogen production by electrolysis is below 16%. It is anticipated that the efficiency of hydrogen production based on renewable energies may be increased in the future, through the use of new types of materials and equipments. (Wang, Z. et al.,2012; Wang, H., et al., 2020; Yadav, D., & Banerjee, R., 2018).

In addition to technological maturity, another major advantage of water electrolysis is that hydrogen production can provide the necessary energy during the night or on days with adverse weather conditions, by using renewable energy stored in hydrogen. (Wang, Z. et al.,2012; Maggio, G., et al., 2019).

Photoelectrolysis and Photoelectrochemical Technologies

The terminology suggests that these technologies are adopted when the electrode uses at least one absorption light and electricity from external power sources to carry out the reduction or oxidation reactions. (Van de Krol, R., & Grätzel, M., 2012; Wang, Z. et al.,2012)

The hydrogen production unit in this case may be an electrochemical electrolyzer or photoelectrolyzer, which consists of an electrode that absorbs sunlight and a counter electrode immersed in the electrolyte. The sunlight absorption electrode must be arranged so as to be incident with the sunlight window to capture sufficient solar radiation in the process of generating electric potential, the electricity being generated by this electrode which is made of materials that can create electric potential. due to its exposure to the sun. *(Wang, Z. et al.,2012; Grimm, A., et al., 2020; Bai, S., et al., 2021)*

Photolysis and Photochemical Technologies

These technologies suggest the use of sunlight in the interaction with a semiconductor-electrolyte surface for the generation of hydrogen from water or solvents, the appearance of which is observed at the semiconductor-electrolyte interface in the phenomenon of photocorrosion. (*Valente, A., et al, 2017).*

The decomposition of water into hydrogen and oxygen by photochemical mechanisms involves a chain of endergonic or endoergic reactions of the "up-hill" type, which are accompanied by a positive variation of Gibbs free energy *(Punekar, N. S., 2018).*Through these reactions, the energy of the photon is converted to chemical energy, identical to the reactions that take place in the process of green plants photosynthesis.

The basic components of a photochemical cell for hydrogen production are: the solar window, the sensitizer and the catalyst, which are immersed and evenly distributed in the water. Briefly, the photocatalytic decomposition process can be summarized in three simple steps: photons with energy greater than 1.23 eV (λ = 1000 nm) are absorbed by the catalyst, photogenerated pairs of electron-gaps migrate separately to the catalyst surface, species absorbed on the surface of the catalyst (water) are reduced and oxidized by electrons and positive voids. *(Dong, Y., et al., 2021; Cornwell, O., et al, 2018)*

Hybrid Technologies

Two or more of the technologies presented in the previous sections can be combined together for hydrogen production, which would mean a hybrid production technology with a higher energy supply potential and thus an improved efficiency of hydrogen production. *(Wang, Z. et al.,2012; Dawood, F., Anda, M., & Shafiullah, G. M., 2020).*

A typical hybrid technology, which uses both thermal and electrical energy, is high temperature- electrolysis (HTE). In HTE, electricity could be the external power supply, and the overall hydrogen production efficiency can reach 40%. *(Wang, Z. et al.,2012; Ayers, K., et al., 2019)*

Other examples of hybrid hydrogen generation technologies using solar energy are: electricity and photonic energy - used by water photoelectrolysis technology, photonic energy and biochemical energy - used by methods such as water biophotolysis or artificial photosynthesis. *(Abdel-Basset, M., et al., 2021)*

Assuming that hydrogen can be produced from a wide variety of raw materials available almost anywhere in nature, using a wide variety of technological processes it can be concluded that each region of the world may be able to produce more of its own energy with minimal impact on the environment *(Holladay, J. D., 2009)*, while increasing energy efficiency through primary energy savings, reducing grid losses, reducing the price and cost of electricity for consumers, all together contributing to the security and sustainability of the regional, national and global energy system *(Felseghi, R. A., & Papp, A., 2014).*

Hydrogen production technologies using renewable energy were presented, innovative technologies that have recently made significant progress in the development of alternative hydrogen production systems. A simplified comparison of the main characteristics and particularities of the types of technologies *(Midilli, A., et al., 2005, a)* developed and currently used in obtaining hydrogen using renewable energy is presented in Table 4.

Table 4. Particularities in obtaining hydrogen based on renewable energy

Method	**Thermochemical**	**Electrolysis**	**Photoelectrolysis**	**Photochemical**
Power supply	**External**	**External**	**Intern**	**Intern**
Basic system components	more than 3 thermal reactors	2 electrodes + electrolyte	2 electrodes, electrolyte + solar lighting window	at least 1 sensitizer + 1 catalyst and sunlight window
Overall efficiency of H_2 production	45%	30%	16%	10%
Energy form at the input	Thermal	Electric	Electric	Photonics
Reaction mechanism	Thermal splitting of water	Electric potential	Electric potential	Photons activate electrons
Production scale	Wide scale	Medium-scale or local fuel stations	Local fuel stations	Local fuel stations
getting H_2 / O_2	Separated	Separated	Mixed	Mixed
Distribution network	Needed	It depends on the production scale	It's not necessary	It's not necessary
prod. H_2 constructive type & solar energy capture	No	No	Yes	Yes

Source: (Badea, G., et al., 2021; Iordache, I., & Stefănescu, I., 2010)

Following a critical analysis of the current state of development of hydrogen production technologies based on renewable energy, it can be stated that the electrolysis of water fed by energy generated by these sources is the most mature of the technologies, and thermochemical cycles offer the possibility of obtaining hydrogen on a scale. high production while minimizing energy losses caused by high-temperature requirements (Wang, Z. et al.,2012; Midilli, A., et al., 2005, b; Wieczorek, A. K., et al., 2006). Hydrogen production technologies based on photochemical and photoelectrolytic methods are limited, in terms of efficiency, significantly by the quality of materials, photocatalysts and photoelectrolytes that are part of the constructive composition of systems.

Sustained research and development programs in the field of hydrogen production systems and technologies will eliminate technical shortcomings while reducing costs, and hydrogen will compete with traditional energy resources. Any form of renewable energy, in order to be a convincing replacement for fossil fuels, must allow for storage and transport. Hydrogen is often proposed as an attractive option to supplement the use of renewable energy, while hydrogen can prove to be the most ubiquitous fuel available. (Badea, G., et al., 2021; Holladay, J. D., 2009; Felseghi, R.A., et al., 2019, a)

Hydrogen Transport and Distribution

The detailed analyzes that can be found in the literature on the distance/cost ratio of hydrogen transport from the place of production to the place of consumption can be summarized as follows: for small consumers with low demand it is preferable to supply hydrogen in the form of gas tablet (CGH_2), for moderate long-distance consumption hydrogen is delivered in liquid form (LH_2), and for crowded urban centers with high demand hydrogen will be delivered through distribution networks. (Wulf, C., et al., 2018; Armaroli, N., & Balzani, V., 2011; Iulianelli, A., & Basile, A., 2014; EHFCTP, 2005).

Land or naval means of transport can be used to transport hydrogen from the manufacturer to distribution stations, in the form of compressed gas, cryogenic liquid or stored in special compounds, carriers that store hydrogen in the form of reversible chemical bonds (metal hydrides, carbon nanofibers).

Worldwide, more than 820 approved hydrogen stations were registered worldwide for 2019, intended to power cars that use fuel cells to generate electricity. Hydrogen is delivered through distribution stations in the form of liquid hydrogen at -253 °C and hydrogen gas at 300-350 bar and 700 bar, and as sources of hydrogen are: pipelines from plants that generate secondary hydrogen, plants local electrolysis or catalytic reforming of methane gas. (E4tech, 2019)

The use of hydrogen transport pipelines is not a novelty, but given the specificity of hydrogen transport, for example, the hydrogenation of steel leads to a deterioration of its quality becoming breakable, the construction of such pipelines requires very large initial investments. An advanced solution to accelerate the transition to a hydrogen economy would be to use the current gas network to transport a mix of 80% natural gas and 20% hydrogen, or a liquid compound (ethanol) from which hydrogen can be easily extracted (by reforming). (Cerniauskas, S., et al., 2020; Abd, A. A., et al., 2020).

Scientific studies analyze the possibility of creating a super-network for the simultaneous transport of liquid hydrogen and electricity, the same pipe/cable in which liquid hydrogen would have the role of coolant necessary to obtain the state of superconductivity and would be used at the exit of the cable as and fuel in hydrogen filling stations. (Al-Breiki, M., & Bicer, Y., 2021).

Hydrogen Storage

A key issue in ensuring the viability of a hydrogen-based economy is the storage of manufactured hydrogen. This problem arose with the development of hydrogen production processes using renewable energy, unpredictable, weather-dependent energy. With classical technologies, hydrogen is difficult to store due to its physicochemical properties. Thus, in the presence of air, an explosive mixture occurs, because hydrogen, mixed with oxygen in a proportion of 5–85%, is flammable. Due to the extremely small size of its molecules, hydrogen diffuses through a large number of materials, the diffusion process being favored by high pressure and heat. In prolonged contact with steel, its hydrogenation occurs, which makes it brittle. Due to its very low density, the volume energy power of hydrogen is three times lower than that of methane gas, which means that, at the same amount stored, a tank with a pressure three times higher than for methane gas. (Zhou, L., 2005; Eberle, U., Felderhoff, M., & Schueth, F., 2009).

All these problems are currently being solved by developing new materials and storage technologies (Ren, J., et al. 2017). Thus, by using new materials, diffusion losses were reduced and special seals were made, which allowed the storage of hydrogen in the form of gas under pressure (CGH_2). Liquid hydrogen tanks (LH_2) are currently used for large quantities. This reduces the reactivity of hydrogen and increases its density by approx. 800 times the gaseous state at normal temperature, but heat loss on evaporation occurs. High-cost issues involve the thermal insulation of tanks and pipes, which is necessary to avoid the occurrence of boiling, but also the formation of ice around them. (Chapelle, D., et al., 2021)

To reduce the pressure of molecular hydrogen, research is being done to dissolve hydrogen in other materials that are either liquid or solid. The principle consists in the storage of molecular hydrogen in the voids of the network of the respective materials. These hydrogenated materials are called hydrides. The problems raised by the storage of hydrogen in hydrides are the temperature and pressure of absorption and release of hydrogen, toxicity in contact with the skin and eyes, but also the weight of the tanks. Among the hydrides possible to be used for hydrogen storage are magnesium, sodium, lithium or calcium hydrides and aluminum or boron. (Dwivedi, S. K., & Vishwakarma, M., 2018)

Another means of storing hydrogen may be ammonia (NH_3), from which it can be released by a catalytic reforming process. Ammonia has the advantage of having a high density of hydrogen storage in liquid form and without special pressure and temperature problems. When mixed with water, it can be stored at normal ambient temperature and pressure. Also, there are no problems with the production, transport and distribution of ammonia, there is already a well-developed infrastructure in this regard, and by catalytic reforming, hydrogen is obtained without toxic waste. The disadvantage, however, is that the manufacture of ammonia is energy-consuming. (Badea, G., et al., 2021; Makepeace, J. W., et al., 2019; Palys, M. J., & Daoutidis, P., 2020; Ishaq, H., & Dincer, I., 2020)

In recent years, however, a new storage technique has emerged, using graphite nanotubes, called fullerenes. Hydrogen is stored between layers of graphite nanotubes with diameters of 5–10 nm. Thus, about 30 liters of hydrogen are stored for every gram of graphite. The disadvantage is that the charging time of a 25-liter tank is 4–24 hours and can only be done 4-5 times. Also, the theory of storage mode has not yet been fully elucidated. Ongoing research is also on the storage of hydrogen in metal-organic networks using a series of synthetic porous materials, polymers with attached titanium atoms, glass microspheres of 10-300 m. (Lyu, J., Kudiiarov, V., & Lider, A., 2020; Ahamed, M. I., et al., 2021)

It is worth mentioning the initiative taken by the European Commission, which financed through the public-private partnership FCH JU (Fuel Cells and Hydrogen Joint Undertaking) within the HyUnder project (Assessment of the potential, the actors and relevant business cases for large scale and seasonal storage of renewable electricity by hydrogen underground storage in Europe). The idea behind the project is to establish a European initiative to implement hydrogen-based energy technologies generated by increasing the percentage of capitalization of renewable resources. This research project focused on large-scale hydrogen storage in underground caverns, related to the energy market and existing storage technologies, and identified and analyzed areas of applicability, possible stakeholders, safety rules, regulatory framework and the societal impact on public acceptance. Within this project, case studies were carried out in several areas of Europe, including Romania. Each case study analyzed the competitiveness of hydrogen storage compared to other types of energy storage, the geological zonal potential of hydrogen storage and how this hypothesis of hydrogen storage can be implemented in the energy market. (Aşchilean, I., et al., 2021; Cernat, A., et al., 2023; Chisalita, D. A., et al., 2023; Felseghi, R. A., et al., 2021; Filote, C., et al, 2020)

FUTURE RESEARCH DIRECTIONS

In the light of the new strategies of global policies on climate neutrality of energy systems that propose plans for the energy system of the future and place clean hydrogen in the spotlight, it is necessary to address as soon as possible the following study directions:

- carrying out a study to outline the socio-economic perception, viability and public acceptance on the use of hydrogen as an energy alternative and the regional transition to sustainable and ecological hydrogen-based energy generation systems;
- identifying and inventorying the availability of hydrogen economy implementation, estimating the potential for hydrogen production, storage, transport, distribution and use at the regional level;
- in the context of the often discussed topic of the circular economy, a series of studies can be developed on how to obtain hydrogen from various wastes, but also from wastewater by using renewable energies in production processes;
- conducting feasibility studies on the implementation of energy systems that integrate hydrogen and fuel cells in the power supply of individual consumers and at the level of local communities.

CONCLUSION

Hydrogen, called the energy vector, is currently designated as the fuel of the future for transportation, for industry, for heating buildings and more. Its most important aspect is that it does not emit CO_2 and does not pollute the atmosphere when used. Currently, in the large world, the amount of hydrogen is produced mainly from fossil fuels with harmful effects on the environment. Hydrogen production is currently based on a variety of processes, but it is known that almost all of them are associated with a wide range of emissions with harmful implications for the environment, but also with a different footprint in terms of production costs.

For an overview, an important part of the processes and methods of hydrogen production can be highlighted, keeping in mind:

- hydrogen based on electricity, when it is produced by water electrolysis - in which case the greenhouse gas emissions belong entirely to the electricity production cycle;
- hydrogen based on fossil fuels - in which case a variety of processes are considered that use natural gas as a raw material by reforming or coal by

gasification. It can be appreciated that currently, this method is the most used, and with the hydrogen production is created the most important amount of harmful greenhouse gases;

- hydrogen based fossil fuels with the use - carbon capture process - a situation in which the effect of greenhouse gas emissions is greatly diminished.

After the review it can be easily highlighted that hydrogen obtained from renewable sources is the most suitable because it is the "cleanest", which is why it is the only one compatible with the objectives of each country and implicitly of the U.E. on climate neutrality. For the future, the differentiation between the hydrogen types depending on the technologies used for its production will be the basis of the strategies adopted, the main objective being to obtain clean hydrogen.

REFERENCES

Abd, A. A., Naji, S. Z., Thian, T. C., & Othman, M. R. (2021). Evaluation of hydrogen concentration effect on the natural gas properties and flow performance. *International Journal of Hydrogen Energy*, *46*(1), 974–983. doi:10.1016/j.ijhydene.2020.09.141

Abdel-Basset, M., Gamal, A., Chakrabortty, R. K., & Ryan, M. J. (2021). Evaluation of sustainable hydrogen production options using an advanced hybrid MCDM approach: A case study. *International Journal of Hydrogen Energy*, *46*(5), 4567–4591. doi:10.1016/j.ijhydene.2020.10.232

Ahamed, M. I., Shakeel, N., Anwar, N., Khan, A., Asiri, A. M., & Dzudzevic-Cancar, H. (2021). Graphene-based nanocomposite for hydrogen storage application. In *Nanomaterials for Hydrogen Storage Applications* (pp. 57–78). Elsevier. doi:10.1016/B978-0-12-819476-8.00006-2

Al-Breiki, M., & Bicer, Y. (2021). Comparative life cycle assessment of sustainable energy carriers including production, storage, overseas transport and utilization. *Journal of Cleaner Production*, *279*, 123481. doi:10.1016/j.jclepro.2020.123481

Armaroli, N., & Balzani, V. (2011). The hydrogen issue. *ChemSusChem*, *4*(1), 21–36. doi:10.1002/cssc.201000182 PMID:21226208

Aşchilean, I., Cobîrzan, N., Bolboaca, A., Boieru, R., & Felseghi, R. A. (2021). Pairing solar power to sustainable energy storage solutions within a residential building: A case study. *International Journal of Energy Research*, *45*(10), 15495–15511. doi:10.1002/er.6982

Atems, B., Mette, J., Lin, G., & Madraki, G. (2023). Estimating and forecasting the impact of nonrenewable energy prices on US renewable energy consumption. *Energy Policy, 173*, 113374. doi:10.1016/j.enpol.2022.113374

Ayers, K., Danilovic, N., Ouimet, R., Carmo, M., Pivovar, B., & Bornstein, M. (2019). Perspectives on low-temperature electrolysis and potential for renewable hydrogen at scale. *Annual Review of Chemical and Biomolecular Engineering, 10*(1), 219–239. doi:10.1146/annurev-chembioeng-060718-030241 PMID:31173524

Azam, W., Khan, I., & Ali, S. A. (2023). Alternative energy and natural resources in determining environmental sustainability: A look at the role of government final consumption expenditures in France. *Environmental Science and Pollution Research International, 30*(1), 1949–1965. doi:10.100711356-022-22334-z PMID:35925458

Badea, G., Felseghi, R., Aschilean, I., Bolboaca, A., Muresan, D., Soimosan, T., & Raboaca, S. (2015, August). Energen system for power supply of passive house: case study. In *2015 Second International Conference on Mathematics and Computers in Sciences and in Industry (MCSI)* (pp. 24-31). IEEE. 10.1109/MCSI.2015.31

Badea, G., Felseghi, R. A., & Aşchilean, I. (2021). Hydrogen-Energy Vector Within a Sustainable Energy System for Stationary Applications. In *Hydrogen Fuel Cell Technology for Stationary Applications* (pp. 1–21). IGI Global. doi:10.4018/978-1-7998-4945-2.ch001

Badea, G., Felseghi, R. A., Aşchilean, I., Răboacă, S. M., & Şoimoşan, T. (2017, December). The role of hydrogen as a future solution to energetic and environmental problems for residential buildings. In AIP Conference Proceedings (Vol. 1917, pp. 030004). AIP Publishing LLC. doi:10.1063/1.5018277

Bai, S., Tian, K., Meng, J. C., Zhao, Y., Sun, J., Zhang, K., Feng, Y., Luo, R., Li, D., & Chen, A. (2021). Reduced graphene oxide decorated SnO2/BiVO4 photoanode for photoelectrochemical water splitting. *Journal of Alloys and Compounds, 855*, 156780. doi:10.1016/j.jallcom.2020.156780

Bellocchi, S., Colbertaldo, P., Manno, M., & Nastasi, B. (2023). Assessing the effectiveness of hydrogen pathways: A techno-economic optimisation within an integrated energy system. *Energy, 263*, 126017. doi:10.1016/j.energy.2022.126017

Carrillo, R. J., & Scheffe, J. R. (2017). Advances and trends in redox materials for solar thermochemical fuel production. *Solar Energy, 156*, 3–20. doi:10.1016/j.solener.2017.05.032

CernatA.PanaC.NegurescuN.NutuC.FuiorescuD.LazaroiuG. Aspects of an Experimental Study of a Hydrogen Use at Automotive Diesel Engine. *Available at* SSRN 4315127. doi:10.1016/j.heliyon.2023.e13889

Cerniauskas, S., Junco, A. J. C., Grube, T., Robinius, M., & Stolten, D. (2020). Options of natural gas pipeline reassignment for hydrogen: Cost assessment for a Germany case study. *International Journal of Hydrogen Energy*, *45*(21), 12095–12107. doi:10.1016/j.ijhydene.2020.02.121

Chapelle, D., Maynadier, A., Bebon, L., & Thiébaud, F. (2021). Hydrogen Storage: Different Technologies, Challenges and Stakes. Focus on TiFe Hydrides. In *Advances in Renewable Hydrogen and Other Sustainable Energy Carriers* (pp. 1–10). Springer. doi:10.1007/978-981-15-6595-3_1

Chisalita, D. A., Petrescu, L., Galusnyak, S. C., & Cormos, C. C. (2023). Environmental evaluation of hydrogen production employing innovative chemical looping technologies–A Romanian case study. *International Journal of Hydrogen Energy*, *48*(32), 12112–12128. doi:10.1016/j.ijhydene.2022.06.029

Cloete, S., Ruhnau, O., & Hirth, L. (2021). On capital utilization in the hydrogen economy: The quest to minimize idle capacity in renewables-rich energy systems. *International Journal of Hydrogen Energy*, *46*(1), 169–188. doi:10.1016/j.ijhydene.2020.09.197

Cobîrzan, N., Thalmaier, G., Balog, A. A., Constantinescu, H., Timiş, I., & Streza, M. (2018). Thermophysical properties of fired clay bricks with waste ceramics and paper pulp as pore-forming agent. *Journal of Thermal Analysis and Calorimetry*, *134*(1), 843–851. doi:10.100710973-018-7092-3

Colbertaldo, P., Agustin, S. B., Campanari, S., & Brouwer, J. (2019). Impact of hydrogen energy storage on California electric power system: Towards 100% renewable electricity. *International Journal of Hydrogen Energy*, *44*(19), 9558–9576. doi:10.1016/j.ijhydene.2018.11.062

Cornwell, O., Radford, S. E., Ashcroft, A. E., & Ault, J. R. (2018). Comparing hydrogen deuterium exchange and fast photochemical oxidation of proteins: A structural characterisation of wild-type and ΔN6 β2-microglobulin. *Journal of the American Society for Mass Spectrometry*, *29*(12), 2413–2426. doi:10.100713361-018-2067-y PMID:30267362

Dawood, F., Anda, M., & Shafiullah, G. M. (2020). Hydrogen production for energy: An overview. *International Journal of Hydrogen Energy*, *45*(7), 3847–3869. doi:10.1016/j.ijhydene.2019.12.059

Dillman, K., & Heinonen, J. (2023). Towards a Safe Hydrogen Economy: An Absolute Climate Sustainability Assessment of Hydrogen Production. *Climate (Basel), 11*(1), 25. doi:10.3390/cli11010025

Dincer, I., & Zamfirescu, C. (2012). Sustainable hydrogen production options and the role of IAHE. *International Journal of Hydrogen Energy, 37*(21), 16266–16286. doi:10.1016/j.ijhydene.2012.02.133

Dong, Y., Peng, W., Liu, Y., & Wang, Z. (2021). Photochemical origin of reactive radicals and halogenated organic substances in natural waters: A review. *Journal of Hazardous Materials, 401*, 123884. doi:10.1016/j.jhazmat.2020.123884 PMID:33113752

Dwivedi, S. K., & Vishwakarma, M. (2018). Hydrogen embrittlement in different materials: A review. *International Journal of Hydrogen Energy, 43*(46), 21603–21616. doi:10.1016/j.ijhydene.2018.09.201

Eberle, U., Felderhoff, M., & Schueth, F. (2009). Chemical and physical solutions for hydrogen storage. *Angewandte Chemie International Edition, 48*(36), 6608–6630. doi:10.1002/anie.200806293 PMID:19598190

Eiler, J. M. (2020). The Once and Future Hydrogen Economy. *Nature Journal.*

European Hydrogen and Fuel Cell Technology Platform. (2005), Strategic Research Agenda. fch. https://www.fch.europa.eu/sites/default/files/documents/hfp-sra004_ v9-2004_sra-report-final_ 22jul2005.pdf

Felseghi, R. A., Aşchilean, I., Cobîrzan, N., Bolboacă, A. M., & Raboaca, M. S. (2021). Optimal Synergy between Photovoltaic Panels and Hydrogen Fuel Cells for Green Power Supply of a Green Building—A Case Study. *Sustainability (Basel), 13*(11), 6304. doi:10.3390u13116304

Felseghi, R. A., Carcadea, E., Raboaca, M. S., Trufin, C. N., & Filote, C. (2019). a). Hydrogen fuel cell technology for the sustainable future of stationary applications. *Energies, 12*(23), 4593. doi:10.3390/en12234593

Felseghi, R. A., & Papp, A. (2014). Considerations regarding hybrid systems of power generator from renewable energy sources. *Studia Universitatis Babes-Bolyai, Ambientum, 59.*

Felseghi, R. A., Şoimoşan, T. M., Filote, C., & Răboaca, M. S. (2019, b). Considerations Regarding the Green Retrofitting of Residential Buildings From Human Wellbeing Perspectives. In Retrofitting for Optimal Energy Performance (pp. 143-175). IGI Global.

Filote, C., Felseghi, R. A., Raboaca, M. S., & Aşchilean, I. (2020). Environmental impact assessment of green energy systems for power supply of electric vehicle charging station. *International Journal of Energy Research, 44*(13), 10471–10494. doi:10.1002/er.5678

Genovese, M., Schlüter, A., Scionti, E., Piraino, F., Corigliano, O., & Fragiacomo, P. (2023). Power-to-hydrogen and hydrogen-to-X energy systems for the industry of the future in Europe. *International Journal of Hydrogen Energy, 48*(44), 16545–16568. doi:10.1016/j.ijhydene.2023.01.194

Grimm, A., de Jong, W. A., & Kramer, G. J. (2020). Renewable hydrogen production: A techno-economic comparison of photoelectrochemical cells and photovoltaic-electrolysis. *International Journal of Hydrogen Energy, 45*(43), 22545–22555. doi:10.1016/j.ijhydene.2020.06.092

Hagen, M., & Azevedo, A. (2022). Climate Changes Consequences from Sun-Earth Connections and Anthropogenic Relationships. *Nature and Science, 14*(2), 24–41.

Holladay, J. D., Hu, J., King, D. L., & Wang, Y. (2009). An overview of hydrogen production technologies. *Catalysis Today, 139*(4), 244–260. doi:10.1016/j.cattod.2008.08.039

Hosseini, S. E., & Wahid, M. A. (2020). Hydrogen from solar energy, a clean energy carrier from a sustainable source of energy. *International Journal of Energy Research, 44*(6), 4110–4131. doi:10.1002/er.4930

Iordache, I. & Stefănescu, I. (2010). Tehnici Avansate De Tratare a Apei ¸Si Separare a Hidrogenului. Editura AGIR: Bucureşti, Romania.

Ishaq, H., & Dincer, I. (2019). A comparative evaluation of three CuCl cycles for hydrogen production. *International Journal of Hydrogen Energy, 44*(16), 7958–7968. doi:10.1016/j.ijhydene.2019.01.249

Ishaq, H., & Dincer, I. (2020). A comprehensive study on using new hydrogen-natural gas and ammonia-natural gas blends for better performance. *Journal of Natural Gas Science and Engineering, 81*, 103362. doi:10.1016/j.jngse.2020.103362

Iulianelli, A., & Basile, A. (Eds.). (2014). *Advances in hydrogen production, storage and distribution*. Elsevier.

Kakoulaki, G., Kougias, I., Taylor, N., Dolci, F., Moya, J., & Jäger-Waldau, A. (2021). Green hydrogen in Europe–A regional assessment: Substituting existing production with electrolysis powered by renewables. *Energy Conversion and Management, 228*, 113649. doi:10.1016/j.enconman.2020.113649

Kiani, R., Mirzaei, F., Ghanbari, F., Feizi, R., & Mehdipour, F. (2020). Real textile wastewater treatment by a sulfate radicals-Advanced Oxidation Process: Peroxydisulfate decomposition using copper oxide (CuO) supported onto activated carbon. *Journal of Water Process Engineering*, *38*, 101623. doi:10.1016/j.jwpe.2020.101623

Lagioia, G., Spinelli, M. P., & Amicarelli, V. (2023). Blue and green hydrogen energy to meet European Union decarbonisation objectives. An overview of perspectives and the current state of affairs. *International Journal of Hydrogen Energy*, *48*(4), 1304–1322. doi:10.1016/j.ijhydene.2022.10.044

Li, Y., Sun, Y., Qin, Y., Zhang, W., Wang, L., Luo, M., Yang, H., & Guo, S. (2020). Recent advances on water-splitting electrocatalysis mediated by noble-metal-based nanostructured materials. *Advanced Energy Materials*, *10*(11), 1903120. doi:10.1002/aenm.201903120

Linssen, J., & Hake, J. F. (2016). Hydrogen research, development, demonstration, and market deployment activities. *Hydrogen Science and Engineering: Materials, Processes, Systems and Technology*, 57-84.

Liu, W., Sun, L., Li, Z., Fujii, M., Geng, Y., Dong, L., & Fujita, T. (2020). Trends and future challenges in hydrogen production and storage research. *Environmental Science and Pollution Research International*, *27*(25), 31092–31104. doi:10.100711356-020-09470-0 PMID:32529621

Lyu, J., Kudiiarov, V., & Lider, A. (2020). An overview of the recent progress in modifications of carbon nanotubes for hydrogen adsorption. *Nanomaterials (Basel, Switzerland)*, *10*(2), 255. doi:10.3390/nano10020255 PMID:32024092

Maggio, G., Nicita, A., & Squadrito, G. (2019). How the hydrogen production from RES could change energy and fuel markets: A review of recent literature. *International Journal of Hydrogen Energy*, *44*(23), 11371–11384. doi:10.1016/j.ijhydene.2019.03.121

Makepeace, J. W., He, T., Weidenthaler, C., Jensen, T. R., Chang, F., Vegge, T., & David, W. I. (2019). Reversible ammonia-based and liquid organic hydrogen carriers for high-density hydrogen storage: Recent progress. *international journal of hydrogen energy*, *44*(15), 7746-7767.

Mandal, S., & Jana, A. K. (2020). Simulating reactive distillation of HIx (HI–H2O–I2) system in Sulphur-Iodine cycle for hydrogen production. *Nuclear Engineering and Technology*, *52*(2), 279–286. doi:10.1016/j.net.2019.07.033

McCarty, R. D., Hord, J., & Roder, H. M. (1981). *Selected properties of hydrogen (engineering design data)* (Vol. 168). US Department of Commerce, National Bureau of Standards. doi:10.6028/NBS.MONO.168

Mehrpooya, M., & Habibi, R. (2020). A review on hydrogen production thermochemical water-splitting cycles. *Journal of Cleaner Production*, *275*, 123836. doi:10.1016/j.jclepro.2020.123836

Midilli, A., Ay, M., Dincer, I., & Rosen, M. A. (2005). b). On hydrogen and hydrogen energy strategies: I: current status and needs. *Renewable & Sustainable Energy Reviews*, *9*(3), 255–271. doi:10.1016/j.rser.2004.05.003

Midilli, A. D. N. A. N., Ay, M., Dincer, I., & Rosen, M. A. (2005). a). On hydrogen and hydrogen energy strategies II: Future projections affecting global stability and unrest. *Renewable & Sustainable Energy Reviews*, *9*(3), 273–287. doi:10.1016/j.rser.2004.05.002

Mneimneh, F., Ghazzawi, H., Abu Hejjeh, M., Manganelli, M., & Ramakrishna, S. (2023). Roadmap to achieving sustainable development via green hydrogen. *Energies*, *16*(3), 1368. doi:10.3390/en16031368

Molnarne, M., & Schroeder, V. (2019). Hazardous properties of hydrogen and hydrogen containing fuel gases. *Process Safety and Environmental Protection*, *130*, 1–5. doi:10.1016/j.psep.2019.07.012

Momirlan, M., & Veziroglu, T. N. (2002). Current status of hydrogen energy. *Renewable & Sustainable Energy Reviews*, *6*(1-2), 141–179. doi:10.1016/S1364-0321(02)00004-7

Noor, M., Khan, D., Khan, A., & Rasheed, N. (2023). The impact of renewable and non-renewable energy on sustainable development in South Asia. *Environment, Development and Sustainability*, •••, 1–18. doi:10.100710668-023-03210-3

Notteboom, T., & Haralambides, H. (2023). Seaports as green hydrogen hubs: Advances, opportunities and challenges in Europe. *Maritime Economics & Logistics*, *25*(1), 1–27. doi:10.105741278-023-00253-1

Nuttall, W. J., & Bakenne, A. T. (2020). Towards a Hydrogen Economy. In *Fossil Fuel Hydrogen* (pp. 43–52). Springer. doi:10.1007/978-3-030-30908-4_4

Oruc, O., & Dincer, I. (2021). Assessing the potential of thermochemical water splitting cycles: A bridge towards for clean and sustainable hydrogen generation. *Fuel*, *286*, 119325. doi:10.1016/j.fuel.2020.119325

Palys, M. J., & Daoutidis, P. (2020). Using hydrogen and ammonia for renewable energy storage: A geographically comprehensive techno-economic study. *Computers & Chemical Engineering*, *136*, 106785. doi:10.1016/j.compchemeng.2020.106785

Pareek, A., Dom, R., Gupta, J., Chandran, J., Adepu, V., & Borse, P. H. (2020). Insights into renewable hydrogen energy: Recent advances and prospects. *Materials Science for Energy Technologies*, *3*, 319–327. doi:10.1016/j.mset.2019.12.002

Pezhmani, Y., Oskouei, M. Z., Rezaei, N., & Mehrjerdi, H. (2023). Optimal operation of active distribution networks hosting hybrid hydrogen-electricity refuelling stations considering water demand under a stochastic-IGDT approach. *IET Generation, Transmission & Distribution*, *17*(7), 1461–1476. doi:10.1049/gtd2.12752

Pingkuo, L., & Xue, H. (2022). Comparative analysis on similarities and differences of hydrogen energy development in the World's top 4 largest economies: A novel framework. *International Journal of Hydrogen Energy*, *47*(16), 9485–9503. doi:10.1016/j.ijhydene.2022.01.038

Posso, F., Pulido, A., & Acevedo-Páez, J. C. (2023). Towards the hydrogen economy: Estimation of green hydrogen production potential and the impact of its uses in Ecuador as a case study. *International Journal of Hydrogen Energy*, *48*(32), 11922–11942. doi:10.1016/j.ijhydene.2022.05.128

Punekar, N. S. (2018). Hallmarks of an Enzyme Catalyst. In *ENZYMES: Catalysis, Kinetics and Mechanisms* (pp. 43–51). Springer. doi:10.1007/978-981-13-0785-0_5

Razi, F., & Dincer, I. (2022). Challenges, opportunities and future directions in hydrogen sector development in Canada. *International Journal of Hydrogen Energy*, *47*(15), 9083–9102. doi:10.1016/j.ijhydene.2022.01.014

Ren, J., Gao, S., Liang, H., Tan, S., & Dong, L. (2023). The role of hydrogen energy: Strengths, weaknesses, opportunities, and threats. In *Hydrogen Economy* (pp. 3–43). Academic Press. doi:10.1016/B978-0-323-99514-6.00014-5

Ren, J., Musyoka, N. M., Langmi, H. W., Mathe, M., & Liao, S. (2017). Current research trends and perspectives on materials-based hydrogen storage solutions: A critical review. *International Journal of Hydrogen Energy*, *42*(1), 289–311. doi:10.1016/j.ijhydene.2016.11.195

Ren, X., Dong, L., Xu, D., & Hu, B. (2020). Challenges towards hydrogen economy in China. *International Journal of Hydrogen Energy*, *45*(59), 34326–34345. doi:10.1016/j.ijhydene.2020.01.163

Safari, F., & Dincer, I. (2020). A review and comparative evaluation of thermochemical water splitting cycles for hydrogen production. *Energy Conversion and Management*, *205*, 112182. doi:10.1016/j.enconman.2019.112182

Sarker, A. K., Azad, A. K., Rasul, M. G., & Doppalapudi, A. T. (2023). Prospect of Green Hydrogen Generation from Hybrid Renewable Energy Sources: A Review. *Energies*, *16*(3), 1556. doi:10.3390/en16031556

Sazali, N. (2020). Emerging technologies by hydrogen: A review. *International Journal of Hydrogen Energy*, *45*(38), 18753–18771. doi:10.1016/j.ijhydene.2020.05.021

Scheller, F., Wald, S., Kondziella, H., Gunkel, P. A., Bruckner, T., & Keles, D. (2023). Future role and economic benefits of hydrogen and synthetic energy carriers in Germany: A review of long-term energy scenarios. *Sustainable Energy Technologies and Assessments*, *56*, 103037. doi:10.1016/j.seta.2023.103037

Shadman, M., Roldan-Carvajal, M., Pierart, F. G., Haim, P. A., Alonso, R., Silva, C., Osorio, A. F., Almonacid, N., Carreras, G., Maali Amiri, M., Arango-Aramburo, S., Rosas, M. A., Pelissero, M., Tula, R., Estefen, S. F., Pastor, M. L., & Saavedra, O. R. (2023). A review of offshore renewable energy in South America: Current status and future perspectives. *Sustainability (Basel)*, *15*(2), 1740. doi:10.3390u15021740

Sillero, L., Solera, R., & Perez, M. (2022). Anaerobic co-digestion of sewage sludge, wine vinasse and poultry manure for bio-hydrogen production. *International Journal of Hydrogen Energy*, *47*(6), 3667–3678. doi:10.1016/j.ijhydene.2021.11.032

Stennikov, V. (2023, January). Transformation challenges faced by energy systems. In. AIP Conference Proceedings: Vol. 2552. *No. 1* (p. 080030). AIP Publishing LLC.

Tiwari, A. (2021). The emerging global trends in hydrogen energy research for achieving the net zero goals. *Advanced Materials Letters*, *12*(10), 1–5. doi:10.5185/aml.2021.15697

Valente, A., Iribarren, D., & Dufour, J. (2017). Life cycle assessment of hydrogen energy systems: A review of methodological choices. *The International Journal of Life Cycle Assessment*, *22*(3), 346–363. doi:10.100711367-016-1156-z

Van de Krol, R., & Grätzel, M. (2012). *Photoelectrochemical hydrogen production* (Vol. 90). Springer. doi:10.1007/978-1-4614-1380-6

Veziroğlu, T. N. (2000). Quarter century of hydrogen movement 1974–2000. *International Journal of Hydrogen Energy*, *25*(12), 1143–1150. doi:10.1016/S0360-3199(00)00038-0

Veziroğlu, T. N. (2004). IJHE thrust toward meeting the needs of implementation. *International Journal of Hydrogen Energy, 1*(29), 15. doi:10.1016/j.ijhydene.2003.09.003

Wang, H., Kong, H., Pu, Z., Li, Y., & Hu, X. (2020). Feasibility of high efficient solar hydrogen generation system integrating photovoltaic cell/photon-enhanced thermionic emission and high-temperature electrolysis cell. *Energy Conversion and Management, 210*, 112699. doi:10.1016/j.enconman.2020.112699

Wang, Z., Roberts, R. R., Naterer, G. F., & Gabriel, K. S. (2012). Comparison of thermochemical, electrolytic, photoelectrolytic and photochemical solar-to-hydrogen production technologies. *International Journal of Hydrogen Energy, 37*(21), 16287–16301. doi:10.1016/j.ijhydene.2012.03.057

Weeda, M., & Segers, R. (2020). *The Dutch hydrogen balance, and the current and future representation of hydrogen in the energy statistics*. Research Gate.

Wieczorek, A. K., Krystian, M., & Zehetbauer, M. (2006). SPD processed alloys as efficient vacancy-hydrogen systems. [). Trans Tech Publications Ltd.]. *Diffusion and Defect Data, Solid State Data. Part B, Solid State Phenomena, 114*, 177–182. doi:10.4028/www.scientific.net/SSP.114.177

Wolniak, R., & Skotnicka-Zasadzień, B. (2022). Development of Photovoltaic Energy in EU Countries as an Alternative to Fossil Fuels. *Energies, 15*(2), 662. doi:10.3390/en15020662

Wulf, C., Reuß, M., Grube, T., Zapp, P., Robinius, M., Hake, J. F., & Stolten, D. (2018). Life Cycle Assessment of hydrogen transport and distribution options. *Journal of Cleaner Production, 199*, 431–443. doi:10.1016/j.jclepro.2018.07.180

Xu, X., Zhou, Q., & Yu, D. (2022). The future of hydrogen energy: Bio-hydrogen production technology. *International Journal of Hydrogen Energy, 47*(79), 33677–33698. doi:10.1016/j.ijhydene.2022.07.261

Yadav, D., & Banerjee, R. (2018). Economic assessment of hydrogen production from solar driven high-temperature steam electrolysis process. *Journal of Cleaner Production, 183*, 1131–1155. doi:10.1016/j.jclepro.2018.01.074

Ying, Z., Wang, Y., Zheng, X., Geng, Z., Dou, B., & Cui, G. (2020). Experimental study and development of an improved sulfur–iodine cycle integrated with HI electrolysis for hydrogen production. *International Journal of Hydrogen Energy, 45*(24), 13176–13188. doi:10.1016/j.ijhydene.2020.03.037

Yuzer, B., Selcuk, H., Chehade, G., Demir, M. E., & Dincer, I. (2020). Evaluation of hydrogen production via electrolysis with ion exchange membranes. *Energy*, *190*, 116420. doi:10.1016/j.energy.2019.116420

Zhiznin, S. Z., Shvets, N. N., Timokhov, V. M., & Gusev, A. L. (2023). Economics of hydrogen energy of green transition in the world and Russia. Part I. *International Journal of Hydrogen Energy*, *48*(57), 21544–21567. doi:10.1016/j.ijhydene.2023.03.069

Zhiznin, S. Z., Vassilev, S., & Gusev, A. L. (2019). Economics of secondary renewable energy sources with hydrogen generation. *International Journal of Hydrogen Energy*, *44*(23), 11385–11393. doi:10.1016/j.ijhydene.2019.03.072

Zhou, L. (2005). Progress and problems in hydrogen storage methods. *Renewable & Sustainable Energy Reviews*, *9*(4), 395–408. doi:10.1016/j.rser.2004.05.005

ADDITIONAL READING

Ball, M., Basile, A., & Veziroglu, T. N. (Eds.). (2015). *Compendium of hydrogen energy: hydrogen use, safety and the hydrogen economy*. Woodhead Publishing.

Barbir, F. (2012). *PEM fuel cells: theory and practice*. Academic press.

Barbir, F., Basile, A., & Veziroglu, T. N. (Eds.). (2015). *Compendium of Hydrogen Energy: Hydrogen Energy Conversion*. Woodhead Publishing.

Graham, T. (2020). *Tomorrow's Energy, Renewables Alone? Complementary Resources for Tomorrow: Proceedings of Energy and Resources for Tomorrow 2019*. University of Windsor.

Gupta, R., Basile, A., & Veziroglu, T. N. (Eds.). (2016). *Compendium of Hydrogen Energy: Hydrogen Storage, Distribution and Infrastructure*. Woodhead Publishing.

Ito, K., Li, H., & Hao, Y. M. (2016). *Hydrogen Energy Engineering: A Japanese Perspective*. Research Gate.

Miranda, P. E. (Ed.). (2018). *Science and Engineering of Hydrogen-Based Energy Technologies: Hydrogen Production and Practical Applications in Energy Generation*. Academic Press.

Scipioni, A., Manzardo, A., & Ren, J. (Eds.). (2017). *Hydrogen economy: supply chain, life cycle analysis and energy transition for sustainability*. Academic Press.

Seelam, P. K., Rathnayake, B., Pitkäaho, S., Turpeinen, E., & Keiski, R. L. (2020). Overview on recent developments on hydrogen energy: Production, catalysis, and sustainability. In Current Trends and Future Developments on (Bio-) Membranes (pp. 3-32). Elsevier. doi:10.1016/B978-0-12-817110-3.00001-1

Sherif, S. A., Goswami, D. Y., Stefanakos, E. L., & Steinfeld, A. (Eds.). (2014). *Handbook of hydrogen energy*. CRC Press. doi:10.1201/b17226

Zohuri, B. (2019). Cryogenics and liquid hydrogen storage. In *Hydrogen Energy* (pp. 121–139). Springer. doi:10.1007/978-3-319-93461-7_4

KEY TERMS AND DEFINITIONS

Alternative fuel: Also known as non-conventional fuel, means any chemical substance that can be used as a fuel, other than conventional fuels, derived from fossil fuels, i.e. from crude oil, coal and natural gas.

Circular economy: In the linear economy, raw natural resources are taken, transformed into products and get disposed of. On the opposite, a circular economy model aims to close the gap between the production and the natural ecosystems' cycles – on which humans ultimately depend upon. This means, on one hand, eliminating waste – composting biodegradable waste or, if it's a transformed and non-biodegradable waste, reusing, remanufacturing and finally recycling it. On the other hand, it also means cutting off the use of chemical substances (a way to help regenerate natural systems) and betting on renewable energy.

Electric Vehicle (EV): A vehicle uses one or more electric motors or traction motors for propulsion. An electric vehicle can be powered by a collection system with electricity from extra-vehicle sources, or it can be self-powered with a battery, solar panels, or an electric generator or fuel cell to convert fuel into electricity. EVs include, but are not limited to, road and rail vehicles, surface and underwater vessels, electric aircraft, and electric spacecraft.

Energy system: A system primarily designed to supply energy services to end-users.

Fuel cell vehicles: A type of electric vehicle that uses fuel cells to power an internal electric motor. Fuel cells produce electricity, usually using oxygen from the air and hydrogen from a tank located in the car.

Fuel cell: A device that continuously changes the chemical energy of a fuel (such as hydrogen) and an oxidant directly into electrical energy.

Hydrogen: A nonmetallic gaseous chemical element with atomic number 1 that is the simplest and lightest of the elements.

Hydrogen fuel: This refers to hydrogen that is burned as a fuel with oxygen. It can be a zero-carbon fuel, provided it is created in a process that does not involve carbon.

Hydrogen economy: An envisioned future in which hydrogen is used as an energy vector for power systems, fuel for heat production, fuel for hydrogen vehicles, medium for energy storage, and for long-distance transport of energy. In order to phase out fossil fuels and limit global warming, hydrogen can be created from water using intermittent renewal sources such as wind and solar, and its combustion only releases water vapor to the atmosphere.

Sustainable development: Economic development that is conducted without depletion of natural resources.

Chapter 2
Clean and Efficient Transportation With Hydrogen Fuel Cell Electric Vehicles

Emine Teke Öner
Ataturk University, Turkey

Ayşe Bayrakçeken Yurtcan
Ataturk University, Turkey

ABSTRACT

The number of disasters caused by global warming has increased in recent years, making it important to accelerate the studies on alternative and clean energy conversion systems to decrease these detrimental effects. Conventional transportation systems mainly based on fossil fuels seem to be the most critical issue for the environment. Electric vehicles are promising clean candidates that can be used for transportation. Electric vehicle technology can be classified into battery electric vehicles (BEVs), hybrid electric vehicles (HEVs), plug-in hybrid electric vehicles (PHEVs), and fuel cell electric vehicles (FCEVs). In recent years, due to the development in nanotechnology and material science engineering, newly designed battery and fuel cell-based technologies draw attention. This chapter aims to give brief information about the electric vehicle types, a comparison of the existing systems, hydrogen fuel cell electric vehicles and drawbacks of these systems, and the current status of the hydrogen economy.

DOI: 10.4018/978-1-6684-6721-3.ch002

INTRODUCTION

One of the biggest problems of our era is global warming. Studies conducted by scientists on this subject show that human beings have dragged the world they live into uninhabitable with their own hands. Today, the use of renewable energy is supported by the governments of the countries, and intensive studies are carried out on increasing energy efficiency. In scientific research, it is estimated that the energy-based CO_2 emission rate will increase by 20% in 2035, which means that the earth's temperature will increase by 3.6 °C on average in the long run. It is seen that this value, which is determined as 2 °C at the international level, is well above the target (Karlsson et al., 2020) (Li & Zhang., 2009). Global warming causes the earth's temperature to increase by 0.3 °C on average every ten years, and this causes an average of 150,000 deaths per year. An average increase of 2 °C in the earth's temperature is a severe problem that will expose millions of people to hunger, malaria, flood disasters, and freshwater scarcity (Kovats & Haines 2005). The severity of this situation has made it necessary to turn to renewable energy sources in order to reduce the number of greenhouse gases released into the atmosphere all over the world. There are many measures that can be taken to reduce greenhouse gases released into the atmosphere. One of them is to reduce the use of fossil fuels. The transportation sector is known as one of the largest fuel consumers in the global energy market. Currently, there are approximately 1 billion vehicles in the world and most of these vehicles are based on fossil fuels. In many families, each individual has his own vehicle. This situation has considerably increased the number of motor vehicles in traffic. The environmental damage caused by fossil fuels used in motor vehicles is too great to be ignored. Nowadays, it has become necessary to use nature-friendly, renewable resources instead of fossil energy resources, whose reserves are decreasing and polluting the environment. In addition, the orientation towards clean and sustainable energy sources is a very effective driving force for the continuation of economic development and improvement of living conditions. Studies show that electric vehicles create less pollution than vehicles using fossil fuels (Verma et al., 2021). This gives even better results if the electric vehicle is charged with electricity produced entirely from renewable energy. For this reason, electric vehicles attract a lot of attention around the world and many improvements are being made in this area. When examining electric vehicle technology, it is seen that it is much more advanced than its counterparts designed 10 years ago. In this context, when the studies on electric vehicles are examined, it is quite remarkable the way fuel cell technology has been covered in automobile applications. The applications of PEM fuel cells in electric vehicles are increasing day by day and are the subject of many R&D studies. Studies conducted today show that among the electric vehicles targeted for widespread use in the future, fuel-cell electric vehicles will effectively exist.

CASE STUDY

Historical Advance of Vehicles

After the first discovery of steam-powered vehicles by Nicolas Cugnot in 1769, Oliver Evans developed and patented high-pressure steam engines in 1790. Later, Richard Trevithick used the high-pressure steam engine in the manufacture of locomotives in 1803. At the same time, Francis and Freelan Stanley brothers built the steam cars. Electric cars, on the other hand, were first made in the late 1800s, but could not be used effectively because sufficient power could not be produced for the vehicle. For example, electric cars produced in the early 1900s could only travel a range of 32 km before being charged. Due to the fact that electric vehicles have a short range and are very costly compared to the time of steam vehicles, the time of internal combustion engines started to not shut down for a long time. The adventure of internal combustion engines, which started with the invention of the 4-piston engine of inventor Nikolas Otto in 1876, continued with the engine that Karl Benz started to try in 1878 and patented as Karl's motorwagen in 1886. Mass production of internal combustion engine vehicles that are replacing electric and steam engines due to their high range and low production cost began in the early 1900s (Bailey, 2015). After using fossil fuels with internal combustion engines for a long time, environmental factors such as global warming, increase in CO_2 gas emissions, and also the decrease in fossil fuel reserves have increased the interest in electric vehicles again. Electric vehicle technology, which continues to develop in this process, has started to be used again with hybrid vehicles in the 2000s (Bae & Kim, 2017). Today, many companies such as Toyota, Ford, Renault, Tesla, and BMW continue to develop electric vehicle technology and launch these cars to the market (Slowik, Pavlenko & Lutsey 2016).

ELECTRIC VEHICLES

Electric vehicles are vehicles that store electrical energy in electrochemical systems, use it when needed, and provide propulsion with electric motors. Electric vehicle technology can be classified into battery electric vehicles (BEVs), hybrid electric vehicles (HEVs), plug-in hybrid electric vehicles (PHEV), and fuel cell electric vehicles (FCEVs). When talking about electric vehicles, the first thing that comes to mind is batteries. Batteries are classified according to their chemical structure. These are primary and secondary batteries. Primary batteries are non-rechargeable batteries; they have a single-use feature. Once discharged, they cannot be recharged. Primary batteries are common in small and portable devices (Mantell, 1982). Secondary batteries are briefly defined as rechargeable batteries. Since the reaction in these batteries is reversible, these batteries can be charged and reused. In the first designed electric vehicles, primary batteries were used, which is not sustainable. With the invention of secondary batteries, electric vehicles have become practically usable (McCullough, 1989). Electric vehicle technology can be used not only with electric motors but also with internal combustion engines, auxiliaries, and sub-systems. Its use of different energy sources and energy management systems, as well as its compatibility with new generation technologies, have made it an important place in the sector compared to traditional (conventional) vehicles. Other than batteries, fuel cell-based electric vehicles are also good candidates for replacing conventional motor vehicles. Fuel cells are devices that convert the chemical energy of the fuel into electricity directly. Among the other fuel cell types, proton exchange membrane (PEM) fuel cells seem to be more promising for mobile and portable applications due to their outstanding properties such as low operating temperature, high power density, and durability.

- **Battery electric vehicles (BEVs)**

BEVs are also known as vehicles that exclusively run on electricity because in BEV types, all of the necessary electricity is provided by the battery and these cars don't utilize a petrol tank. A sizable rechargeable battery that doesn't release hazardous chemicals into the atmosphere makes up the battery system in use (Wolfram & Lutsey, 2016). Researchers have concentrated on the manufacture of the batteries used in these cars as well as the environmental harm caused by discharged batteries. Similar to mobile applications, these cars use a charging connector to allow the battery to be charged from mains power or any other external energy source (Mishra et al., 2021). Issues with charging are another area where BEVs are being improved. It takes 6 to 8 hours to fully charge normally, and 20 to 40 minutes to fully charge quickly. These timeframes are substantially longer than the time it takes for an internal combustion engine car to fill up with fuel. The amount of time it takes to charge a car varies based on the battery's capacity, how it's charged, and whether a serial or parallel connection is being utilized. Another area that needs improvement is the shorter battery life in fast charging applications compared to conventional charging. The performance of BEVs is largely influenced by two things. The heat management system is one of them, and battery capacity is the other. In order to determine the battery's performance and lifespan, temperature control is crucial. The range of the car is between 100 and 400 kilometers, depending on the battery that is installed. These vehicles are suited for short distances and stop-start situations because they efficiently transfer electrical energy into mechanical energy (Thomas, 2009).

- **Hybrid electric vehicles (HEVs)**

Hybrid electric vehicles (HEVs) incorporate multiple power and energy sources, making them a combination of different technologies. These vehicles utilize energy sources such as batteries, flywheels, regenerative braking, and power sources like fuel cells, ultra-capacitors, and internal combustion engines in complex ways depending on the technology used. By employing various energy management strategies, longer distances can be covered while reducing emissions and fuel consumption. HEVs provide a backup system for the primary fuel, such as diesel and gasoline, through their secondary source. They are classified into series, parallel, and dual HEVs based on the energy sources applied to the driveline (Chan & Wong, 2004).

The electric motor and IEC immediately supply the electricity needed to start the engine and charge the battery in series HEVs. The electric motor serves as the only source of traction and driving force in this setup. These cars feature a modest internal combustion engine (ICE), a sizable electric motor, and a sizable backup battery (Zhang et al., 2014). As the ICE and transmission are also not mechanically connected in serial HEVs, the ICE can function as efficiently as possible. The number of conversion stages between the ICE and gearbox in this model lowers overall efficiency and increases the cost of the battery and its parts. The lack which arises in this design is its essential drawback (Chau &Wong 2002).

To provide traction in parallel HEVs, the rear driveshaft is connected in parallel to both the electric motor and the ICE. The ICE serves as the main driveline source in these cars. The design's electric motor functions as a support to boost traction by maintaining optimal fuel efficiency. The most notable benefits of this car type are that parallel HEVs have a modest battery backup that lowers overall costs as well as charging the battery when in regenerate brakes and propulsion function (He et al., 2017).

Models of dual HEVs include a combination of series and parallel hybrids. In this configuration, there are typically two different powertrain arrangements. One is an electric propulsion system made up of an electric motor, battery, and generator, and the other is an internal combustion engine (ICE) and a generator coupled by a gear assembly (Yang et al., 2016) (Fredriksson, Weiefors & Egardt, 2002). The ability to function in both serial and parallel modes is the main benefit of dual-mode HEVs. Yet, in practice, this feature makes it the most complex and expensive sort of HEV model. As a result, parallel HEV configurations are the most common.

- **Plug-in hybrid electric vehicles (PHEVs)**

Plug-in hybrid electric vehicles (PHEVs) can run on both batteries and gasoline and are designed with batteries that can be charged at a charging station or home, with an ordinary plug, or using regenerative braking (Fernández, Cilleruelo &

Martinez, 2016). PHEVs use battery power for short-distance driving, producing no exhaust emissions, but rely on gasoline power for longer distances once battery power is depleted. PHEVs have greater range compared to battery electric vehicles (BEVs), but their high initial cost and non-environmentally friendly emissions at the end of production make them less popular. Researchers are conducting various studies to address these challenges (Hao et al., 2020).

- **Fuel cell electric vehicles (FCEVs)**

Fuel cell electric vehicles (FCEVs) use fuel cells instead of batteries to power the electric motor for traction, with hydrogen being the fuel source (Das, Tan &Yatim, 2017). Fuel cells are more efficient than conventional internal combustion engines, and FCEVs have higher efficiency than ICEs (Emadi, Williamson & Khaligh, 2006). The proton exchange membrane (PEM) fuel cell is the most widely used fuel cell type in FCEVs, due to its zero emissions, quiet operation, high power density, and flexible operating range characteristics. However, FCEVs are dependent on hydrogen production, require complex storage technologies, and have high installation costs (Offer et al., 2010).

COMPARISON OF ELECTRIC VEHICLES

Environmental problems such as global warming and greenhouse gas emissions are currently on the agenda of the world, and one of the biggest causes of socioeconomic difficulties arising from these is fossil fuel-based energy consumption. One of the areas where energy consumption from this fossil fuel is the highest is in the transportation sector. The use of greener technologies in this area has become mandatory to control this situation. In this context, the carbon emission of all EV types produced is lower than that of internal combustion engine technology. This is the basic requirement for improving EVs and making their use more widespread by eliminating their shortcomings.

BEVs attract a lot of attention from researchers in this context. These vehicles are seen as environmentally friendly technology with zero carbon emissions (Sudjoko et al., 2021). However, some researchers are working on the environmental effects of the batteries used in these vehicles and are focusing on the effects of this situation. Although no waste product is formed during the use phase, the environmental effects of the battery system in the vehicle architecture after the end of its life and during the production phase have been another research topic. In these vehicles, which are generally referred to as clean technology during use, battery charging times and the range covered are the main limiting factors in making use of them more common.

Hybrid mechanisms used in HEVs have improved the range distances, but although these technologies have low emissions, they have more emissions compared to BEVs. Since the vehicle is designed as a combination of different technologies in hybrid structures, their costs are high (Zeng et al., 2019). Among the series, parallel, and dual HEVs types, parallel HEVs are the most common.

PHEVs have a longer range than the BEV model, but these vehicles have higher initial costs and are not as environmentally friendly as BEVs as they produce emissions after use. Intensive studies are carried out on battery technologies that can be used more efficiently in order to eliminate these problems (Hoffert et al., 2002). It is also noteworthy that the fuel consumption amount of PHEVs is above the value mentioned by automobile manufacturers.

Fuel cells are relatively more efficient than ICE, which is used extensively today, thus making the use of fuel more economical (Bauman & Kazerani, 2008). PEM fuel cell, which is frequently used in FCEVs, is an environmentally friendly technology with zero emission features (Alaswad et al., 2020). The main problem with FCEVs is the complex storage technology for fuel and high installation costs. *Table 1.* summarizes the different characteristics of EV types (Chan, 2002).

Table 1. Different characteristics of EV types

	BEV	**HEV**	**FCEV**	**PHEV**
Energy system	Battery	Battery	FC	Battery
	UC	ICE	Battery	ICE
		UC	UC	UC
Energy source	Charging power station	Fuel pump	Hydrogen fuel	Charging power station
		Gasoline		Fuel pump
				Gasoline
Traction power	Electric motor	Electric motor	Electric motor	Electric motor
Advantage	Zero emissions	Less emissions	Zero emissions	Less emissions
	Independence on fossil fuels	Long- distance covered	High efficiency	Long- distance covered
Disadvantage	High initial cost Battery replacement	Dependence on fossil fuels Higher cost	Hydrogen fuel storage High cost of fuel	Dependence on fossil fuels Higher cost
Major issue	Battery management Charging time and charging facilities Cost	Control and management of multiple energy systems utilization Battery sizing	Control and management of multiple energy systems utilization Battery management	Fuel cell cost Hydrogen production and storage Battery life

HYDROGEN FUEL CELL ELECTRIC VEHICLE

The world is in search of clean energy sources and seeks to diversify its energy supply and find cleaner sources in the transportation sector, like many other sectors. However, alternative fuels were not available everywhere, and different engine technologies are required for the efficient use of fuels such as ethanol, gasoline, biodiesel, or methane. Hydrogen, which is mostly mentioned today and seen as the energy source of the future, can be produced from all these raw materials as well as from many other sources (Kalamaras & Efstathiou, 2013). Hydrogen is not easily found in nature like fossil fuels. Although it is the most abundant element in the universe, it forms compounds with other elements due to its high energy, and therefore it can be produced from any primary energy source, in which case it brings the ways of obtaining hydrogen. The hydrogen obtained by various processes can then be used as fuel for direct combustion in an internal combustion engine or a fuel cell, and only water is formed as a by-product (Edwards et al., 2008) (Ren, Zhou & Ou 2020). Hydrogen has a very high energy content and globally it is considered an alternative to fossil fuels as well as an environmentally harmless secondary form of renewable energy (Nicoletti et al., 2015) (Veziroğlu & Şahin, 2008). In this context, cheap production methods are needed to obtain hydrogen and much research is carried out. When H2 production is examined, many methods and classifications are seen in the literature. One of the areas of use of hydrogen today is EV types that enter our lives as nature-friendly technologies. Hydrogen used in FCEVs, one of the EV types, provides zero emissions in these vehicles (Choi et al., 2020). When we look at the methods of obtaining hydrogen in today's technology, we can divide the general title into two according to the source we obtained. It can first be obtained from fossil fuels by hydrocarbon reforming and pyrolysis methods. In the hydrocarbon reforming process, the relevant chemical techniques are steam reforming, partial oxidation, and auto-thermal steam reforming.

Hydrogen production from renewable sources such as biomass or water is the second method. When we examine the literature, methods using biomass as raw materials can be classified as thermochemical and biological processes (Das & Veziroğlu, 2001). Thermochemical technology mainly includes pyrolysis, gasification, combustion, and liquefaction, while biological processes mainly include direct and indirect bio-photolysis, dark fermentation, photo-fermentation, and sequential dark and photo-fermentation. When water is used as raw material, methods such as electrolysis, thermolysis, and photo-electrolysis that can produce H_2 through the decomposition of water are encountered in the literature. Today, there are many studies on the more efficient and affordable production of hydrogen from renewable sources, which is seen as an alternative to fossil sources and is also described as clean energy. In this context, green hydrogen studies on the basis of countries also draw attention (Nikolaidis & Poullikkas, 2017).

Another remarkable issue regarding hydrogen technologies is the storage of hydrogen. Hydrogen can be stored by different methods such as gas storage, liquid storage, solid storage, metal hydride storage, metal-organic framework, and carbon nanotubes (Ross, 2006). Today, the storage of gaseous forms of hydrogen is widely used. Storage in gaseous form is carried out at a pressure of 700 bar, and storage in liquid form at a very low temperature of 5K (−268.15 °C) (Verma et al., 2021). Solid-state storage requires absorption in the carrier material to form hydride or surface absorption (Rusman & Dahari, 2016). It is more effective when storing large amounts of hydrogen at moderate temperature and pressure. In an adsorption process, a gas molecule interacts with several atoms on the surface of a solid, where it is bound and reversibly released as needed. In general, the subject of hydrogen storage is open to development and many researchers are working on this subject (Verma et al., 2021). Studies on these environmentally friendly technologies continue in the literature, gaining momentum. Hydrogen comes to the forefront as the most efficient and useful energy source among alternative fuels in research to reduce the use of fossil fuel-based energy (Bicer & Dincer, 2018). Hydrogen is seen worldwide as a solution to future energy problems. Hydrogen energy and fuel cells are seen as the most promising clean, alternative energy source of the future, as they have low or zero carbon emissions and environmental pollution, depending on whether hydrogen is produced from non-renewable or renewable energy sources. In this context, PEM fuel cells, also known as hydrogen fuel cells, have attracted great interest in recent years.

FCEV is a type of electric vehicle that uses a fuel cell to power the electric motor (İnci et al., 2021). Global pollution will almost never occur when hydrogen was used as a fuel in a vehicle with FCEV (Fayaz et al., 2012) (Desideri & Yan, 2013) (Xiaoluan et al., 2022). Studies on hydrogen fuel cell electric motors have focused on the best features of battery-powered electric vehicles such as zero emissions, high efficiency, quiet operation, and long life, with long range and fast refueling time of internal combustion engine vehicles and about aforementioned combining features. Studies on this environmentally friendly approach are continuously accelerating in the literature (Najjar, 2013). A FCEV consists of a fuel cell and an electric motor. The stationary fuel cell converts chemical energy directly into electrical energy (Wang et al., 2011). A hydrogen fuel cell is the most popular type used in this type of vehicle (Alaswad et al., 2016) (Cunanan et al., 2021). It consumes hydrogen and oxygen as fuel and produces only water vapor and heat as exhaust leftovers (Schwarze et al., 2017). Therefore, the hydrogen fuel cell vehicle produces no greenhouse gas (GHG) emissions (Liu et al., 2018). Because hydrogen can be produced by a variety of methods, such as steam methane reforming and electrolysis of water, the adoption of FCEVs will significantly reduce a country's dependence on foreign oil (Choi et al., 2020). Unlike battery electric vehicles that have limited range and require long charging times, FCEVs can be recharged in minutes and have a range of up to several hundred miles (Forrest et al., 2020). FCEVs have been the focus of attention several times in the automobile industry, and this technology has been considered by many researchers as "technology of the distant future"fuel cell was first adapted to motor vehicles by General Motors (GM) in 1966. The project resulted having been called; as a result of two years of research led by Dr. Crai Marks. In the system, 32 fuel cell modules with a continuous output of 32 kW were used and a power of 160 kW was obtained. Pure liquid hydrogen and liquid oxygen were used as fuel in the vehicle. The vehicle produced has reached a maximum speed of 70 MPH and had a range of 120 miles. After this attempt, many car companies have worked on motor vehicles working with these systems in certain periods of history (Ohno, 2019). In 1994, Daimler-Benz introduced the vehicle produced under the name of New electric car number 1 (NECAR-1) to the public (Oin, Brooker & Raissi, 2017). The fuel cell used in the vehicle was manufactured by Ballard, one of the leading PEM fuel cell manufacturers. The NECAR 1 used a 50 kW PEM fuel cell and used a compressed hydrogen tank in the vehicle, which stores gas at 300 bar and reaches speeds of 56 mph and a range of 81 miles. In the following years, many major automakers such as Toyota, GM, Mazda, Ford, Honda, Nissan, and Volkswagen have also implemented fuel cell vehicle projects(Sierzchula et al., 2012). They used fuel cell systems with power ranging from 10 to 75 kW in the vehicles they produced. In these research and development studies, a vehicle range of up to 310 miles was achieved. In the early 2000s, Toyota started to lease the FCHV it produced for the first time in the USA and Japan on a limited basis (Aso, Kizaki & Mizuno, 2009). The powertrain of these vehicles consisted of a 90 kW fuel cell and a nickel-metal hybrid battery. When the vehicles were driven at low speed, they were powered only by the FCHV battery, but when higher performance was required, the fuel cell and battery powered the vehicle together. The combined range of the fuel cell and battery in the FCHVs was 155 miles. Since then, major automakers Toyota, Daimler, Ford, GM, Nissan, Kia, Honda, and Hyundai have undertaken significant studies to test the real-world performance of fuel cell vehicles. The first fuel cell vehicle announcement to the market was made by Toyota and Hyundai in 2015 (Budde, Alkemade & Hekkert, 2015). The limited rental program for these vehicles started in 2014. This process first started in California. This is because of the fact that California has public hydrogen fueling stations. Both models released (Tesla Model S-85 with a range of 265 miles) showed similar ranges, surpassing the best battery-electric car on the current market. The refueling time for both models is comparable to petrol vehicles. Toyota plans to decrease the price to achieve market penetration of these vehicles in the 2020s (Asif & Schmidt, 2021). Despite the potential benefits of FCEVs, there are technical challenges that need to be addressed before they are widely adopted, and it is necessary to have sustainability in that these issues should be addressed before they become widespread (Saritas, Meissner & Sokolov, 2019). These challenges are fuel cell durability, fuel cell efficiency, fuel cell cost, and performance, as well as the storage of hydrogen used as fuel (Miller et al., 2020). In the current situation, a maximum endurance of 3,900 hours has been achieved on road with the fuel cell. However, this value must reach 5000 hours of endurance, which can travel approximately 150000 miles. For this, it is necessary to improve the catalyst and electrode efficiency, increase the stability of the proton exchange membrane used in the system and improve the adaptation process to the commonly used vehicle structure. Today, the combustion efficiency of vehicles using fossil fuel-based is around 25%, while it is around 60% in FCEVs (Wang et al., 2021) (Dash, et al., 2022). However, while 65% of efficiencies have been achieved in some studies, the scientific world is working to increase this value even more in the near future (Qin, Raissi & Brooker, 2014). For the widespread use of FCEVs, the amount of charge of the high-cost platinum group elements used in the electrodes should be reduced. In addition, the hydrogen used as fuel in these vehicles must be stored, transported, and loaded into vehicles safely. When these conditions are improved, the use of FCEVs will be expanded in a sustainable manner (Muthukumar et al., 2021).

DRAWBACKS OF FCEVS

Fuel cells have the technology that is constantly being researched as clean energy systems and is being developed a little more every day with many prototypes (Conte et al., 2001). In recent years, Europe and the USA have been working to advance this technology and make it more usable. However, some difficulties need to be overcome in order to commercialize this technology. Although the main problem here is considered to be high fuel cell costs and cell durability, one of the biggest problems encountered in this technology is hydrogen infrastructure and storage of hydrogen (Fan, Tu & Chan, 2021). All these problems in front of this technology are the main factors limiting its commercialization and widespread use today.

One of the most promising technological developments of our time is fuel cell studies. However, when the cost item, which is one of the problems in front of the widespread use of this technology, is evaluated, the cost of a fuel cell; the cost of the manufacturing plant and its equipment can be classified as the cost of workers and the procurement costs of fuel cell components (Frenette & Forthoffer, 2009). When these topics are evaluated, labor and capital costs can be reduced by mass production, but the costs arising from fuel cell components are technological developments and expenses related to the market. A fuel cell is obtained by combining different materials such as a catalyst, bipolar plate, and membrane in optimum conditions and order. PEMFCs are generally used in vehicles due to priorities such as operating temperature and usage advantages (Acar, Beskese & Temur, 2021). Pt catalysts are generally preferred in PEMFCs. Because the best performance in PEMFCs is obtained in the presence of these catalysts. Platinum (Pt), used as a catalyst, is a rare metal with an annual production rate of approximately 250 tons and is mined in South Africa, Russia, and North America (Smith et al.,2022). World Pt reserves are estimated at more than 30,000 tons. Platinum is a very valuable element and therefore Pt can be recycled at the end of the vehicle's useful life. Many researchers are conducting technological studies that can reduce the recovery of Pt or the amount of Pt used with different hybrid or bimetallic catalyst trials. The positive developments to be experienced as a result of these studies will ultimately affect the cost of fuel cells (Öner, Öztürk & Yurtcan, 2020) (Daş & Yurtcan, 2016) (Çelik, et al., 2020). Also, the fuel cell is a complex system and there are many parameters that can be worked on in this structure. These studies aim to design the optimum fuel cell and improve its durability by increasing its power density. Scientific studies on this subject will have a positive impact on the cost. In particular, the reduction of platinum in the catalyst layer will affect the overall cost of the fuel cell and contribute to mass production in the transportation industry (Thompson & Papageorgopoulos, 2019). Nowadays, platinum-free catalyst development studies are carried out in line with the aim of reducing fuel cell costs. In these technologies, different metals such as cobalt, iron, or hybrid structures are tried instead of Pt. In fuel cells, the catalyst layer is the heart of the system. Because the electrochemical reactions that produce the energy take place on these surfaces. In this context, it is aimed to reduce fuel cell cost and increase fuel cell performance with many different studies.The FlowCath fuel cell design, developed by ACAL energy in England and patented by ACAL Energy, is a different work. In this study, a liquid polymer cathode solution was used to replace the platinum-based solid cathode used in standard PEM fuel cells (Wilberforce et al., 2017). In addition, many studies are carried out on the proton-permeable membrane used in the system. Although perfluorosulfonic acid membranes are generally used, different trials are also carried out to increase the performance (Suzuki, Shoji & Tsuru, 2021). ITM Power, an energy storage and clean fuel company founded in the UK in 2001, has received another patent to demonstrate the high power densities they have achieved by replacing perfluorosulfonic acid membranes, which are also commonly used in industry, with a membrane produced using ionic polymers. While the fuel cell proposed by the ITM Power fuel cell was $35/kW, Imperial College patented a new fuel cell stack design. This new patented Flexi-planar design is an array of layered printed circuit board materials glued on top of each other to form a fuel cell stack with internal fuel, water and air channels (Wilberforce et al., 2017).

Another issue to be considered in fuel cell systems is durability. The first problem that comes to mind in terms of durability in fuel cells is water management in the cell (Vincent, Lee & Kim, 2020). Water management that cannot be provided at the optimum level adversely affects battery performance in all conditions. For example, when there are floods in the system, excess water cannot be removed from the cell and may cause the channels on the bipolar plate to become clogged. In this case, the inability of the reacting gases to reach the catalyst surface effectively limits the performance or vice versa, when the humidification in the system is insufficient, the proton permeability limits the performance, so it is very important to be at the optimum level in both cases. In addition, controlling the system conditions such as the flow rates of the gases fed to the system and the temperature can affect the performance and life of the fuel cells (Wu et al. 2020). Durable and affordable bipolar plates can be developed to increase the mechanical durability of PEM fuel cells. Most bipolar plates commonly used are composed of graphite due to their ability to resist corrosion and their high surface conductivity in the PEM fuel cell environment (Wlodarczyk, 2022). However, due to the brittle nature of graphite and the intense vibration and load occurring in the transportation sector, cost-effective and durable bipolar plates that can replace graphite from metals, metal alloys, and carbon-based composites can be developed recently (Baroutaji et al., 2014). The advantages of carbon-based composites over metals for fuel cells used in the transportation industry are numerous. Because these structures can be made thinner for higher power density and mechanical strength (Alaswad et al., 2016). It has been found that metals such as stainless steel, titanium, nickel and aluminum are susceptible to corrosion in their uncoated use (Tawfik, Hung & Mahajan, 2007). When the corrosion behavior of metallic flow plates was examined, it was seen that the fuel cell increased the electrical resistance, and decreased its efficiency and power output. In recent years, carbon-based and metal-based coatings have been researched. Studies have been carried out on conductive polymers, graphite, and composite coatings in carbon-based coatings (De Las Heras et al., 2009) (Tawfik, Hung & Mahajan, 2007) (Yang et al., 2010). In addition, many studies have been carried out on metal-based coatings such as metal carbides, metal nitrides, and noble metals (Dinh et al., 2019) (Song et al., 2020) (Ren, Zhang & Wang, 2012)

One of the durability issues is carbon erosion in the catalyst layer. The most promising electro-catalyst applied for PEMFC is carbon black (Pt/C) supported Pt catalysts. However, when operating at high-temperature values in high current density regions, Pt nanoparticles can be separated from the carbon support and performance drops can be observed (Tan et al., 2021). Can provide higher durability than graphene carbon black, which has been mentioned a lot lately and attracts attention with its properties (Xu et al., 2020).

When we evaluate the membrane, which is another fuel cell component, in terms of durability, the studies emphasize the importance of optimum hydration. Overhydration will cause safety issues such as voltage loss at high current density, voltage instability at low current density, and unreliable starting under freezing conditions. The durability of fuel cells to be used in automotive applications against different environmental conditions is important. In this context, alternatives to sulfonated copolymers can be developed. In a study, it was stated that the SPP-b-PAES(6.5K)-3.0 membrane is an alternative with its high performance, durability, and proton conductivity (Lee et al., 2015).

CURRENT STATUS OF HYDROGEN ECONOMY

The concept of the 'hydrogen economy' was first introduced by John Bockris in the 1970s. Bockris put forward the idea that hydrogen would be produced through the electrolysis of water and distributed to factories, homes and fuel stations, where it would be converted back into electricity in fuel cells (Bockris, 2013). Japan has included hydrogen in its strategic energy plans and has presented several plans on hydrogen as the main component of secondary energy sources. Researchers at the U.S. Department of Energy's National Renewable Energy Laboratory presented a view involving different sectors (Sinclair, 2001). In a green economy for a renewable energy society, 96% of hydrogen is produced from fossil fuels by the hydrocarbon reforming method. The most common processes include steam methane reformation of hydrogen (48%), naphtha and heavy oil reformation (30%), and coal gasification (18%), all of which release large amounts of CO_2 80-90% of the emitted CO_2 can be captured by combining the steam methane reforming process and the CO_2 capture process, but this process is costly. Low-carbon (or zero-carbon) hydrogen production techniques are required to achieve an environmentally friendly hydrogen economy (Dawood, Anda & Shafiullah). Water electrolysis with renewable energy is the best way to produce green hydrogen but is currently more costly than steam methane reforming with CO_2 capture processes (Guoxing et al., 2022)

Although the transition to hydrogen energy is a process that will progress depending on the results of heuristics, supporting this transition with country policies provides convenience in the acceptance and implementation of the process (Kakodkar et al.,2022). In this context, some countries are pursuing policies that support hydrogen energy in various ways. For example; in countries such as Belgium, Germany and Italy, hydrogen use is encouraged in the transportation sector, some countries such as France, Norway, and Spain encourage vehicle purchases, and some countries such as Norway and the Netherlands encourage hydrogen production (Lebrouhi et al., 2022). There is very little supportive policy for the use of hydrogen based on renewable energy in the industry. Hydrogen strategies developed in order to meet the energy needs of many countries (Kakoulaki et al.,2021). After Europe, China is seen as the world's largest producer of hydrogen and is developing strong actions and strategies for hydrogen and fuel cell applications with the target of achieving a 10% share of domestic energy consumption by 2040 (Lebrouhi et al., 2022). Australia focuses on large-scale hydrogen centers and the long-term perspective to export large volumes of renewable hydrogen to other countries (Terlouw et al., 2022). An industry association called Hydrogen Europe is promoting a massive increase in electrolyzer production in the EU and North Africa. The project has targeted topics such as the development of high-efficiency technologies, the development of hydrogen transport technologies over long distances, and the evaluation and monitoring of the safety of existing technologies used (Falcone, Hiete & Sapio, 2021). When evaluated on the basis of countries, the support of hydrogen energy and technological developments related to this energy opens the doors to a clean and efficient technology era in the transportation sector, as in many other fields (Falcone, Hiete & Sapio, 2021).

CONCLUSION

Problems such as environmental concerns, increasing costs, and the inability of resources to meet the supply due to fossil fuel use. It leads to the search for new openings in the transportation sector, which is one of the most intense areas of use. In this context, the interest in electric vehicle technologies has increased considerably. In addition to many researchers, many companies continue their R&D studies on this subject and introduce new products to the market. Electric vehicles can be classified as BEV, HEV, PHEV and FCEVs. All of these vehicles are cleaner technologies than fossil fuel-based ICEs.

FCEVs do not use a battery system like other electric vehicles; instead, they have a fuel tank and an FC system like ICEs in their designs. In general, these technologies, in which PEMFCs are used, are characterized as environmentally friendly with zero carbon emissions. Although hydrogen, which is used as fuel in these vehicles, is the most abundant element in the universe, it always bonds with other elements due to its high energy and therefore it must be obtained by various production methods.

When the methods of obtaining hydrogen are examined in general, it is seen that the classification is made according to the source from which it will be obtained. Today, a very large percentage of hydrogen is obtained by hydrocarbon reforming processes, in which high-purity carbon dioxide is released. Many countries are working on the supply of hydrogen, which is seen as an alternative to fossil fuels, from renewable sources. The best-known way to obtain clean hydrogen is the electrolysis of water. Many countries have added renewable energy sources, clean hydrogen and electric vehicle technology-themed R&D items to their energy policies. Projects and support policies carried out in this context will facilitate the adoption and dissemination of these new technologies. Although the use of these technologies has been started, speeding up the R&D studies for the widespread use and solution-oriented elimination of the identified problems will accelerate this process.

ACKNOWLEDGMENT

This research received no specific grant from any funding agency in the public, commercial, or not-for-profit sectors.

REFERENCES

Acar, C., Beskese, A., & Temur, G. T. (2022). Comparative Fuel Cell Sustainability Assessment With A Novel Approach. *International Journal of Hydrogen Energy*, *47*(1), 575–594. doi:10.1016/j.ijhydene.2021.10.034

Alaswad, A., Baroutaji, A., Achour, H., Carton, J., Al Makky, A., & Olabi, A. G. (2016). Developments In Fuel Cell Technologies In The Transport Sector. *International Journal Of Hydrogen Energy*, *41*(37), 16499–1650. doi:10.1016/j.ijhydene.2016.03.164

Alaswad, A., Omran, A., Sodre, J. R., Wilberforce, T., Pignatelli, G., Dassisti, M., & Olabi, A. G. (2020). Technical And Commercial Challenges Of Proton-Exchange Membrane (Pem) Fuel Cells. *Energies*, *14*(1), 144. doi:10.3390/en14010144

Asif, U., & Schmidt, K. (2021). Fuel Cell Electric Vehicles (Fcev): Policy Advances To Enhance Commercial Success. *Sustainability (Basel)*, *13*(9), 5149. doi:10.3390u13095149

Aso, S., Kizaki, M., & Mizuno, H. (2009). Development Progress Of The Toyota Fuel Cell Hybrid Vehicle. *Sae International Journal Of Engines*, *1*(1), 296–303. doi:10.4271/2008-01-0420

Bae, C., & Kim, J. (2017). Alternative Fuels For Internal Combustion Engines. *Proceedings Of The Combustion Institute*, *36*(3), 3389–3413. doi:10.1016/j.proci.2016.09.009

Bailey, D. (2015). *How The Automobile Changed History*. Abdo.

Baroutaji, A., Carton, J., Stokes, J., & Olabi, A. G. (2014). Design And Development Of Proton Exchange Membrane Fuel Cell Using Open Pore Cellular Foam As Flow Plate Material. *Journal Of Energy Challenges And Mechanics, 1*(7).

Bauman, J., & Kazerani, M. (2008). A Comparative Study Of Fuel-Cell–Battery, Fuel-Cell–Ultracapacitor, And Fuel-Cell–Battery–Ultracapacitor Vehicles. *Ieee Transactions On Vehicular Technology, 57*(2), 760–769. doi:10.1109/TVT.2007.906379

Bicer, Y., & Dincer, I. (2018). Life Cycle Environmental Impact Assessments And Comparisons Of Alternative Fuels For Clean Vehicles. *Resources, Conservation And Recycling, 132*, 141–157. doi:10.1016/j.resconrec.2018.01.036

Bockris, J. O. M. (2013). The Hydrogen Economy: Its History. *International Journal Of Hydrogen Energy, 38*(6), 2579–2588. doi:10.1016/j.ijhydene.2012.12.026

Budde, B., Alkemade, F., & Hekkert, M. (2015). On The Relation Between Communication And Innovation Activities: A Comparison Of Hybrid Electric And Fuel Cell Vehicles. *Environmental Innovation And Societal Transitions, 14*, 45–59. doi:10.1016/j.eist.2013.11.003

Chan, C. C. (2002). The State Of The Art Of Electric And Hybrid Vehicles. *Proceedings Of The Ieee, 90*(2), 247–275. doi:10.1109/5.989873

Chan, C. C., & Wong, Y. S. (2004). Electric Vehicles Charge Forward. *Ieee Power & Energy Magazine, 2*(6), 24–33. doi:10.1109/MPAE.2004.1359010

Chau, K. T., & Wong, Y. S. (2002). Overview Of Power Management In Hybrid Electric Vehicles. *Energy Conversion And Management, 43*(15), 1953–1968. doi:10.1016/S0196-8904(01)00148-0

Choi, W., Yoo, E., Seol, E., Kim, M., & Song, H. H. (2020). Greenhouse Gas Emissions Of Conventional And Alternative Vehicles: Predictions Based On Energy Policy Analysis In South Korea. *Applied Energy, 265*, 114754. doi:10.1016/j.apenergy.2020.114754

Choi, W., Yoo, E., Seol, E., Kim, M., & Song, H. H. (2020). Greenhouse Gas Emissions Of Conventional And Alternative Vehicles: Predictions Based On Energy Policy Analysis In South Korea. *Applied Energy, 265*, 114754. doi:10.1016/j.apenergy.2020.114754

Conte, M., Iacobazzi, A., Ronchetti, M., & Vellone, R. (2001). Hydrogen Economy For A Sustainable Development: State-Of-The-Art And Technological Perspectives. *Journal Of Power Sources, 100*(1-2), 171–187. doi:10.1016/S0378-7753(01)00893-X

Cunanan, C., Tran, M. K., Lee, Y., Kwok, S., Leung, V., & Fowler, M. (2021). A Review Of Heavy-Duty Vehicle Powertrain Technologies: Diesel Engine Vehicles, Battery Electric Vehicles, And Hydrogen Fuel Cell Electric Vehicles. *Clean Technologies*, *3*(2), 474–489. doi:10.3390/cleantechnol3020028

Das, D., & Veziroğlu, T. N. (2001). Hydrogen Production By Biological Processes: A Survey Of Literature. *International Journal Of Hydrogen Energy*, *26*(1), 13–28. doi:10.1016/S0360-3199(00)00058-6

Daş, E., & Yurtcan, A. B. (2016). Effect Of Carbon Ratio In The Polypyrrole/Carbon Composite Catalyst Support On Pem Fuel Cell Performance. *International Journal Of Hydrogen Energy*, *41*(30), 13171–13179. doi:10.1016/j.ijhydene.2016.05.167

Das, H. S., Tan, C. W., & Yatim, A. H. M. (2017). Fuel Cell Hybrid Electric Vehicles: A Review On Power Conditioning Units And Topologies. *Renewable & Sustainable Energy Reviews*, *76*, 268–291. doi:10.1016/j.rser.2017.03.056

Dash, S. K., Chakraborty, S., Roccotelli, M., & Sahu, U. K. (2022). Hydrogen Fuel For Future Mobility: Challenges And Future Aspects. *Sustainability (Basel)*, *14*(14), 8285. doi:10.3390u14148285

Dawood, F., Anda, M., & Shafiullah, G. M. (2020). Hydrogen Production For Energy: An Overview. *International Journal Of Hydrogen Energy*, *45*(7), 3847–3869. doi:10.1016/j.ijhydene.2019.12.059

De Las Heras, N., Roberts, E. P. L., Langton, R., & Hodgson, D. R. (2009). A Review Of Metal Separator Plate Materials Suitable For Automotive Pem Fuel Cells. *Energy & Environmental Science*, *2*(2), 206–214. doi:10.1039/B813231N

Desideri, U., & Yan, J. (2012). *Clean Energy Technologies And Systems For A Sustainable World.*

Dinh, K. N., Liang, Q., Du, C. F., Zhao, J., Tok, A. I. Y., Mao, H., & Yan, Q. (2019). Nanostructured Metallic Transition Metal Carbides, Nitrides, Phosphides, And Borides For Energy Storage And Conversion. *Nano Today*, *25*, 99–121. doi:10.1016/j.nantod.2019.02.008

Edwards, P. P., Kuznetsov, V. L., David, W. I., & Brandon, N. P. (2008). Hydrogen And Fuel Cells: Towards A Sustainable Energy Future. *Energy Policy*, *36*(12), 4356–4362. doi:10.1016/j.enpol.2008.09.036

Emadi, A., Williamson, S. S., & Khaligh, A. (2006). Power Electronics Intensive Solutions For Advanced Electric, Hybrid Electric, And Fuel Cell Vehicular Power Systems. *Ieee Transactions On Power Electronics*, *21*(3), 567–577. doi:10.1109/TPEL.2006.872378

Falcone, P. M., Hiete, M., & Sapio, A. (2021). Hydrogen Economy And Sustainable Development Goals: Review And Policy Insights. *Current Opinion In Green And Sustainable Chemistry*, *31*, 100506. doi:10.1016/j.cogsc.2021.100506

Fan, L., Tu, Z., & Chan, S. H. (2021). Recent Development Of Hydrogen And Fuel Cell Technologies: A Review. *Energy Reports*, *7*, 8421–8446. doi:10.1016/j.egyr.2021.08.003

Fayaz, H., Saidur, R., Razali, N., Anuar, F. S., Saleman, A. R., & Islam, M. R. (2012). An Overview Of Hydrogen As A Vehicle Fuel. *Renewable & Sustainable Energy Reviews*, *16*(8), 5511–5528. doi:10.1016/j.rser.2012.06.012

Fernández, R. Á., Cilleruelo, F. B., & Martínez, I. V. (2016). A New Approach To Battery Powered Electric Vehicles: A Hydrogen Fuel-Cell-Based Range Extender System. *International Journal Of Hydrogen Energy*, *41*(8), 4808–4819. doi:10.1016/j.ijhydene.2016.01.035

Forrest, K., Mac Kinnon, M., Tarroja, B., & Samuelsen, S. (2020). Estimating The Technical Feasibility Of Fuel Cell And Battery Electric Vehicles For The Medium And Heavy Duty Sectors In California. *Applied Energy*, *276*, 115439. doi:10.1016/j.apenergy.2020.115439

Fredriksson, J., Weiefors, H., & Egardt, B. (2002). Powertrain Control For Active Damping Of Driveline Oscillations. *Vehicle System Dynamics*, *37*(5), 359–376. doi:10.1076/vesd.37.5.359.3527

Frenette, G., & Forthoffer, D. (2009). Economic & Commercial Viability Of Hydrogen Fuel Cell Vehicles From An Automotive Manufacturer Perspective. *International Journal Of Hydrogen Energy*, *34*(9), 3578–3588. doi:10.1016/j.ijhydene.2009.02.072

Guoxing, C., Tu, X., Homm, G., & Weidenkaff, A. (2022). Plasma Pyrolysis For A Sustainable Hydrogen Economy. *Nature Reviews. Materials*.

Hao, X., Zhou, Y., Wang, H., & Ouyang, M. (2020). Plug-In Electric Vehicles In China And The Usa: A Technology And Market Comparison. *Mitigation And Adaptation Strategies For Global Change*, *25*(3), 329–353. doi:10.100711027-019-09907-z

He, R., Tian, X., Ni, Y., & Xu, Y. (2017). Mode Transition Coordination Control For Parallel Hybrid Electric Vehicle Based On Switched System. *Advances In Mechanical Engineering*, *9*(8), 1687814017715564. doi:10.1177/1687814017715564

Hoffert, M. I., Caldeira, K., Benford, G., Criswell, D. R., Green, C., Herzog, H., & Wigley, T. M. (2002). Advanced Technology Paths To Global Climate Stability: Energy For A Greenhouse Planet. *Science*, *298*(5595), 981–987. doi:10.1126cience.1072357 PMID:12411695

İnci, M., Büyük, M., Demir, M. H., & İlbey, G. (2021). A Review And Research On Fuel Cell Electric Vehicles: Topologies, Power Electronic Converters, Energy Management Methods, Technical Challenges, Marketing And Future Aspects. *Renewable & Sustainable Energy Reviews*, *137*, 110648. doi:10.1016/j.rser.2020.110648

Kakodkar, R., He, G., Demirhan, C. D., Arbabzadeh, M., Baratsas, S. G., Avraamidou, S., & Pistikopoulos, E. N. (2022). A Review Of Analytical And Optimization Methodologies For Transitions In Multi-Scale Energy Systems. *Renewable & Sustainable Energy Reviews*, *160*, 112277. doi:10.1016/j.rser.2022.112277

Kakoulaki, G., Kougias, I., Taylor, N., Dolci, F., Moya, J., & Jäger-Waldau, A. (2021). Green Hydrogen In Europe–A Regional Assessment: Substituting Existing Production With Electrolysis Powered By Renewables. *Energy Conversion And Management*, *228*, 113649. doi:10.1016/j.enconman.2020.113649

Kalamaras, C. M., & Efstathiou, A. M. (2013). Hydrogen Production Technologies: Current State And Future Developments. In *Conference Papers In Science*. Hindawi.

Karlsson, M., Alfredsson, E., & Westling, N. (2020). Climate Policy Co-Benefits: A Review. *Climate Policy*, *20*(3), 292–316. doi:10.1080/14693062.2020.1724070

Kovats, R. S., & Haines, A. (2005). Global Climate Change And Health: Recent Findings And Future Steps. *Canadian Medical Association Journal*, *172*(4), 501–502. doi:10.1503/cmaj.050020 PMID:15710942

Lebrouhi, B. E., Djoupo, J. J., Lamrani, B., Benabdelaziz, K., & Kousksou, T. (2022). Global Hydrogen Development-A Technological And Geopolitical Overview. *International Journal Of Hydrogen Energy*, *47*(11), 7016–7048. doi:10.1016/j.ijhydene.2021.12.076

Lee, J. Y., Yu, D. M., Kim, T. H., Yoon, S. J., & Hong, Y. T. (2015). Multi-Block Copolymers Based On Poly (P-Phenylene) S With Excellent Durability And Fuel Cell Performance. *Journal Of Membrane Science*, *492*, 209–219. doi:10.1016/j.memsci.2015.04.013

Li, S., & Zhang, C. (2009, March). Study On Battery Management System And Lithium-Ion Battery. In 2009 *International Conference On Computer And Automation Engineering* (Pp. 218-222). IEEE.

Liu, F., Zhao, F., Liu, Z., & Hao, H. (2018). The Impact Of Fuel Cell Vehicle Deployment On Road Transport Greenhouse Gas Emissions: The China Case. *International Journal Of Hydrogen Energy*, *43*(50), 22604–22621. doi:10.1016/j.ijhydene.2018.10.088

Mantell, C. L. (1982). *Batteries And Energy Systems*.

Mccullough, F. P., Levine, C. A., & Snelgrove, R. V. (1989). Secondary Battery. *Noûs (Detroit, Mich.)*, 4830938.

Miller, E. L., Thompson, S. T., Randolph, K., Hulvey, Z., Rustagi, N., & Satyapal, S. (2020). Us Department Of Energy Hydrogen And Fuel Cell Technologies Perspectives. *Mrs Bulletin*, *45*(1), 57–64. doi:10.1557/mrs.2019.312

Mishra, S., Verma, S., Chowdhury, S., Gaur, A., Mohapatra, S., Dwivedi, G., & Verma, P. (2021). A Comprehensive Review On Developments In Electric Vehicle Charging Station Infrastructure And Present Scenario Of India. *Sustainability (Basel)*, *13*(4), 2396. doi:10.3390u13042396

Muthukumar, M., Rengarajan, N., Velliyangiri, B., Omprakas, M. A., Rohit, C. B., & Raja, U. K. (2021). The Development Of Fuel Cell Electric Vehicles–A Review. *Materials Today: Proceedings*, *45*, 1181–1187. doi:10.1016/j.matpr.2020.03.679

Najjar, Y. S. (2013). Hydrogen Safety: The Road Toward Green Technology. *International Journal Of Hydrogen Energy*, *38*(25), 10716–10728. doi:10.1016/j.ijhydene.2013.05.126

Nicoletti, G., Arcuri, N., Nicoletti, G., & Bruno, R. (2015). A Technical And Environmental Comparison Between Hydrogen And Some Fossil Fuels. *Energy Conversion And Management*, *89*, 205–213. doi:10.1016/j.enconman.2014.09.057

Nikolaidis, P., & Poullikkas, A. (2017). A Comparative Overview Of Hydrogen Production Processes. *Renewable & Sustainable Energy Reviews*, *67*, 597–611. doi:10.1016/j.rser.2016.09.044

Offer, G. J., Howey, D., Contestabile, M., Clague, R., & Brandon, N. P. (2010). Comparative Analysis Of Battery Electric, Hydrogen Fuel Cell And Hybrid Vehicles In A Future Sustainable Road Transport System. *Energy Policy*, *38*(1), 24–29. doi:10.1016/j.enpol.2009.08.040

Ohno, T. (2019). *Toyota Production System: Beyond Large-Scale Production*. Productivity Press. .

Öner, E., Öztürk, A., & Yurtcan, A. B. (2020). Utilization Of The Graphene Aerogel As Pem Fuel Cell Catalyst Support: Effect Of Polypyrrole (Ppy) And Polydimethylsiloxane (Pdms) Addition. *International Journal Of Hydrogen Energy*, *45*(60), 34818–34836. doi:10.1016/j.ijhydene.2020.05.053

Qin, N., Brooker, P., & Raissi, A. (2017). *Fuel Cell Vehicle Technologies, Infrastructure And Requirements* (No. Evtc Report Number: Fsec-Cr-2059-17). University Of Central Florida. Electric Vehicle Transportation Center (Evtc)

Qin, N., Raissi, A., & Brooker, P. (2014). Analysis Of Fuel Cell Vehicle Developments. The Florida Solar Energy Center (Fsec): Cocoa, Fl, Usa.

Ren, L., Zhou, S., & Ou, X. (2020). Life-Cycle Energy Consumption And Greenhouse-Gas Emissions Of Hydrogen Supply Chains For Fuel-Cell Vehicles In China. *Energy*, *209*, 118482. doi:10.1016/j.energy.2020.118482

Ren, Z., Zhang, D., & Wang, Z. (2012). Stacks With Tin/Titanium As The Bipolar Plate For Pemfcs. *Energy*, *48*(1), 577–581. doi:10.1016/j.energy.2012.10.020

Ross, D. K. (2006). Hydrogen Storage: The Major Technological Barrier To The Development Of Hydrogen Fuel Cell Cars. *Vacuum*, *80*(10), 1084–1089. doi:10.1016/j.vacuum.2006.03.030

Rusman, N. A. A., & Dahari, M. (2016). A Review On The Current Progress Of Metal Hydrides Material For Solid-State Hydrogen Storage Applications. *International Journal Of Hydrogen Energy*, *41*(28), 12108–12126. doi:10.1016/j.ijhydene.2016.05.244

Saritas, O., Meissner, D., & Sokolov, A. (2019). A Transition Management Roadmap For Fuel Cell Electric Vehicles (Fcevs). *Journal Of The Knowledge Economy*, *10*(3), 1183–1203. doi:10.100713132-018-0523-3

Sartbaeva, A., Kuznetsov, V. L., Wells, S. A., & Edwards, P. P. (2008). Hydrogen Nexus In A Sustainable Energy Future. *Energy & Environmental Science*, *1*(1), 79–85. doi:10.1039/b810104n

Schwarze, K., Posdziech, O., Kroop, S., Lapeña-Rey, N., & Mermelstein, J. (2017). Green Industrial Hydrogen Via Reversible High-Temperature Electrolysis. *Ecs Transactions*, *78*(1), 2943–2952. doi:10.1149/07801.2943ecst

Sierzchula, W., Bakker, S., Maat, K., & Van Wee, B. (2012). Technological Diversity Of Emerging Eco-Innovations: A Case Study Of The Automobile Industry. *Journal Of Cleaner Production*, *37*, 211–220. doi:10.1016/j.jclepro.2012.07.011

Sinclair, K. (2001). *Status Of Avian Research At The National Renewable Energy Laboratory (No. Nrel/Cp-500-30546). National Renewable Energy Lab.(Nrel), Golden*. United States.

Slowik, P., Pavlenko, N., & Lutsey, N. (2016). Assessment Of Next-Generation Electric Vehicle Technologies. *International Council On Clean Transportation*, 1225.

Smith, B. J., Graziano, D. J., Riddle, M. E., Liu, D. J., Sun, P., Iloeje, C., & Diamond, D. (2022). Platinum Group Metal Catalysts-Supply Chain Deep Dive Assessment. *Usdoe Office Of Policy* (Po).

Song, Y., Zhang, C., Ling, C. Y., Han, M., Yong, R. Y., Sun, D., & Chen, J. (2020). Review On Current Research Of Materials, Fabrication And Application For Bipolar Plate In Proton Exchange Membrane Fuel Cell. *International Journal Of Hydrogen Energy*, *45*(54), 29832–29847. doi:10.1016/j.ijhydene.2019.07.231

Sudjoko, C., Sasongko, N. A., Utami, I., & Maghfuri, A. (2021, November). Utilization Of Electric Vehicles As An Energy Alternative To Reduce Carbon Emissions. In I*op Conference Series: Earth And Environmental Science* (Vol. 926, No. 1, P. 012094). Iop Publishing.

Suzuki, S., Shoji, N., & Tsuru, T. (2021). Performance Evaluation Of Water Vapor Permeation Through Perfluorosulfonic Acid Capillary Membranes. *Separation And Purification Technology*, *266*, 118508. doi:10.1016/j.seppur.2021.118508

Tawfik, H., Hung, Y., & Mahajan, D. (2007). Metal Bipolar Plates For Pem Fuel Cell—A Review. *Journal Of Power Sources*, *163*(2), 755–767. doi:10.1016/j.jpowsour.2006.09.088

Terlouw, T., Bauer, C., Mckenna, R., & Mazzotti, M. (2022). Large-Scale Hydrogen Production Via Water Electrolysis: A Techno-Economic And Environmental Assessment. *Energy & Environmental Science*, *15*(9), 3583–3602. doi:10.1039/D2EE01023B

Thomas, C. E. (2009). Fuel Cell And Battery Electric Vehicles Compared. *International Journal Of Hydrogen Energy*, *34*(15), 6005–6020. doi:10.1016/j.ijhydene.2009.06.003

Thompson, S. T., & Papageorgopoulos, D. (2019). Platinum Group Metal-Free Catalysts Boost Cost Competitiveness Of Fuel Cell Vehicles. *Nature Catalysis*, *2*(7), 558–561. doi:10.103841929-019-0291-x

Verma, S., Mishra, S., Gaur, A., Chowdhury, S., Mohapatra, S., Dwivedi, G., & Verma, P. (2021). A Comprehensive Review On Energy Storage In Hybrid Electric Vehicle. [English Edition]. *Journal Of Traffic And Transportation Engineering*, *8*(5), 621–637. doi:10.1016/j.jtte.2021.09.001

Veziroğlu, T. N., & Şahin, S. (2008). 21st Century's Energy: Hydrogen Energy System. *Energy Conversion And Management*, *49*(7), 1820–1831. doi:10.1016/j.enconman.2007.08.015

Vincent, I., Lee, E. C., & Kim, H. M. (2020). Solutions To The Water Flooding Problem For Unitized Regenerative Fuel Cells: Status And Perspectives. *Rsc Advances*, *10*(29), 16844–16860. doi:10.1039/D0RA00434K PMID:35521448

Wang, Y., Chen, K. S., Mishler, J., Cho, S. C., & Adroher, X. C. (2011). A Review Of Polymer Electrolyte Membrane Fuel Cells: Technology, Applications, And Needs On Fundamental Research. *Applied Energy*, *88*(4), 981–1007. doi:10.1016/j.apenergy.2010.09.030

Wang, Y., Yuan, H., Martinez, A., Hong, P., Xu, H., & Bockmiller, F. R. (2021). Polymer Electrolyte Membrane Fuel Cell And Hydrogen Station Networks For Automobiles: Status, Technology, And Perspectives. *Advances In Applied Energy*, *2*, 100011. doi:10.1016/j.adapen.2021.100011

Wilberforce, T., El-Hassan, Z., Khatib, F. N., Al Makky, A., Baroutaji, A., Carton, J. G., & Olabi, A. G. (2017). Developments Of Electric Cars And Fuel Cell Hydrogen Electric Cars. *International Journal Of Hydrogen Energy*, *42*(40), 25695–25734. doi:10.1016/j.ijhydene.2017.07.054

Włodarczyk, R. (2022). Corrosion Analysis Of Graphite Sinter As Bipolar Plates In The Low-Temperature Pem Fuel Cell Simulated Environments. *Journal Of Solid State Electrochemistry*, *26*(1), 39–47. doi:10.100710008-021-05015-8

Wu, D., Peng, C., Yin, C., & Tang, H. (2020). Review Of System Integration And Control Of Proton Exchange Membrane Fuel Cells. *Electrochemical Energy Reviews*, *3*(3), 466–505. doi:10.100741918-020-00068-1

Xiaoluan, Z., Farajian, H., Xifeng, W., & Ohshima, K. (2022). Scheduling Of Renewable Energy And Plug-In Hybrid Electric Vehicles Based Microgrid Using Hybrid Crow—Pattern Search Method. *Journal Of Energy Storage*, *47*, 103605. doi:10.1016/j.est.2021.103605

Xu, J., Yu, J., Xu, J., Sun, C., He, W., Huang, J., & Li, G. (2020). High-Value Utilization Of Waste Tires: A Review With Focus On Modified Carbon Black From Pyrolysis. *The Science Of The Total Environment*, *742*, 140235. doi:10.1016/j.scitotenv.2020.140235 PMID:32629243

Yang, L., Yu, H., Jiang, L., Zhu, L., Jian, X., & Wang, Z. (2010). Improved Anticorrosion Properties And Electrical Conductivity Of 316l Stainless Steel As Bipolar Plate For Proton Exchange Membrane Fuel Cell By Lower Temperature Chromizing Treatment. *Journal Of Power Sources*, *195*(9), 2810–2814. doi:10.1016/j.jpowsour.2009.11.018

Yang, Y., Hu, X., Pei, H., & Peng, Z. (2016). Comparison Of Power-Split And Parallel Hybrid Powertrain Architectures With A Single Electric Machine: Dynamic Programming Approach. *Applied Energy*, *168*, 683–690. doi:10.1016/j.apenergy.2016.02.023

Zeng, X., Li, M., Abd El-Hady, D., Alshitari, W., Al-Bogami, A. S., Lu, J., & Amine, K. (2019). Commercialization Of Lithium Battery Technologies For Electric Vehicles. *Advanced Energy Materials*, *9*(27), 1900161. doi:10.1002/aenm.201900161

KEY TERMS AND DEFINITIONS

Battery electric vehicle: Battery-powered electric vehicles are electric vehicle models that take energy from previously stored systems. The vehicle consists of a battery, electric motor and controller to store energy.

Electric vehicle: They are vehicles that use electrical energy obtained from batteries/fuel cells/hybrid systems, use it when necessary, and provide propulsion with electric motors. These vehicles do not have an internal combustion engine.

Fuel cell electric vehicle: It is a type of electric car that uses a fuel cell to power the electric motor. When the vehicle is used, only water and heat by-products are formed, it is a zero-emission technology.

Fuel cell: It is an energy conversion device that directly converts the energy of the fuel into electrical energy by means of an electrochemical reaction.

Hybrid electric vehicle: A type of electric vehicle that combines a conventional internal combustion engine (ICE) system with an electric propulsion system.

Hydrogen: Hydrogen (H_2) is a colorless, odorless and highly flammable gas with atomic number 1 at standard temperature and pressure. It is the most abundant element in nature, and due to its high energy, it is often found in combination with another element.

Plug-in hybrid electric vehicle: A hybrid electric vehicle that can charge rechargeable batteries or another energy storage device by plugging it into an external electrical source. A PHEV combines the characteristics of a conventional hybrid electric vehicle with both an electric motor and an internal combustion engine, and an all-electric vehicle with a plug to connect to the electric grid.

Seconder Battery: Electrochemical devices that store electrical energy as chemical energy and give it as electrical energy when requested. It is also a source of electrical energy found in vehicles.

APPENDIX

Abbreviations

BEV Battery electric vehicle
FCEV Fuel cell electric vehicle
HEV Hybrid electric vehicle
ICE Internal combustion engine
PHEV Plug-in hybrid electric vehicle
PEM Proton exchange membrane
FC Fuel cell
EV Electric vehicle

Chapter 3
Hydrogen Fuel Cell Technologies for Mobile Applications:
A Perspective Towards Sustainable Development Goals

Rakesh Ashok Afre
https://orcid.org/0000-0002-7412-6437
Assam down town University, Guwahati, India

Nallin Sharma
https://orcid.org/0000-0002-0845-4467
Indian Institute of Sciences, Bangalore, India

ABSTRACT

A high-quality energy carrier like hydrogen can be created by thermochemical processes involving the conversion of hydrocarbons, such as coal or natural gas. It can also be obtained by water electrolysis using renewable energy sources, such as wind and solar power. Due to climate change, air pollution, and energy deficiency, the exploration of new transportation fuels is very important. Despite the various challenges that hydrogen faces, it is still regarded as a clean energy source that can help solve the environmental and energy problems of the future. It can be used in fuel cells to provide high-quality electricity and heat. This chapter aims to discuss the various techniques involved in the production of hydrogen and their applications in fuel cells. It also explores the different types of fuel cells, such as PEMFCs, MCFCs, and SOFCs as well as to understand the effect towards sustainable development goals.

DOI: 10.4018/978-1-6684-6721-3.ch003

INTRODUCTION

The world's population has increased dramatically over the past few decades, which has resulted in the need for energy being more prevalent. Fossil fuels are the most common source of energy, though other sources are also available. Although they are cheaper than renewable energy sources, they have drawbacks. The rapid emergence and evolution of hydrogen as a fuel source for cars has been studied. The current state of the art in this area, as well as the future potential of this technology in road transportation, have been discussed (Apostolou and Xydis, 2019). The hydrogen economy is a multi-sector view of the decarbonization of the process sectors and the low-cost clean energy. Hydrogen's ability to store and transport derivatives or hydrogen has been regarded as a game-changer in the decarbonization of the energy sector. It can help lower greenhouse gas emissions and air pollution (Staffell et al., 2019). In addition to vehicles, hydrogen is also one of the most important low-carbon transportation options. Unlike other fuel sources, hydrogen does not contribute to air pollution and land usage issues. It also provides a low-emissions alternative to electric cars. In addition, it has been shown that hydrogen cars are three times more likely to reduce global warming than other vehicles (Falcone et al., 2021).

The goal of this study was to analyze the effects of various types of alternative vehicles on the environment and people's health. It was done by comparing the life cycle of these vehicles to that of traditional cars. Compared to methanol, hydrogen-powered vehicles have a higher energy density 60000 Wh kg^{-1}, which makes them more sustainable when it comes to reducing greenhouse gas emissions and depletion of the ozone layer (Bicer and Dincer, 2017). The COVID-19 coronavirus has also made people more aware of the importance of consuming clean water, air, and food. Until the coronavirus pandemic occurred in 2020, the world was in a carbon age with respect to the use of hydrocarbon fuels. This is the start of a new hydrogen age, wherein the use of fossil fuels will decrease dramatically. The COVID-19 pandemic has caused the transition from the C age to the H_2 age to be chaotic (Dincer, 2020; Apostolou et al., 2018) Renewable resources such as hydrogen are the most promising ways to meet the energy demands of the future. Although it is rare in nature, hydrogen is found in plant materials and can provide more energy than gasoline when consumed in a unit of mass. However, a significant study is needed to design and implement onboard applications of this fuel (Abdalla et al., 2018).

The need for new technologies that can help address the increasing energy demands of the world's population and provide sustainable solutions has become more critical as the use of fossil fuels continues to rise. This is why the development of fuel cell technology is becoming more important. Compared to traditional combustion-based fuel cells, hydrogen-powered fuel cells offer numerous advantages (Hart, 2000 & Rady et. al. 2012). They're far more eco-friendly than their conventional counterparts, and they don't release harmful air pollutants. Also, they're 96% more energy-efficient possessing in terms of internal combustion-based Carnot cycle. Fuel cells and hydrogen can be used in a wide range of applications. They can power cars, portable electronic devices, and backup power systems. Because of their grid-independent nature, fuel cells are ideal for various critical load functions, such as data centers and hospitals. A fuel cell is like batteries that can produce electricity without emissions or combustion. Unlike batteries, they do not need to recharge as long as there is a continuous source of oxygen and fuel. Vehicles powered by fuel cells can reduce carbon dioxide emissions by up to 50% if the hydrogen comes from natural gas or 90% if it comes from renewable sources such as solar and wind (Pehnt, 2003; Midilli et. al., 2005; Xu et. al 2020; Abdelkareem et. al. 2020).

The concept of sustainable mobility refers to a system that is both ecologically beneficial and ubiquitous. While there are no specific goals for transportation, it is still important to reach other sustainable development objectives. The top-performing countries have robust long-term policies for mobility, while the ones with the lowest scores suffer from inadequate infrastructure. Table 2 shows the various SDGs that are indirectly or directly affected by sustainable mobility. The concept of sustainable development refers to a broader framework. The goal of sustainable development is to meet current needs while ensuring that the future generations have the necessary resources to meet their needs. An infographic shows the various benefits of mobility, such as economic development, energy security, and environmental sustainability.

METHODOLOGY ADOPTED

Over the past couple of decades, various approaches have been attempted to encourage the use of alternative fuels in mobile devices. Unfortunately, most of these efforts have failed due to various factors. Various lessons learned from the past efforts should be used to develop new hydrogen technologies and improve the market's acceptance of alternative fuels. This study aims to analyze the various strategies that were used in the past and present mobile systems to promote the use of hydrogen and also reviews the literature on the environmental and technological aspects of hydrogen in fuel cell technology.

FUEL CELL AND ITS TYPES:

Alkaline Fuel Cells

In fuel cells, the electrolyte known as KOH is usually used. For high-temperature applications, the majority of it is used, while the remaining 35% to 50% is utilized for low-temperature operations. A typical fuel cell can operate at temperatures ranging from 50 to 200 °C. This fuel cell typically operates at a temperature of 50 °C. On the other hand, it can operate at 200 °C. The electrolyte and the platinum-based material are the cell's primary components. In fuel cells, H_2 is the fuel. AFCs generally have an average efficiency of around 50% to 60%. The first one was developed in 1960 and was used in the Apollo space vehicle. In terms of their performance, they are better than other fuel cells that operate at O_2 and H_2 (Koscher and Kordesch 2003).

Direct Methanol Fuel Cells

This type of fuel cell is known as DMFC. It is a special type of fuel cell that's based on the proton exchange membrane (PEMFC). The two end plates that are used in this fuel cell are anode and cathode. The other components of this fuel cell include an electrolyte membrane and a catalyst. The H^+ ions are the carrier and the platinum-based material is used as the catalyst in this type of fuel cell (Arico et al. 2009). The primary fuel for this type of fuel cell is methanol. Its average cell efficiency is around 30 to 40%. During the 1960s to the 1980s, the use of conventional electrolytes changed, and during the 1990s, the use of polymer-based electrolytes was introduced (Goor, Menkin,, & Peled 2018; Munjewaret. Al. 2017).

Molten Carbonate Fuel Cells

During the 1960s, a new type of cell was introduced, which is known as the MCFC. This is due to the electrolyte being made from alkali carbonate. The cell's operating temperature is around 700 °C. Nickel is the catalyst, and the carbonate ions are the charge carriers. This type of fuel cell features end plates made from stainless steel, which helps prevent corrosion. It can run on various fuels such as CO_2, CH_4, and H_2. The MCFC's efficiency ranges from 55 to 65% percent, which is higher than other types of fuel cells. This is the second generation of fuel cells after the PAFC, and it's commonly used for commercial purposes as well (Ermete 2012; Salehi et. al. 2019; Barckholtz et. al. 2022; Cui et. al. 2021; Abdollahi et. al. 2019).

Figure 1. Types of fuel cells and their year of inventions

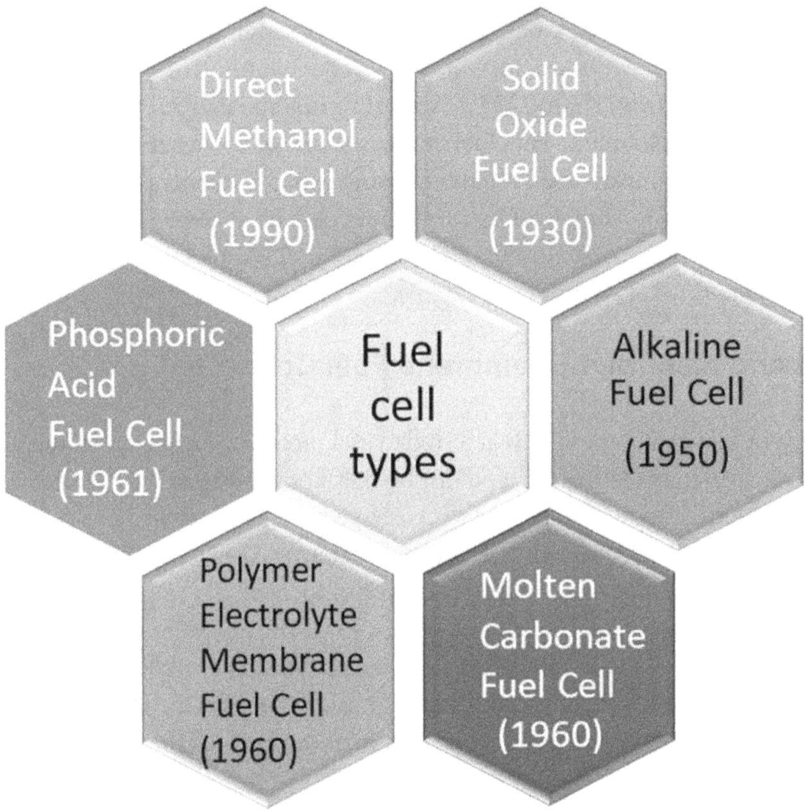

Solid Oxide Fuel Cells

This fuel cell is made up of a solid, nonporous metal oxide electrolyte. Its operating temperature ranges from 700 to 1,000 °C. This makes it the most extreme fuel cell in terms of its temperature. The O^{-} ions are the carriers, and the catalyst is made up of a type of inorganic compound known as perovskite. Ceramic is the cell's primary component. Its efficiency is similar to that of a fuel cell known as MCFC. Its power density is about 1.5 to 2.6 kW/m^3. The flat and tabular configurations of solid oxide fuel cells are different. These were developed during the 1950s, but their mechanical properties have improved significantly during the 2000s (Bessette and Wepfer 1996; Patel et al. 2004, Ni 2018).

Phosphoric Acid Fuel Cells

The development of this type of cell was first observed in 1965. It utilizes a type of electrolyte called phosphoric acid, which has a concentration of 100%. Its operating temperature ranges from 150 to 220 °C. According to some sources, the electrocatalyst utilized is usually platinum. This fuel cell has a charge carrier that is composed of H^+ ions. Since it requires an external reforming process to operate, it usually takes several hours for it to start working. Its power density is lower than that of other fuel cells, but it is the first commercialized fuel cell (Appleby and Foulkes 1993; Bhaskaran et. al. 2021; Nimir et. al. 2023).

Polymeric Electrolyte Membrane Fuel Cell

This type of fuel cell is commonly referred to as a proton exchange membrane. It has a solid electrolyte made from a type of polymer known as perfluoro-sulfonic acid. The membrane electrolyte is utilized to exchange protons between the anode and the cathode of the PEMFC. The simple and quick start and shutdown time of PEMFC fuel cells make them the most widely used type of fuel cell. Its operating temperature can range from 50 to 100 °C, and its efficiency is estimated to be between 40% and 60%. It's divided into two categories: low temperature and high temperature. The temperature range of fuel cells varies from 100 to 200 °C. On the other hand, its low temperature can vary from 50 to 100 °C (Shamardina et al. 2010; Aili et. al. 2020).

BACKGROUND OF HYDROGEN TECHNOLOGIES AND ITS PRODUCTION

Although fossil fuels are the main source of energy for the world, they contribute to the environment's degradation. Renewable energy sources can help minimize this issue. One promising technology for reducing environmental impact is hydrogen. However, this technology faces various challenges. Here we aims to provide an overview of the current state of hydrogen-based energy technologies and their potential to reduce environmental impact. When hydrogen is used in fuel cells, it produces no local pollution since the byproduct is water. It has a high specific density and is made locally. Various substances, such as water, oil, gas, sewage sludge, and biofuels, can be used to produce hydrogen. The availability of water allows the process to be performed in a sustainable manner. Timing plays a crucial role when it comes to deciding which technology and resources are ideal for producing hydrogen for transportation. This is why it is important to thoroughly study the various factors that will affect the supply of the fuel to the automotive industry.

Table 1. Fuel cells with chemical reactions occur at anode, cathode and the overall reaction

Fuel cell	Chemical reaction	Energy (ΔE)
AFC	At anode: $2H_2 + 4OH^- \rightarrow 4H_2O + 4e^-$ At cathode: $O_2 + 2H_2O + 4e^- \rightarrow 4OH^-$ Overall reaction: $2H_2 + O_2 \rightarrow 2H_2O$	$E^- = 0.4009$ V $E^+ = -0.8279$ V $E = 1.23$ V
MCFC	At anode: $H_2 + CO_3^- \rightarrow H_2O + CO_2 + 2e^-$ At cathode: $1/2O_2 + CO_2 + 2e^- \rightarrow CO_3^-$ Overall reaction: $H_2 + 1/2O_2 + CO_2 \rightarrow H_2O + CO_2$	$E^- = 0.86$ V $E^+ = 0.42$ V $E = 1.22$ V
SOFC	At anode: $H_2 + O^- \rightarrow H_2O + 2e^-$ At cathode: $1/2O_2 + 2e^- \rightarrow O^=$ Overall reaction: $H_2 + 1/2O_2 \rightarrow H_2O$	
PAFC	At anode: $H_2 \rightarrow 2H^+ + 2e^-$ At cathode: $1/2O_2 + 2H^+ + 2e^- \rightarrow H_2O$ Overall reaction: $H_2 + 1/2O_2 \rightarrow H_2O$	$E^- = 0.86$ V $E^+ = 0.42$ V $E = 1.23$ V
DMFC	At anode: $CH_3OH + H_2O \rightarrow CO_2 + 6H^+ + 6e^-$ At cathode: $3/2O_2 + 6H^+ + 6e^- \rightarrow 3H_2O$ Overall reaction: $CH_3OH + 3/2O_2 \rightarrow CO_2 + 2H_2O$	$E^- = -0.25$ V $E^+ = 1.46$ V $E = 1.21$ V
PEMFC	At anode: $H_2 \rightarrow 2H^+ + 2e^-$ At cathode: $1/2O_2 + 2H^+ + 2e^- \rightarrow H_2O$ Overall reaction: $H_2 + 1/2O_2 \rightarrow H_2O$	$E^- = 0.78$ V $E^+ = 0.46$ V $E = 1.18$ V

From Fossil Fuels

At commercial scale, hydrogen can be extracted from various sources such as fossil fuels, such as coal, natural gas, and methane. The extraction process for hydrogen is generally performed at larger scale. Some of the technologies used to produce hydrogen include fuel oxidation, gasification, and steam reformation. The main reason why carbon dioxide emissions are rising is due to the depletion of fossil fuel resources. As a result, the production of hydrogen from these sources is not sustainable. Fossil fuels such as natural gas and coal are commonly used to produce hydrogen. Steam reforming is a process that involves using oxygen instead of steam. Different types of reforming techniques are used depending on the oxidization agent used. For instance, steam reforming is used for light hydrocarbons, while partial oxidation is carried out for heavy hydrocarbons, such as coal and oil. After the reforming process, the gas mixture must be thoroughly cleaned to remove any impurities (Hydrogen Production Energy Conversion, 2020).

Production of Hydrogen

Figure 2. Production of hydrogen from different sources

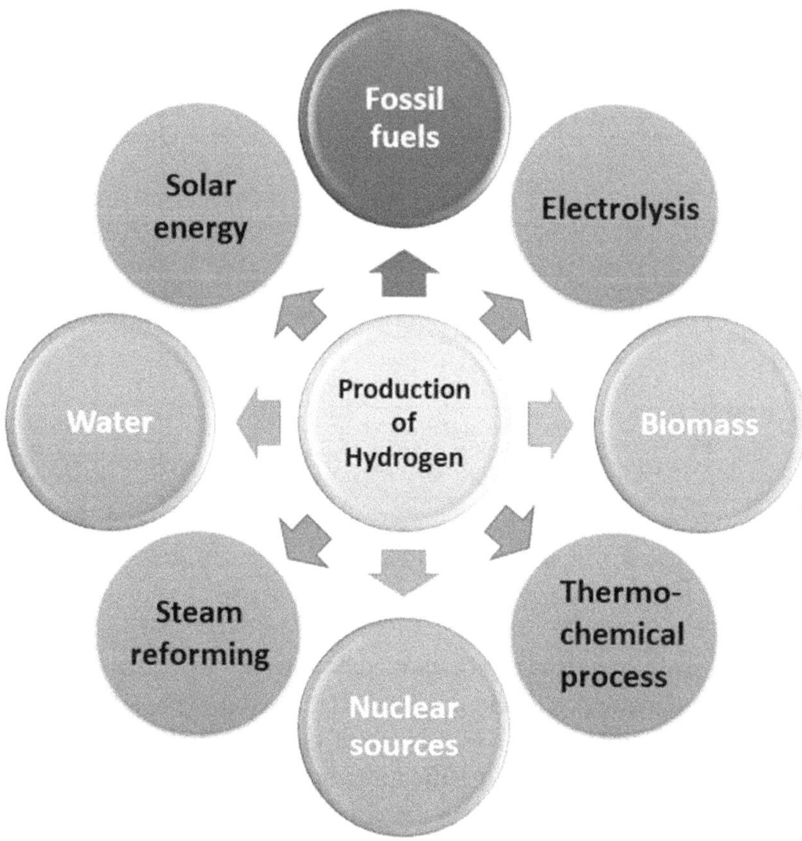

From Biomass

Biomass gasification is a process that can convert biomass into hydrogen. This method has low carbon dioxide emissions and can be used to produce renewable fuel. The most common biomass sources for hydrogen production are municipal solid waste and agriculture. Biogasification, biological reactions, and pyrolysis are the various steps that can be utilized to extract hydrogen from biomass fuels (Xuan et.al., 2009). The most common method of producing hydrogen is through biomass feedstock, as its gasifiers are readily available. Further studies are required to improve the efficiency of these processes and increase the hydrogen production rate. In addition to hydrogen, high temperature gasification can also be used to produce other types of fuel. This process can be used to produce hydrogen without the need for CO_2. The heated light hydrocarbons are then converted into hydrogen and carbon through a process known as plasma arc gasification. Another method is to produce hydrogen by using algae or bacterial decomposition of water (Demirbas & Leung, 2004).

From Nuclear Sources

Collisions between high-energy particles created by a nuclear reactor and water molecules are the process of water splitting called as Radiolysis is not very promising, as it can only produce around 1% efficiency. Thermal energy from a nuclear reactor can be used to produce hydrogen. This process involves splitting molecules by exposing them to high temperature (Verfondern, 2007). If no catalyst is used, a steam of 2500 °C will turn into oxygen and hydrogen after being dissociated. High temperature can affect the design of certain materials. This process produces by-product heat, which is used in the production of hydrogen. Various reactor technologies are utilized to process this waste. The process is considered to be economically attractive, and it does not produce greenhouse gas emissions. Nuclear energy is used to produce hydrogen in four different ways. These include steam reforming, thermochemical water splitting, electrolysis, and high temperature steam. The waste heat from a nuclear reactor is utilized in these two methods, which are more efficient than those that utilize low temperature electrolysis (El-Emam and Khamis, 2019). Various hydrogen production methods can be utilized with nuclear power to produce hydrogen. However, one of the main issues that can affect the safety of this process is the presence of a flammable substance in the system. This issue can lead to the failure of the facility to operate safely.

From Water

The process of electrolysis involves separating water into oxygen and hydrogen. It is regarded as a green route when the electricity comes from sources that are not dependent on fossil fuels. For instance, wind energy doesn't produce CO_2, but fossil fuels do. On the other hand, photo-electrolysis relies on sunlight to power its operation. An electrolyser is used in the process of separating water into hydrogen and oxygen. It can be divided into three types: solid oxide, proton exchange membrane, and an alkaline electrolyser. The three components of an electrolyser are water, an electrolyte, and two electrodes. Free ions are used to make substances electrically conductive. Depending on the type of electrolyte used, an electrolyser can be classified as a low temperature electrolyser, a proton exchange membrane, or a solid oxide electrolyser

(Bhandari et. al., 2014). The former refers to devices that operate at less than 100 °C, while the latter is regarded as high temperature electrolyser. Water electrolysis is known to produce high-purity hydrogen.

From Steam Reforming of Methane and Methanol

A process known as steam reforming involves the conversion of hydrocarbons, which are usually fossil fuels, into hydrogen. When combined with steam, the two components are then subjected to high pressure and temperature that produces carbon dioxide and hydrogen. Although the energy content of hydrogen produced by steam reforming is higher than that of natural gas, the process requires a considerable amount of energy to operate. This process involves the use of natural gas, which is composed of methane. It is mixed with water at high temperature. During this process, the heat is absorbed and the reaction is endothermic. The reaction should be performed at a temperature of around 850 °C. The second step involves a heat release, which is an exothermic reaction. The combined effect of these two reactions adds a significant amount of heat to the process (Milner-Elkharouf, 2020). The process is characterized by a cyclic process that involves the partial oxidation or steam reformation of methane. During the partial oxidation, the use of oxygen instead of steam is used to perform the reaction. This process is known as an auto-thermal reforming technique.

In order to produce hydrogen, methanol (CH_3OH) is used during the reformation process, which is significantly less demanding than the methane steam reforming method. The reformation process can yield up to 75% of the hydrogen concentration. Methanol can be used in stationary hydrogen applications due to its low temperature reformation process (Nancy et. al., 2012). In addition, it can be produced using renewable resources, which have high carbon to hydrogen ratio and low sulfur content. The process requires the use of expensive methanol since the hydrogen produced has impurities. Currently, methanol is produced from natural gas and coal. The world's total demand for this product is around 41 million metric tons. Besides these, other bio-based products such as pig manure, paper mill black liquor, and landfill gas are also used to produce methanol (Palo et. al., 2007). Methanol can be produced using natural gas and coal as the fuel. The process is attractive since it requires low temperature reformation. This method can be performed at a fast start-up and low cost.

From Electrolysis

Through electrolysis, water molecules can be separated into their constituent oxygen and hydrogen by passing an electrical current through it. This process generally results in a modest energy loss. Large electrolysers can achieve efficiency of up to 85%. Despite the small size of its operation, electrolysis has gained widespread attention due to its ability to produce clean hydrogen. However, it is currently only being used by small plants, which can cost around two to three dollars for each kilogram of hydrogen produced. Due to the high cost, it is currently not feasible for large industrial plants to use electrolysis. In the long run, it could become more competitive. Currently, reforming natural gas is more energy-efficient than electrolysis (Nancy et. al., 2012).

From Thermochemical Process

This process involves heating water to a certain temperature, which then turns it into oxygen and hydrogen. The simplest version of this technique is called direct thermal conversion. However, since high temperatures are required, this method is not practical outside of laboratories. Various chemical reactions can be used to reduce the temperature. Their use has been studied in complex multistep processes, and hybrid methods that combine electrolysis with several reactions are being investigated. There has been a lack of recent work on thermochemical methods.

From Solar Energy to Hydrogen

A simple and original form of hydrogen production is called solar hydrogen, which involves using sunlight to generate electricity and electrolyzing water to produce hydrogen. This method would replace fossil fuels such as oil in the future. This concept has gained widespread attention due to its potential to reduce the environmental impact of burning fossil fuels. This concept addresses two of the main barriers preventing the widespread use of solar energy. In photovoltaic systems, which are used to generate electricity, solar energy is used, however it can also be used to collect thermal power. Since it can be stored and terraform into other energy carriers, it can meet the energy requirements of the world. Unfortunately, the availability of sunlight during the daytime makes solar energy less attractive for the production of hydrogen. The cost of implementing solar energy systems and the space required for their operation are some of the drawbacks of this technology. However, it can be used to provide thermal energy to various hydrogen production facilities (Kabir et. al., 2018). These include thermolysis, steam reformation, and thermochemical processes. In addition to photovoltaics, photon energy can also be generated by using various processes such as photo-electrolysis and photo-water splitting. In terms of temperature, solar photovoltaics are known to produce oxygen and hydrogen. In addition, it can be utilized to produce hydrogen through photo-biolysis. In terms of temperature, solar thermal systems have high-temperature requirements for steam reforming. They can also be utilized to produce hydrogen through thermolysis and cracking.

HFC TECHNOLOGIES: PERSPECTIVES TOWARDS SUSTAINABLE DEVELOPMENT

The primary energy source used in the world is fossil fuels. Despite the various advantages of these fuels, they have negative environmental impacts. One way to address these issues is by utilizing fuel cells. Although these types of cells exhibit promising characteristics, they also have some environmental aspects that need to be discussed and effective measures to be addressed. Despite the various advantages of FCs, there are still areas where they can experience environmental and operational setbacks. These setbacks are compared in order to provide a clear and fair comparison between the various types of FCs. Although FCs are generally considered to be more efficient and environmentally friendly, they still face challenges when it comes to their fuel sourcing. For instance, when it comes to obtaining hydrogen, the production of hydrogen can have a negative impact on the environment.

Importance of Sustainable Development

The concept of sustainable development is a framework that aims to meet the various needs of the society while maintaining the ability to provide adequate resources for the people. It is based on the principle of environmental stability. Without undermining its integrity, the concept of sustainable development provides the necessary energy to meet the needs of the society. This is done through the use of renewable energy sources such as wind energy. Since energy systems are abundant in nature, we should look after them for a long time. These systems do not have a negative impact on the environment. One of the most important factors that we should consider is the environmental impact of the energy that we consume. In addition to ensuring that the systems do not emit harmful pollutants, it is also important to consider that they do not affect the atmosphere. There are various types of hydrogen systems that are compatible with each other. The development of sustainable technology is the best way to ensure that the system operates efficiently. Various factors can be considered in order to develop a stronger hydrogen fuel cell and sustainability strategy.

Some of the important factors are:

Table 2. SDGs addressed by sustainable mobility

SDGs	Important factors for sustainable mobility
Quality education (4)	• To raise awareness about the various advantages of hydrogen fuel and its potential to contribute to sustainable development.
Decent work and economic growth (8)	• There is a wide range of training and education programs that are focused on sustainable practices. • Various energy efficiency techniques and production methods can help you achieve better energy performance. • The government and private sectors work together to develop programs and initiatives that benefit both the public and private sectors. • Developing policies and programs that promote sustainable energy use is a vital part of any organization's efforts to reach its goals. • The evaluation and monitoring of various tools should be done in a proper manner.
Reduced inequalities (10)	• The availability of renewable resources and their utilization are two of the most critical factors that affect the energy supply equation. • A sufficient supply of financing and attractive incentives are some of the factors that can help a buyer make a successful purchase. • To make sure that people are aware of its application, road maps will be created for future implementations.
Sustainable cities and communications (11)	• Hydrogen fuel cells and other such systems for mobility are potentially reliable and commercially viable. • Various processes, such as transportation, storage, distribution, and production, can benefit from utilizing cleaner technologies.
Climate action (13)	• Promoting sustainable technologies that do not harm the environment is very important. • Various clean hydrogen production methods have been identified that are economically viable. • The development of sustainable hydrogen infrastructure will provide various areas with sound economic benefits.
Life on land (15)	• Clean technologies can help with the transportation, storage, and production of goods and services.

The sustainable quality of life that people desire is the basis of providing a safe and reliable energy supply. To guarantee a competitive environment, energy systems have to meet the needs of various societal groups at a low cost. The link between environmental impact and efficiency is strong, as it can be observed that for the same products or services, lower pollution and resource utilization are often associated with increased efficiency. Most energy efficiency improvements can lead to direct environmental benefits, such as reduced energy losses and the reduction of pollutants. Analyzing the life cycle of an energy resource or technology helps identify areas where efficiency can reduce the environmental impact. Understanding the various energy consumption and emissions of hydrogen technologies during their lifetimes is very important to ensure that they are competitive with existing technologies. This can help develop future scenarios that are more sustainable.

The increasing recognition of people's relationship with the environment has led to an increase in the scope of sustainable development. Fossil fuels have been known to have significantly affected the planet. With energy being a vital part of the world's economy, the widespread use of these fuels may have had a larger impact than previously believed. Aside from the relatively minor impacts of drilling and mining for fossil fuels, burning these fuels has been regarded as one of the main factors that contributes to climate change. This environmental issue has been linked to various negative effects, such as flooding and food chain disruption. With the help of fuel cells and hydrogen technologies, we can potentially achieve a more sustainable and efficient energy system. This alternative to fossil-fuel-based technologies can be particularly beneficial for the environment. Fuel cell and hydrogen technologies can help provide a more sustainable and efficient alternative to traditional energy sources, such as fossil fuels.

The development and use of fuel cell and hydrogen technologies are crucial for meeting the energy needs of both urban and rural regions. Due to the increasing public awareness about the harmful effects of fossil fuels, the development of these technologies should be prioritized. The rapid emergence and growth of the energy sector is a major concern for the world's developing and industrialized nations. The widespread use of fuel cell and hydrogen technologies can help achieve sustainable energy levels.

Environmental Aspects and Effective Measures

Hydrogen's impact on the stratosphere is important as it can potentially help in the destruction of ozone. It is not possible to study this issue without first carrying out detailed research. Although it is possible to substitute hydrogen for fossil fuels in reducing greenhouse gas emissions, it is also possible that the use of hydrogen in aviation fuel could negatively affect the ozone layer. Due to the complexity of chemical reactions occurring in the atmosphere, it is not possible to predict with certainty the impact of hydrogen on the ozone layer.

The transportation sector is considered to be one of the most energy-intensive industries in the world, and it is therefore facing various challenges when it comes to the burning of energy. One of these is the increasing number of road transportations that are contributing to the greenhouse gas emissions (Zhao et al., 2020). According to the regulations issued by the International Agency for Energy, the global carbon dioxide emission should be reduced to limit the effects of climate change. To achieve this, the automotive industry is promoting the use of hydrogen fuel cells (Meng et al., 2021). Since 2018, the governments of several countries, including the USA, Japan, and South Korea, have urged the use hydrogen in their transportation sectors, and in the next sections, we will talk about the technical and environmental elements of this technology. The transportation sector is one of the biggest sources of greenhouse gas emissions, and it poses a threat to human health. In 2021, the European Union adopted a road map that aims to improve the efficiency of the transport system. For most of its fleet, petroleum-based fuels will continue to be used. Renewable sources, on the other hand, will only account for about 10% of the transportation sector's energy requirements by 2050 (Le Quéré et al., 2020).

Aside from releasing harmful greenhouse gases (GHGs), power generation plants can also contribute to the environment's degradation. Some of these include the emission of carbon dioxide, volatile organic compounds, nitrogen oxides, carbon monoxide, and sulfur dioxide. In addition, they can release particulate matter into the air and water, which can contribute to the pollution of ecosystems. The cooling effect of a power plant is achieved by using vast amounts of water. This water is then treated with various chemicals, which can be toxic to aquatic life (Oetari et al., 2019; Pan et al., 2018). The emissions produced by power plants are classified into two categories: GHGs and aerosols. The former comprises compounds that are produced in the atmosphere by natural and man-made activities, while the latter are particles that reflect solar radiation (Dones and Heck, 2004). The emission of greenhouse gases into the atmosphere causes a decrease in Earth's temperature, which is referred to as the Albedo effect. The balance between the two effects is very important in regulating the planet's climate (Zhou et al., 2020). Global GHG emissions are similar to those that were reported for primary energy supply. According to the International Energy Agency, the main sources of CO_2 emissions are the transportation sector, power and heat, and industry (Wang et al., 2018; US-EPA, 2020). Despite being less than a third of the global power supply, coal is still the primary source of CO_2 emissions for both power and heat. About 70% of the global emissions of sulfur oxides are caused by coal, which is known to cause respiratory issues and worsen the environment due to acidic rains. Another harmful pollutant is nitrogen oxides, which can cause respiratory problems (Shcheklein and Dubinin, 2020). Acid rain is a result of the release of nitrogen oxides and sulfur dioxide. The resulting acidification effect has a negative impact on the environment, including the degradation of forests and crops, the corrosion of exposed structures, and the elevated acidity in bodies of water, such as rivers.

TECHNOLOGICAL CHALLENGES OF FUEL CELLS

The global demand for energy will increase threefold by 2050, which means that the oil and gas supply will not be able to meet the needs of the future. Due to the lack of supply, fuel cells and hydrogen have become an integral part of various energy systems (Edwards et al. 2008; Ni 2018; Rady et. al. 2012). Improvements in the storage, production, and reliability of hydrogen are needed to make it more suitable for future energy systems. There are various technical issues that can affect the hydrogen economy as follows:

- The cost of producing hydrogen is reduced in comparing to gasoline, which is comparable to energy costs. As a result, it is highly relieving for conventionally available fuels. Estimation of overall cost consisting, labor, equipment, supervisions, testing, etc., are draining the economic sustainability. FCs provide a bulkier prospect and improving sustainable development altogether.
- The goal of CO_2 -free route is to create a sustainable hydrogen production system that is competitive with current technology. In order to gain healthy and breathable environment, preventing the increasing exponential rate of CO_2, GHGs, and atmospheric temperature is vital aspect. In general understanding to control releasing of such gases by fuel using FCs is commendable approach.
- These systems are designed to store hydrogen for stationery and mobile applications. A durable mechanistic approach is need of future, able to hold strong energy stored and inhibits any possibility of leaks or drop through components. The reliability of such system needs regulatory measures to withstand huge surplus demands at current rate. Storing also requires protection against physical damages to containing body, also needs to be weatherproof, shock proof, radiation proof; to avoid any minor chances of temporal drudge.
- The goal is to develop a safe and efficient hydrogen distribution and delivery system. Energy delivering is one most critical challenge, ample of studies are required to calibrate any standard operations. Energy delivery not only limits to human operations, rather is dependent upon installed components, fail-safe benchmarks, and adverse condition trainings. FCs behold potential in many different forms such as chemicals and gases, raw energy is projected to be delivered by either of two in near future.

During the transition to a new energy source, such as fuel cells, various processes related to gasification, carbon dioxide capture, and maintenance are required. These include the development of new economical and efficient electrochemical technologies. The replacement of batteries and power systems for cars and distributed generation is also feasible. The fuel cells being offered today are, which a very compelling type of technology. With the availability of hydrogen in the future, the efficiency of hydrocarbons could be increased. However, the challenges that fuel cells face are the increasing cost and the durability of their components.

DISCUSSION AND FUTURE EXPECTATIONS

Hydrogen fuel cell technology is a promising alternative to fossil fuels for mobility. It can be used in a variety of applications, such as electric cars, bicycles, and sport-related vehicles. Its low cost and efficient means it can be a valuable alternative to gasoline. The rapid emergence and evolution of fuel cells has made them an eco-friendly power generation technology. Today, they are commonly used for backup and primary power in commercial, industrial, residential, and automotive buildings. However, they can also be utilized to power various types of vehicles such as boats, forklifts, and cars. Due to the improving performance of fuel cells, the importance of utilizing high-purity materials such as graphene and carbon has become more apparent.

Due to the increasing number of electric cars, driving range has become a major challenge. Although fuel cells can be used in various applications, they require a refueling station. On the other hand, standard electric cars are still limited by their batteries. This limitation is not ideal for commercial vehicles, especially those that are used for long distances. In certain regions, such as Germany, Japan and North America, fuel cells are commonly used in buses and mass transportation. They can provide electrical power comparable to diesel or gas engines. The global goal of market is to provide safe, reliable, and sustainable automobiles that can run on low-emission fuels. Unfortunately, the current technology does not meet these needs. Besides the lack of infrastructure, other factors such as the production of hydrogen from fossil fuels are also hindering the development of green mobility. The current production of hydrogen is less than what's needed to meet the needs of the automobile industry. This means that additional funds are needed to increase production.

Conventional fuels are considered harmful to the environment, and hydrogen is an eco-friendly alternative. It can be used as an energy source and carrier, and it can help reduce the greenhouse gas emissions caused by the combustion of fossil fuels. The immense energy content and low atomic weight of hydrogen make it an ideal candidate to help decrease the global fossil-oil reservoir's burden and improve the environment's sustainability. The extensive study on the various aspects of hydrogen production and storage provides a wide range of information that can be used by researchers/scientists and non-scientists to understand the technology. One of the drawbacks of hybrid cars is the time it takes to refuel twice. With the addition of a second battery pack, which can charge while driving, hybrid cars can extend the time between refueling procedures. One of the most common types of hybrid vehicles is the proton-exchange membrane and hydrogen fuel cell. This type of fuel cell can store chemical energy and convert it into electricity. The challenge of incorporating fuel cell and hydrogen technologies into mainstream cars has been raised by major carmakers. Despite the various efforts being made by carmakers to incorporate fuel cell technologies into mainstream cars, more studies are needed to develop the necessary infrastructure and applications for the use of hydrogen-based fuel cells. The goal of hydrogen fuel is to provide a clean energy source that is ideal for transportation. Compared to electric cars, hydrogen fuel-powered vehicles offer lower costs and a shorter refueling time.

Due to the high prices and limited power density of hydrogen, the widespread adoption of hydrogen cars has become a major issue. This is largely due to the lack of infrastructure that can support the production and distribution of hydrogen. In the short and medium term, an onboard fuel processor can help Fuel Cell Vehicles (FCVs) gain a competitive advantage by enabling direct hydrogen conversion. This concept is ideal for safety reasons as it eliminates the need for a large tank and allows the vehicle to refill using existing infrastructure. An onboard fuel processor can also be used to operate on different fuels, which is unlike the type of internal combustion engine that can only operate on bi-fuels. This eliminates the need for modifications to the operating conditions. Although hydrogen has long been regarded as a potential low-carbon fuel, it has been challenging to integrate it into the transportation mix. In addition to being pollution-free, fuel cells can also deliver better performance than traditional combustion methods. For instance, a traditional power plant can generate electricity at an efficiency of around 33 to 35%. On the other hand, fuel cell systems can achieve an efficiency of up to 60%. Conventional cars are less efficient when it comes to converting the energy from gasoline into power. This means that they consume more fuel than fuel cells, which are more energy-efficient.

The zero environmental impact of hydrogen fuel is attractive to the automobile industry as it allows for the development of hydrogen vehicles. However, it is important to note that the design of hydrogen engines is not ideal for producing combustion. Because of this, the efficiency of hydrogen engines, NOx emissions, and power production can be significantly improved. The cost of producing hydrogen can be influenced by the sources of this fuel. Ideally, a hydrogen plant should be located near its natural resources. Hydrogen technology can be adopted by people if they are aware of its positive effects on the environment. Increasing public support for the growth of this promising fuel can be achieved through education. Due to its zero emission and immense energy content, hydrogen has been regarded as the fuel of the future. However, despite its promising potential, its storage, generation, and usage are still not yet up to standard. The US Department of Energy has set a goal for hydrogen energy storage, which is aimed at ensuring that the fuel can be used to replace fossil-based fuels such as gasoline and diesel. The cost of producing and storing hydrogen is higher than that of fossil fuels. One of the biggest challenges that hydrogen power systems face is achieving a higher efficiency as compared to other types of power systems. This refers to the difference between how much electricity is delivered to and used by the power system. Despite the various technological advancements that have been made in the field of hydrogen, it would still take a long time for majority of cars to run on it.

One of the most promising ways to solve the energy crisis is by utilizing renewable sources such as water, wind, and solar energy to produce hydrogen. This can be done through the use of fuel cells and porous materials. In the future, research will focus on hybrid systems that combine the advantages of both the fuel cell and renewable energy technologies. A car with a stack of fuel cells, hydrogen storage, and solar rooftop is a classic example of how sustainable transportation will be in the future (Oberoi 2015). Unlike traditional fuels, which tend to emit harmful gases, this system would not. In another example, a vehicle equipped with wind turbines on its side panels can use the vehicle's wind draft to power its motors. The system could save the country millions of dollars in the financial investment needed to build hydrogen and electricity-charging stations.

Here we have provided a comprehensive analysis of the current state of hydrogen and its potential role in the future sustainable energy framework. It also aims to highlight the various challenges and opportunities that face the hydrogen industry and its perspective towards sustainable development goals.

CONCLUSION

The world's energy demand is becoming more difficult to meet due to the dependence on fossil fuels, which are known to cause various environmental impacts. To lessen the impact of this issue, the world should gradually transition to alternative energy sources like wind, solar, and hydrogen. As an energy carrier, hydrogen has various characteristics that make it an ideal choice for the future. It can be used in various applications such as power generation and transportation. Due to its clean and safe environment, it is becoming more prevalent in the energy industry. Various commercial methods are available for producing hydrogen. Some of these include steam methane reforming, coal gasification, thermochemical method, and biomass gasification. In the future, solar and electrolysis are expected to be the most prominent methods. There are still a lot of research and development activities that need to be carried out in order to develop efficient hydrogen storage methods. Solid state storage is expected to be a promising technology.

REFERENCES

Abdelkareem, M. A., Elsaid, K., Wilberforce, T., Kamil, M., Sayed, E. T., & Olabi, A. (2020). Environmental aspects of fuel cells: A review. *The Science of the Total Environment, 141803*. Advance online publication. doi:10.1016/j.scitotenv.2020.1418 PMID:32889267

Abdollahi Haghghi, M., Shamsaiee, M., Ghazanfari Holagh, S., Chitsaz, A., & Rosen, M. A. (2019). Thermodynamic, exergoeconomic, and environmental evaluation of a new multi-generation system driven by a molten carbonate fuel cell for production of cooling, heating, electricity, and freshwater. *Energy Conversion and Management, 199*, 112040. doi:10.1016/j.enconman.2019.112040

Aili, D., Henkensmeier, D., Martin, S., Singh, B., Hu, Y., Jensen, J. O., ... Li, Q. (2020). *Polybenzimidazole-Based High-Temperature Polymer Electrolyte Membrane Fuel Cells: New Insights and Recent Progress*. Electrochemical Energy Reviews., doi:10.100741918-020-00080-5

Apostolou, D., Enevoldsen, P., & Xydis, G. (2019). Supporting green urban mobility – the case of a small-scale autonomous hydrogen refuelling station. *International Journal of Hydrogen Energy, 44*(20), 9675–9689. doi:10.1016/j.ijhydene.2018.11.197

Apostolou, D., & Xydis, G. (2019). A literature review on hydrogen refuelling stations and infrastructure. current status and future prospects. *Renewable & Sustainable Energy Reviews, 113*, 109292. doi:10.1016/j.rser.2019.109292

Appleby, A. J., & Foulkes, F. R. (1993). *Fuel cell handbook. Van Norstand Reinhold, New York, NY, (1989)*. Republished by Krieger Publishing Company.

Arico, A. S., Baglio, V., & Antonucci, V. (2009). *Direct methanol fuel cells: history, status and prespectives*. Wiley-VCH.

Barckholtz, T. A., Taylor, K. M., Narayanan, S., Jolly, S., & Ghezel-Ayagh, H. (2022). Molten carbonate fuel cells for simultaneous CO2 capture, Power Generation, and H2 generation. *Applied Energy, 313*, 118553. doi:10.1016/j.apenergy.2022.118553

Bessette, N. F. II, & Wepfer, W. J. (1996). Prediction of on-design and off-design performance for a solid oxide fuel cell power module. *Energy Conversion and Management, 37*(3), 281–293. doi:10.1016/0196-8904(95)00188-3

Bhandari, R., Trudewind, C. A., & Zapp, P. (2014). Life cycle assessment of hydrogen production via electrolysis – A Review. *Journal of Cleaner Production, 85*, 151–163. doi:10.1016/j.jclepro.2013.07.048

Bhaskaran, R., Abraham, B. G., & Chetty, R. (2021). Recent advances in electrocatalysts, mechanism, and cell architecture for direct formic acid fuel cells. *WIREs Energy and Environment*, *11*(2). Advance online publication. doi:10.1002/wene.419

Bicer, Y., & Dincer, I. (2017). Comparative life cycle assessment of hydrogen, methanol and electric vehicles from well to wheel. *International Journal of Hydrogen Energy*, *42*(6), 3767–3777. doi:10.1016/j.ijhydene.2016.07.252

Cui, C., Li, S., Gong, J., Wei, K., Hou, X., Jiang, C., & Ma, J. (2021). Review of molten carbonate-based direct carbon fuel cells. *Materials for Renewable and Sustainable Energy*, *10*(2), 12. Advance online publication. doi:10.100740243-021-00197-7

Demirbas, A. Y. H. A. N., & Arin, G. O. N. E. N. C. (2004). Hydrogen from biomass via pyrolysis: Relationships between yield of hydrogen and temperature. *Energy Sources*, *26*(11), 1061–1069. doi:10.1080/00908310490494568

Dincer, I. (2020). Covid-19 coronavirus: Closing carbon age, but opening hydrogen age. *International Journal of Energy Research*, *44*(8), 6093–6097. doi:10.1002/er.5569 PMID:34173427

Dones, R., Heck, T., & Hirschberg, S. (2004). Greenhouse gas emissions from energy systems, comparison and Overview. Encyclopedia of Energy, 77–95. doi:10.1016/B0-12-176480-X/00397-1

Edwards, P. P., Kuznetsov, V. L., David, W. I. F., & Brandon, N. P. (2008). Hydrogen and fuel cells: Towards a sustainable energy future. *Energy Policy*, *36*(12), 4356–4362. doi:10.1016/j.enpol.2008.09.036

El-Emam, R. S., & Khamis, I. (2019). Advances in nuclear hydrogen production: Results from an IAEA International Collaborative Research Project. *International Journal of Hydrogen Energy*, *44*(35), 19080–19088. doi:10.1016/j.ijhydene.2018.04.012

Falcone, P. M., Hiete, M., & Sapio, A. (2021). Hydrogen Economy and Sustainable Development Goals: Review and Policy Insights. *Current Opinion in Green and Sustainable Chemistry*, *31*, 100506. doi:10.1016/j.cogsc.2021.100506

Garland, N. L., Papageorgopoulos, D. C., & Stanford, J. M. (2012). Hydrogen and fuel cell technology: Progress, challenges, and Future Directions. *Energy Procedia*, *28*, 2–11. doi:10.1016/j.egypro.2012.08.034

Goor, M., Menkin, S., & Peled, E. (2018). High power direct methanol fuel cell for mobility and portable applications. *International Journal of Hydrogen Energy*, *44*(5), 3138–3143. doi:10.1016/j.ijhydene.2018.12.019

Hart, D. (2000). Sustainable energy conversion: Fuel cells — the competitive option? *Journal of Power Sources*, *86*(1-2), 23–27. doi:10.1016/S0378-7753(99)00455-3

Kabir, E., Kumar, P., Kumar, S., Adelodun, A. A., & Kim, K.-H. (2018). Solar energy: Potential and future prospects. *Renewable & Sustainable Energy Reviews*, *82*, 894–900. doi:10.1016/j.rser.2017.09.094

Koscher, G. A., & Kordesch, K. (2003). *Handbook of fuel cells- fundamentals, technology and applications* (Vol. 4). John Wiley and Sons.

Le Quéré, C., Jackson, R. B., Jones, M. W., Smith, A. J., Abernethy, S., Andrew, R. M., De-Gol, A. J., Willis, D. R., Shan, Y., Canadell, J. G., Friedlingstein, P., Creutzig, F., & Peters, G. P. (2020). Temporary reduction in daily global CO2 emissions during the COVID-19 forced confinement. *Nature Climate Change*, *10*(7), 647–653. doi:10.103841558-020-0797-x

Meng, X., Gu, A., Wu, X., Zhou, L., Zhou, J., Liu, B., & Mao, Z. (2021). Status quo of China hydrogen strategy in the field of Transportation and International Comparisons. *International Journal of Hydrogen Energy*, *46*(57), 28887–28899. doi:10.1016/j.ijhydene.2020.11.049

Midilli, A., Ay, M., Dincer, I., & Rosen, M. A. (2005). On hydrogen and Hydrogen Energy Strategies. *Renewable & Sustainable Energy Reviews*, *9*(3), 255–271. doi:10.1016/j.rser.2004.05.003

Milner-Elkharouf, L., Khzouz, M., & Steinberger-Wilckens, R. (2020). Catalyst development for indirect internal reforming (IIR) of methane by partial oxidation. *International Journal of Hydrogen Energy*, *45*(8), 5285–5296. doi:10.1016/j.ijhydene.2019.05.026

Munjewar, S. S., Thombre, S. B., & Mallick, R. K. (2017). Approaches to overcome the barrier issues of passive direct methanol fuel cell – Review. *Renewable & Sustainable Energy Reviews*, *67*, 1087–1104. doi:10.1016/j.rser.2016.09.002

Ni, M. (2018). New developments and challenges of solid oxide fuel cell (SOFC)-based technologies. *International Journal of Energy Research*, *42*(15), 1–6. doi:10.1002/er.4213

Nimir, W., Al-Othman, A., Tawalbeh, M., Al Makky, A., Ali, A., Karimi-Maleh, H., Karimi, F., & Karaman, C. (2023). Approaches towards the development of heteropolyacid-based high temperature membranes for PEM fuel cells. *International Journal of Hydrogen Energy, 48*(17), 6638–6656. doi:10.1016/j.ijhydene.2021.11.174

Oberoi, A. S. (2015). *Reversible electrochemical storage of hydrogen in activated carbons from Victorian brown coal and other precursors.* [PhD thesis, RMIT University Australia]

Oetari, P. S., Hadi, S. P., & Huboyo, H. S. (2019). Trace elements in fine and coarse particles emitted from coal-fired power plants with different air pollution control systems. *Journal of Environmental Management, 250*, 109497. doi:10.1016/j.jenvman.2019.109497 PMID:31491717

Palo, D. R., Dagle, R. A., & Holladay, J. D. (2007). Methanol steam reforming for hydrogen production. *Chemical Reviews, 107*(10), 3992–4021. doi:10.1021/cr050198b PMID:17845061

Pan, S.-Y., Snyder, S. W., Packman, A. I., Lin, Y. J., & Chiang, P.-C. (2018). Cooling water use in thermoelectric power generation and its associated challenges for addressing water-energy nexus. *Water-Energy Nexus, 1*(1), 26–41. doi:10.1016/j.wen.2018.04.002

Patel, P., Maru, H. C., Borglum, B., Stokes, R. A., Petri, R. J., Sishtla, C., Krist, K., Armstrong, T., & Virkar, A. (2004). *Thermally integrated power systems, high power density SOFC generator. Proceedings of the Fuel Cell Seminar*, San Antonio, TX.

Pehnt, M. (2003). Assessing future energy and transport systems: The case of Fuel Cells. *The International Journal of Life Cycle Assessment, 8*(5), 283–289. doi:10.1007/BF02978920

Rady, A. C., Giddey, S., Badwal, S. P. S., Ladewig, B. P., & Bhattacharya, S. (2012). Review of Fuels for Direct Carbon Fuel Cells. *Energy & Fuels, 26*(3), 1471–1488. doi:10.1021/ef201694y

Salehi, A., Mousavi, S. M., Fasihfar, A., & Ravanbakhsh, M. (2019). Energy, exergy, and environmental (3E) assessments of an integrated molten carbonate fuel cell (MCFC), Stirling engine and organic Rankine cycle (ORC) cogeneration system fed by a biomass-fueled gasifier. *International Journal of Hydrogen Energy, 44*(59), 31488–31505. doi:10.1016/j.ijhydene.2019.10.038

Shamardina, O., Chertovich, A., Kulikovsky, A. A., & Khokhlov, A. R. (2010). A simple model of a high temperature PEM fuel cell. *International Journal of Hydrogen Energy, 35*(18), 9954–9962. doi:10.1016/j.ijhydene.2009.11.012

Staffell, I., Scamman, D., Abad, A. V., Balcombe, P., Dodds, P. E., Ekins, P., Shah, N., & Ward, K. R. (2018). The role of hydrogen and fuel cells in the global energy system. doi:10.31224/osf.io/rzm4gosf.io/rzm4g

Verfondern, K. (2007). *Nuclear energy for hydrogen production*. Forschungszentrum, Zentralbibliothek.

Xu, Z., Qiu, D., Yi, P., Peng, L., & Lai, X. (2020). Towards mass applications: A review on the challenges and developments in metallic bipolar plates for PEMFC. *Progress in Natural Science*, *30*(6), 815–824. doi:10.1016/j.pnsc.2020.10.015

Xuan, J., Leung, M. K. H., Leung, D. Y. C., & Ni, M. (2009). A review of biomass-derived fuel processors for fuel cell systems. *Renewable & Sustainable Energy Reviews*, *13*(6-7), 1301–1313. doi:10.1016/j.rser.2008.09.027

Zhao, F., Mu, Z., Hao, H., Liu, Z., He, X., Victor Przesmitzki, S., & Ahmad Amer, A. (2020). Hydrogen fuel cell vehicle development in China: An industry chain perspective. *Energy Technology (Weinheim)*, *8*(11), 2000179. doi:10.1002/ente.202000179

Zhou, M., Chen, G., Dong, Z., Xie, B., Gu, S., & Shi, P. (2020). Estimation of surface albedo from meteorological observations across China. *Agricultural and Forest Meteorology*, *281*, 107848. doi:10.1016/j.agrformet.2019.107848

APPENDIX

Acronyms

AFC - Alkaline Fuel Cell
DMFC - Direct Methanol Fuel Cell
FCV - Fuel cell vehicle
GHG - Green House Gases
HFC - Hydrogen Fuel Cell
PEMFC - Proton Exchange Membrane Fuel Cell
MCFC - Molten Carbonate Fuel Cell
PAFC - Phosphoric Acid Fuel Cell
PEMFC - Proton Exchange Membrane Fuel Cell
SOFC - Solid Oxide Fuel Cell
SDG - Sustainable Development Goal

Chapter 4
Hydrogen Economy and Fuel Cell Technology:
A South African Context

Athule Ngqalakwezi
https://orcid.org/0000-0003-2024-0389
Catalysis Group-Advanced Materials Division, Mintek, South Africa

ABSTRACT

The improvement in the standard of living due to urbanization and industrialization can without a doubt be accredited to the discovery and use of fossil fuels over the centuries. Although this energy source has had a huge positive impact on the human population, the negatives are just as huge. The continuous use of fossil has been detrimental to the environment and has pushed the world into a paradigm shift where green technology has to be implemented. Hydrogen as an alternative source has intrigued many researchers due to its zero emission characteristic and higher energy content compared to traditional energy sources. Fuel cells which use hydrogen as a fuel source, are the fastest growing clean technologies. Fuel cells are used in both stationary and on-board applications. In this chapter, the hydrogen economy and fuel cell technology are discussed in the South African context and whether or not the complete replacement of fossil fuel with hydrogen is feasible or not.

INTRODUCTION

Access to clean energy has become one the most important global needs that have to be addressed with urgency due to the prevalent climate issues. As a rehabilitation effort, many governmental laws have been set in place to drive the economy towards clean

DOI: 10.4018/978-1-6684-6721-3.ch004

energy sources (Gallagher, 2014; Gerrard, 2011; Martin & Rice, 2014; Mormann, 2015; Msimanga & Sebitosi, 2014; Sovacool, 2021; Winkler, 2005). Hydrogen, a clean source of energy with energy twice as much as traditional sources, has garnered a lot of attention for application in both stationary and mobile application (Hwang & Varma, 2014; Ngqalakwezi et al., 2020; Schlapbach & Züttel, 2001). As such, the hydrogen economy has been established in most countries globally and with Southern Africa being no exception. Since this establishment the political and environmental momentum for advanced hydrogen energy research and market has grown tremendously.

Hydrogen roadmaps have been drafted and hydrogen strategies have been set in place as initial step towards developing hydrogen economy, deploying hydrogen energy systems and technologies in South Africa (Pollet et al., 2015; Santhanam et al., 2017). Under the Paris climate agreement, South Africa has progressed in trying to establish a green economy and meeting emission targets for current technology. The establishment of the Hydrogen South Africa (HySA) program by the Department of Science and Technology was one the key steps taken to grow the scale of research, innovation and the development hydrogen technology in SA (Barrett, 2014; Pollet et al., 2014a). The key mandate of the program being to attain goals of the national and global strategy towards a greener environment. Under the program, three centers of competency were established with three primary focus areas HySA catalysis, HySA infrastructure and HySA systems. Mintek, a metals and mining consulting company in SA, co-hosts the HySA catalysis R&D with the University of Cape Town. HySA infrastructure and HySA systems are hosted the Council for Scientific and Industrial Research (CSIR), North West University and the University of the Western Cape (Barrett, 2014).

The primary aim and focus of this research and of the centers is to beneficiate the natural resources available, moving from merely manufacturing platinum ores and processed metals to producing fuel cell catalyst and components in large scale (Barrett, 2014; Pollet et al., 2015). This is a deliberate strategy that is promoting the use of our local resources to grow our research capacity while also creating knowledge and expanding the human resource capacity (Bessarabov et al., 2017; Irena, 2011; Santhanam et al., 2017; Technology, 2007). The initial goal/mandate of the program was, a). To create a base for hydrogen processes, storage technologies and hydrogen production, b). Using our PGM resources, establish a base for creating Pt/PGM based electrocatalyst for fuel cells and c). Address local developmental challenges through developing fuel cell technology and building on the already existing global knowledge (Pollet et al., 2015).

A lot of these goals have been achieved with the program being extended for the next 10 coming years. In this chapter, a background in fuel cells and fuel cell technology is discussed, the impact and progress the centers have made on a global scale, and the progress the HySA organisation has made and prototypes it has developed thus far are also discussed.

HYDROGEN ECONOMY

PGMs in South Africa

Platinum group metals (PGMs) are precious metals which include platinum (Pt), palladium (Pd), rhodium (Rh), Iridium (Ir), Osmium (Os) and ruthenium (Ru) (Online-Content, 2023). These metals are commonly known for their catalytic properties, purity and high melting points (Online-Content, 2023).South Africa, alongside Russia possess the largest reserves that bear platinum group metals (PGMs) and are amongst the major producers of PGMs (O'Connor & Alexandrova, 2021). The South African reserves are estimated to hold around 85 000 metric tons (about 77 110 703 kg) (Cawthorn, 2010). These reserves are found in the Bushveld complex. The Bushveld complex is located in the northern region of South Africa and is subdivided into three limbs i.e. western, eastern and northern limbs (Cawthorn, 2010; O'Connor & Alexandrova, 2021). The mineral and geological properties of these limbs have been studied a lot (Cramer, 2001; Engelbrecht, 2012; Schouwstra et al., 2000).

The largest concentrations of PGMs are found in the upper critical zone of the Bushveld complex. The map shown in figure 1 shows the zoning and different limbs of the Bushveld complex. The upper critical zone consists of the Merensky Reef, UG2 and the Platreef. The major platinum group metals are found in the sperrylite (PaAs2), cooperate (PtS) and braggite ((Pt.Pd)NiS) (Engelbrecht, 2012). The most commonly used PGMs out of the six, is platinum (Pt) and palladium (Pd). The graph below shows how the production of platinum in South Africa from the year 2010-2021. These largest production rates of platinum and reserves have prompted the need to utilize and beneficiate the PGMs within the country.

Figure 1. The map of the South African Bushveld complex with the different limbs and zone shown.
Reprinted with permission (Latypov et al., 2022) Open Access

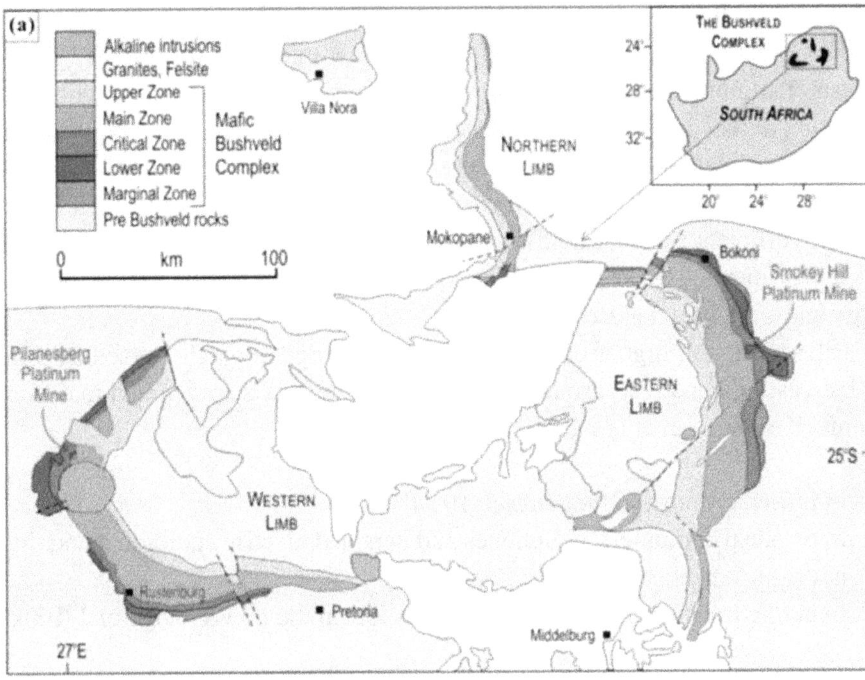

Figure 2. South Africa's Production of platinum from 2010 to 2021
(Garside, 2023)

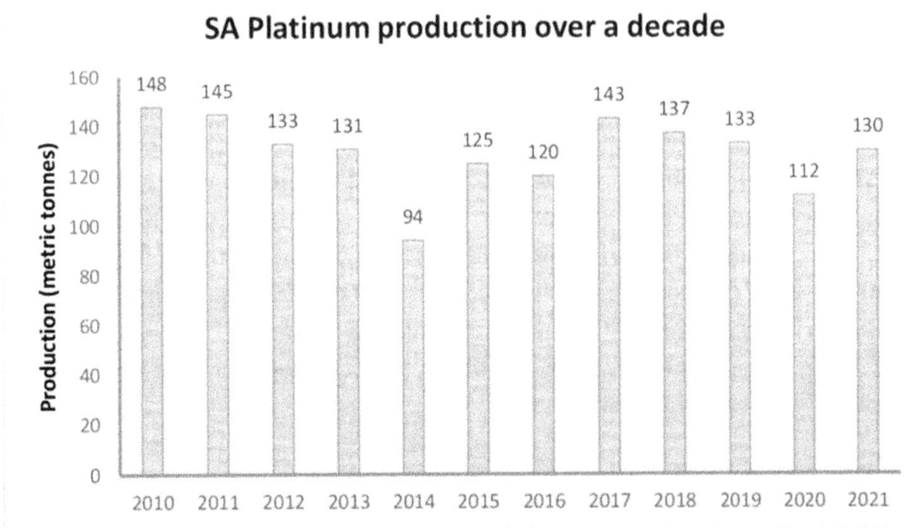

Properties of Platinum Group Metals

PGMs have excellent reduction and oxidations properties which make them applicable in different industrial processes, commercial and technological applications (Online-Content, 2023). They are excellent as raw material and are very good as catalysts. Platinum (Pt) and palladium (Pd) based catalysts have been chosen for application in fuel cells. Fuel cells are contraptions that produce electricity through an electrochemical reaction, where oxygen and hydrogen are combined to produce water, heat and electricity. The primary interest in fuel cells was sparked by their reliability, efficiency and cleanness in terms of environment emissions. This interest has grew tremendously because of its potential for commercial application in the energy and stationary landscape.

Furthermore, the higher energy of density hydrogen, which is used as a fuel, also increases the interests around these devices. Fuel cells are applied in different systems (Kirubakaran et al., 2009):

1. In power systems, in the range 1-10 MW.
2. In personal computers, cellphones and personal electric equipment and this is in the power range of 1W-10 Kw
3. In public transportation and power vehicles: in the power range of 1-100kW

A basic fuel cell comprises of the electrolyte membrane, an anode and a cathode. The cell functions by passing through hydrogen via the anode side of the fuel cell and passing oxygen on the cathode side. The hydrogen molecules are split into electrons and protons at the anode side using platinum/carbon electrocatalysts. The separated protons pass through the perforated electrolyte membrane and the electrons are forced through the circuit creating heat and electric current. When the electrons and protons combine with oxygen they produce pristine water molecules. There different types of fuel cells such as direct methanol fuel cells (DMFCs), polymer electrolyte membrane fuel cells (PEMFCs), solid oxide fuel cell (SOFCs), reversible fuel cells, phosphoric acid fuel cell (PAFCs) and alkaline fuel cells (AFCs) (Blomen & Mugerwa, 2013; Carrette et al., 2000; Mekhilef et al., 2012; O'hayre et al., 2016; Wang & Jiang, 2017). Amongst these, PEM fuel cells are the most commonly used devices. PEMFCs have two distinct types the high temperature PEMFCs (HTPEMFCs) and the low temperature PEMFCs (LTPEMFCs) (Carrette et al., 2000; Mekhilef et al., 2012). The difference between the HTPEMFC and the LTPEMFC is listed on the table below:

Table 1. Properties of different fuel cells

Fuel Cell Type	Pt Loading	Power density	Electrolyte	Operating Temp	CO tolerance
HTPEMFCs	1– 2 mg/cm^2	Lower	Mineral acid based	Up to 200 °C	1-5% by volume
LTPEMFCs	0.2-0.8 mg/cm^2	Higher	Water based	80-100 °C	<50 ppm

The research surrounding HT-PEMFC has gained a lot of traction over the past two decades and is slowly becoming an alternative to the common LT-PEMFC. The high operating temperature (150–180 °C) of HT-PEMFCs has advantages compared to LT-PEMFCs. The main advantage being that at higher temperatures the system is more CO tolerant compared to LT-PEMFCs. Li et al, reported that up to 5% CO tolerance was observed on the anode gas supply for a single cell (Q. Li et al., 2003). In the space of high temperature fuel cell technology, HySA primarily looks at developing the technology and its application with a greater goal of integrating and completing CHP systems.

South African 'Platinum Valley'

The South African Platinum is an initiative that forms part of the government plan for economic recovery in South Africa. This initiative consolidates to other national policies in South Africa such as the national climate change response white paper and the minerals beneficiation strategy (DSI, 2020). The development of the platinum valley will align South Africa to global strategies that support sustainable extraction and processing and recycling of platinum group metals using low carbon technologies.

The key mandate for developing the South African Platinum valley is to bring different hydrogen applications and form an integrated South African hydrogen ecosystem (DSI, 2020). The key objective of this initiative is to reduce our carbon footprint, increase the valorisation of our PGM reserves, develop new industries and subsequently create new job opportunities to boost the economy. This will be done through the identification of opportunities to initiate hydrogen activities in promising areas and a lot of this has been kick-started (DSI, 2020). In the North West, a province in South Africa, fuel cell were deployed in Poelano High School which is situated in an off-grid area to supply power. More recently, more fuel cell systems were installed in 1 Military hospital as primary power suppliers as part of COVID 19 response. A recent project of installing fuel cells as a demonstration of technology is also currently underway at Mintek, a mining and metals consulting company in Randburg. More projects are set to start in PGM mining regions/areas such as Limpopo and Gauteng.

FUEL CELL TECHNOLOGY

HySA Centers of Competence

The HySA centers of competence integrate to form hydrogen research and they collaborate with industry and institutions in the R&D sector(Pollet et al., 2014a). The HySA program has five key programs which are distributed and shared amongst the five centers of competence. The table below shows the centers and the programs they run:

Table 2. The centers and the key programmes they ran

Centers of Competence	Institution/s in charge	Key Programme
HySA Catalysis	Mintek & University of Cape Town (UCT)	Portable power systems
HySA Systems	University of the Western Cape (UWC)	Combined heat and power
HySA Systems	University of the Western Cape (UWC)	Hydrogen fuelled vehicles
HySA Infrastructure	CSIR & North West University (NWU)	Hydrogen filling stations
HySA Infrastructure	CSIR & North West University (NWU)	Renewable hydrogen production

Hydrogen Fueled Vehicles

The clean fuel strategy launched in January 2013 by the European Commission that's aimed at reducing oil-dependent vehicles in the transport industry. The strategy states that member countries are to deploy about 9 million electric vehicles (EVs) by 2020 (Asif & Schmidt, 2021; Ciura, 2019; Pollet et al., 2014b). This quota includes Fuel cell vehicles (FCVs), to which the International energy agency (IEA) estimates 10 million fuel cell vehicles will be sold around the world by 2050. Fuel cell vehicles are powered using hydrogen which produces electricity using a fuel cell rather than get power from a battery (Department-of-Energy-US, 2022). Figure 3 below shows a typical fuel cell vehicles and its main components.

Predictions in the demand fuel cell vehicles is projected to larger for passenger vehicles with quite a number of models being exhibited globally (Pollet et al., 2014b). A lot of government investments and stringent policies in terms of emissions have pushed the more R&D work to be done in the field. As much as fuel cells vehicles are promising, however the fuel cell durability, infrastructure and the overall remain critical challenges that impede market uptake of fuel cell vehicles. In this regard, Hydrogen South Africa's approach in the FCVs market is to find opportunities and offer solutions to some of the impeding challenges thereby having a stance in the mainstream automotive industry. The mobile application of hydrogen opens a major opportunity for the utilization of hydrogen fuel cells not only on the South African market but globally (HySA_Systems, 2014). Some demonstration have been made already by mainstream automotive giants.

One of HySA's key programme is to remodel and assimilate systems and modules that can be utilized in fuel cell vehicles globally based on HySA developed technology. These technologies included lithium-ion battery systems, on-board hydrogen storage, fuel cell stack (stack and its controller) and power electronics. Early demonstrations using forklifts and special purpose vehicles both which are primarily used in the mining sector, were shown to prove the concept (Asif & Schmidt, 2021; Pollet et al., 2014b). Figure 4 below shows the HySA systems R&D work being done towards the development of niche fuel cell vehicles.

Figure 3. Fuel cell vehicle.
Reprinted with permission from (Department-of-Energy-US, 2022) Open Access

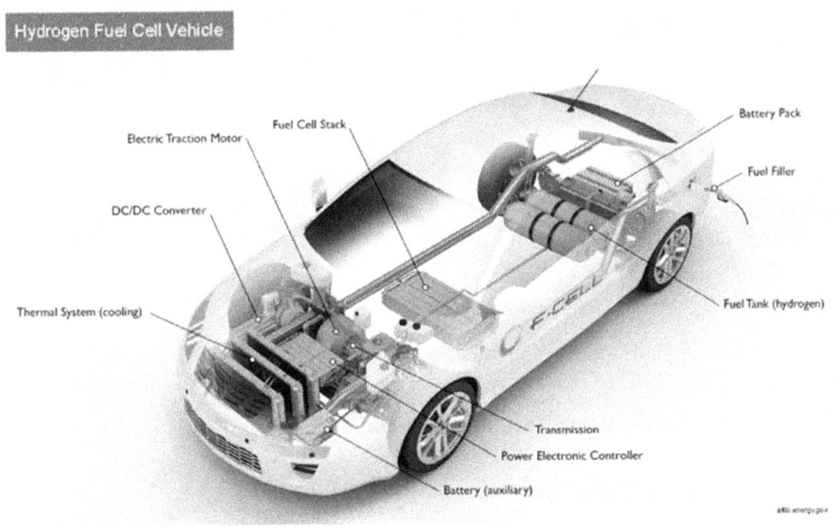

Fuel Cell Stack

The fuel cell stack is the most important part of the fuel cell system. It produces electricity from electrochemical reaction in the form of direct current. Typically, a single fuel cell stack produces small amounts of voltage usually less than 1 volts, and these small amounts are not good for any applications. As such, stacking these cells is mandatory for any practical application. A fuel cell stack may be stacked with hundreds of fuel cells. Various parameters such as cell size, operating temperature, fuel cell type, pressure of the gas supplied to the cell and cell size all affect the amount of power produced by the cell. The typical fuel cell used is the polymer membrane fuel cell (PEMFCs). This fuel cell uses perfluorosulfonic acid (known as Nafion) as a polymer membrane. This membrane is acidic and it allows the flow of hydrogen ions or protons.

Figure 4. HySA systems R&D work towards the development of niche fuel cell vehicles. Reprinted with permission from (Pollet et al., 2015) under Open Access license agreement

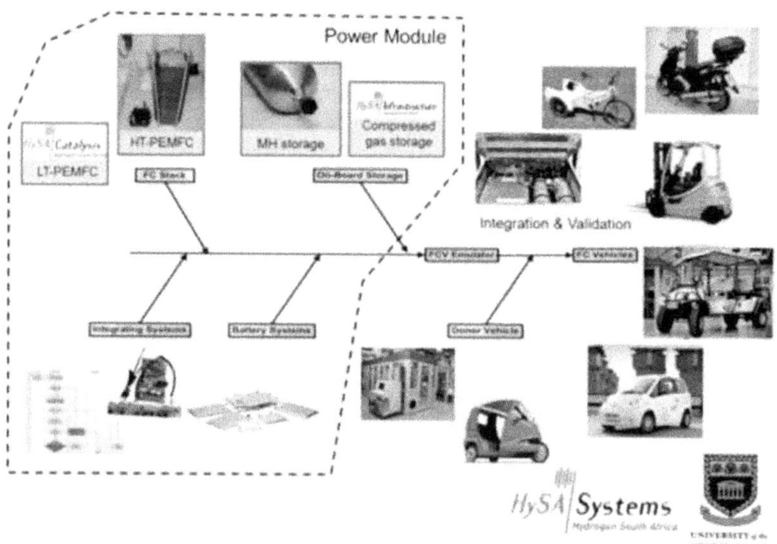

The fuel cell chemical reaction is given as:

$H_2 \rightarrow 2H^+ + 2e - H_2 \rightarrow 2H^+ + 2e^-$

The oxidation reaction is given as:

$H_2 \rightarrow 2H^+ + 2e^-$

The reduction reaction is given as:

$H_2 + O^{2-} \rightarrow H_2O + 2e^-$

The figure below shows an example of a typical fuel cell stack.

Figure 5. A fuel cell stack
(EDN, 2006)

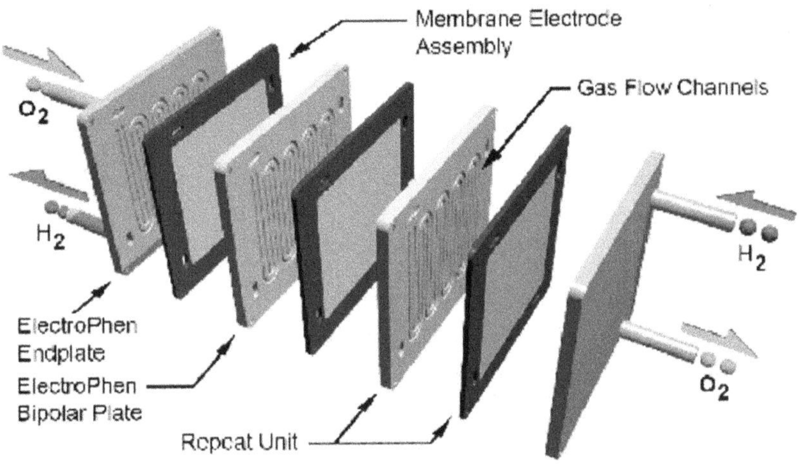

Heat Removal and Heat Generation in Fuel Cell Stacks

One of the major issues in fuel cell stacks is the heat removal and heat generation form the stacks. The issue is due to the fact that fuel cells produce heat that is equated to the electric power output and the conversion efficiency is about 50% (Bargal et al., 2020a; Ju et al., 2005; Kandlikar & Lu, 2009). The heat generated occurs because of ohmic resistance, irreversibility of reaction, entropic heat (reversible heat of the electrochemical reaction) and heat from water vapour condensation. This heat is produced both on the cathode and anode because of the separation of the fuel cell reaction into two electrode reactions. Furthermore the heat due to ohmic heat occurs because of the resistance of electrons to flow through fuel cell components due to conductivity and ion flow resistance (Bargal et al., 2020b). The sum of ohmic heat, entropic heat and irreversible heat account for10%, 55% and 35% of the total heat generated. The heat generated in one fuel cell can be calculated using an energy balance equation given as follows:

$$Q \text{ generate} = \left(\frac{1}{2}F\right)H_{HHVn_{cell}} - IV_{celln_{cell}}$$

H_{HHV} is the high heating value with the assumption that all the water produced leaves the fuel cell stack as liquid. F is the Faraday constant (96485C/mol), I is the current density, Vcell is the cell voltage and ncell is the number of fuel stack cells. A conceptualized image of the heat generated and heat lost in fuel cells is depicted in the figure below:

Figure 6. Conceptualized demonstration of the heat generated and heat removed from HT-PEMFC stacks
Reprinted with permission from (Bargal et al., 2020b) under Open Access license agreement

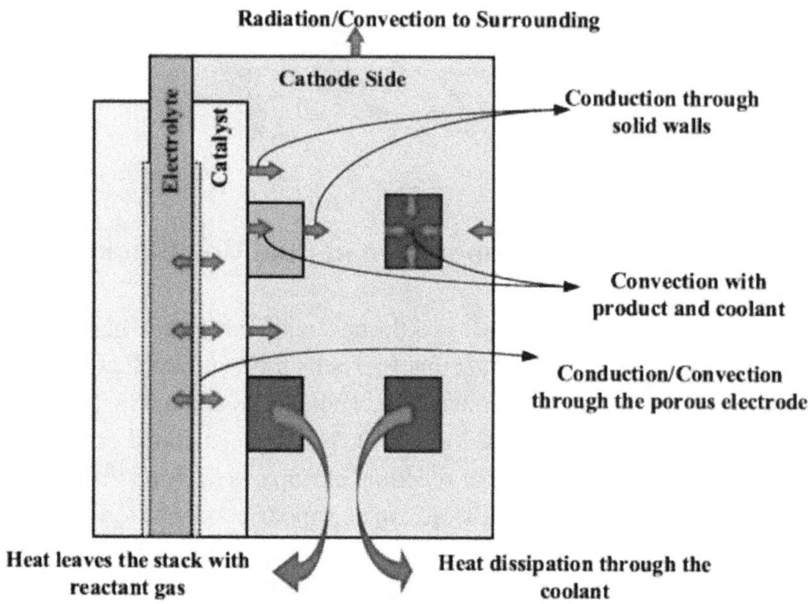

Increasing current density and decreasing cell voltage increases the heat generated. To address some of these issues HySA systems together with ZSW develop and produce stacks for laboratory testing and validation. The center produces carbon based bipolar plates which require high precision tools because the machining tolerance is in the micrometrics range. The validation and test work of the HT-PEMFC stacks is done with the help of the high end commercial testing stations. In these stations, the EIS method is applied to study the electrochemical process that transpires when the stacks are operated. A 5-cell stack has been assembled and validated at the center and the preliminary results of conditioning and performance are promising (for a current density of 0.78 A cm^{-2}, 0.35 W cm^{-2} was measured).

In 2012, a 2kW 96- cell HT-PEMFC stack was manufactured by HySA. This stacks was then integrated with a breadboard system prototype from CHP. The figure below shows an image of the HT-PEMFC integrated with a CHP prototype.

Figure 7. Assembled 2 kW stack on the left and on the right the stack is integrated in a CHP prototype breadboard system
Reprinted with permission from (Pollet et al., 2015) under Open Access license agreement

MEMBRANE ELECTRODE ASSEMBLY (MEA's)

Understanding the core and underlying principle of MEA's is very important in overcoming the current challenges we have in fuel cells such costs and overall durability of the catalysts. A lot of developmental research has been focused and done on low temperature MEA's which are used in low temperature polymer membrane fuel cells and less work has been done on high temperature MEA's (Bose et al., 2011; Scherer, 1990; Shao et al., 2007; Steen et al., 2002). In addition, a lot of R&D work has been focused on making polymer electrolyte membranes as well as making suitable catalyst and gas diffusion layers (Cindrella et al., 2009; Holdcroft, 2014; Park et al., 2012; Zhang, 2008). The HySA systems center focuses on perfecting the MEA architecture by enhancing the durability and reducing the catalyst loading. The center has developed different coating method for use in catalyst coated membrane and catalyst coated substrate for MEA preparation method. The MEA performances have been benchmarked to some of the best results obtained under the same parameters. In this work, the gas diffusion layer was prepared by a novel automatic catalyst spurting under irradiation technique (Pollet et al., 2015). This was done at optimum operating conditions, the current density was at 0.38 A cm^{-2} at 0.6V and the peak power density was 0.61 W cm^{-2} for a single cell.

Improvement and optimization of the GDEs has also been one of the key focus area where parameters such as acid controlling layers, binders and proton conductors are being studied. The effect of different binders (PVDF, PBI-PVDF blend, Nafion, PTFE and PBI) was studied. Nafion® showed in stability for HT-MEA's because of its proton conductivity restrictions at operating temperature of 160 °C (Pollet et al., 2015). The PTFE binder was found to be suitable at high current densities while the PVDF binder was found to be suitable at low current densities. This performance can be observed on the polarization curves depicted in the figure 5 below. They also benchmarked the HySA prepared HT-MEA's with commercial HT-MEAs with the HySA HT-MEA (Pollet et al., 2015). The polarisation curves on the right of figure 5 show the difference in performance.

Figure 8. (a) Polarisation curves of the different binders 8(b) Polarisation curves of the commercial MEA's vs the commercial bought MEA's.
Reprinted with permission from (Pollet et al., 2015) under Open Access license agreement

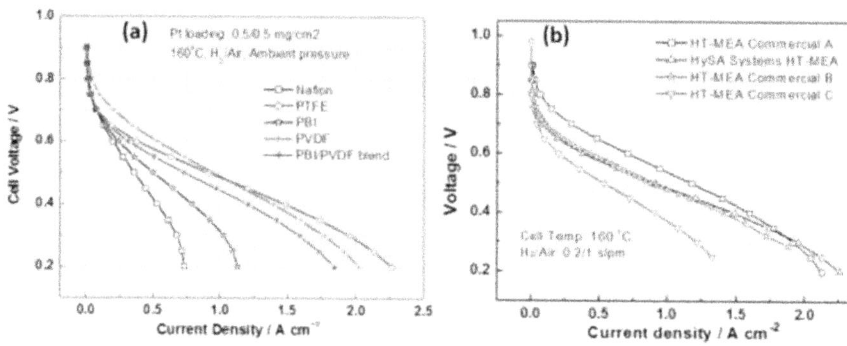

HySA on-board storage

The ongoing development of hydrogen as a transportation fuel is garnering a lot of attention for vehicle applications. In vehicle applications, the hydrogen can be stored in several parts of the production chain, i.e. from manufacture to utilization (Andersson & Grönkvist, 2019; Schlapbach & Züttel, 2001). The hindering challenge in the use of hydrogen for mobile applications is the storage. Hydrogen as the lightest and smallest molecule in the periodic table has a very low density, where 1 kg of hydrogen occupies over 11m^3 at standard pressure and temperature (Schlapbach & Züttel, 2001). For hydrogen to be a viable candidate for mobile applications its density must be increased and there's various methods in which hydrogen can be stored at increased density(Carpetis, 1982; Durbin & Malardier-Jugroot, 2013; Eberle et al., 2009; Luo et al., 2015). These methods however require a lot of energy either in the form of heat, hydrogen-binding material and work (Eberle et al., 2009;

Hwang & Varma, 2014; Ngqalakwezi et al., 2020; Ngqalakwezi & Nkazi, 2020; Paster et al., 2011). A promising hydrogen storage method is the solid state storage of hydrogen into materials. The mechanism is the hydrogen will be stored into any material that has a high hydrogen storage capacity and then it will be released at the desorption temperature of that particular material. The biggest challenge with solid state storage of hydrogen is that material with high hydrogen storage capacity usually have high desorption temperatures which are above operational temperatures of fuel cells (Ngqalakwezi et al., 2020; Ngqalakwezi & Nkazi, 2020). Various materials such as metal hydrides, intermetallic hydrides, complex hydrides, borohydrides and chemical hydrides have been tested for practical mobile applications (Orimo et al., 2007; Özcan & Akın, 2019; Puszkiel et al., 2017; Sandrock, 1999; Von Colbe et al., 2019).

The HySA center has done a substantial amount of work in solid state storage of hydrogen particularly using metal hydrides (light weight metal hydrides) and metal organic frameworks (MOFs). The primary focal activity in the center is the develop a metal hydride hydrogen storage system for a hydrogen powered forklift, with a 10 kW nominal fuel cell output and with a capacity of 3 tonnes (Pollet et al., 2014a). They calculated that a forklift operating for 8 hours at fuel cell efficiency of 50% will require 54 Nm^3 H_2 and this amount will require 300 kg of metal hydride with a hydrogen storage capacity of 180 L of hydrogen per kg. The center then proposed a novel system for the fuel cell powered forklift with subsystems (Pollet et al., 2014a). The following figure shows the overall concept for the development of the fuel cell powered forklift.

Metal Organic Frameworks

The ultra-high porosity and tunabilty of porous framework materials makes them ideal candidates for a wide range applications (Evans et al., 2020). Metal organic frameworks, highly porous material, have been applied in catalysis, hydrogen storage, sensing, carbon dioxide sequestration and optics (Cui et al., 2018; Ding et al., 2019; Furukawa et al., 2013; Stassen et al., 2017). These materials are physisorption based due to the weak binding energy and interaction which is in the range of 5-9 kJ mol^{-1}. MOF's are commonly synthesized using the solvothermal method using DMF and DEF. These solvents are used to control the rate of product crystallization, solvent-ligand exchange kinetics and solution polarity (Suh et al., 2012). Furthermore MOFs can synthesized using different combinations of ligands and metal ions. MOFs with larger surface area have high hydrogen storage capacity however the hydrogen take up is usually at cryogenic temperatures at about 77K (Suh et al., 2012). According to the US standards for hydrogen storage materials, the minimum hydrogen uptake is 5.5wt% for the system at standard temperature and pressure and to date no MOF has met the DOE requirements. The hydrogen storage of these materials goes down with increasing temperature and at high pressures the hydrogen storage capacity increases. The pore volume of MOFs is proportional to the specific surface area and therefore to increase the hydrogen uptake, a high pore volume and specific surface area MOF should be synthesized. Furthermore, to increase the hydrogen storage uptake in MOFs the heat of hydrogen adsorption can be increased to approximately 15 kJ mol^{-1}, an amount effective for adequate hydrogen uptake (Bhatia & Myers, 2006; Lochan & Head-Gordon, 2006). Han and Goddard simulated Li based metal organic frameworks for reverse hydrogen storage at ambient temperatures (Han & Goddard, 2007b). The lithium as a dopant on the framework was to enhance its hydrogen storage uptake at standard conditions. The MOFs were built using metal oxide (ZnO) using carboxylate links. The figure below shows the zinc oxide and different carboxylate links that were used to build Li based MOFs.

Figure 9. The block flow diagram for the fuel cell powered forklift
Reprinted with permission from (Pollet et al., 2015) under Open Access license agreement

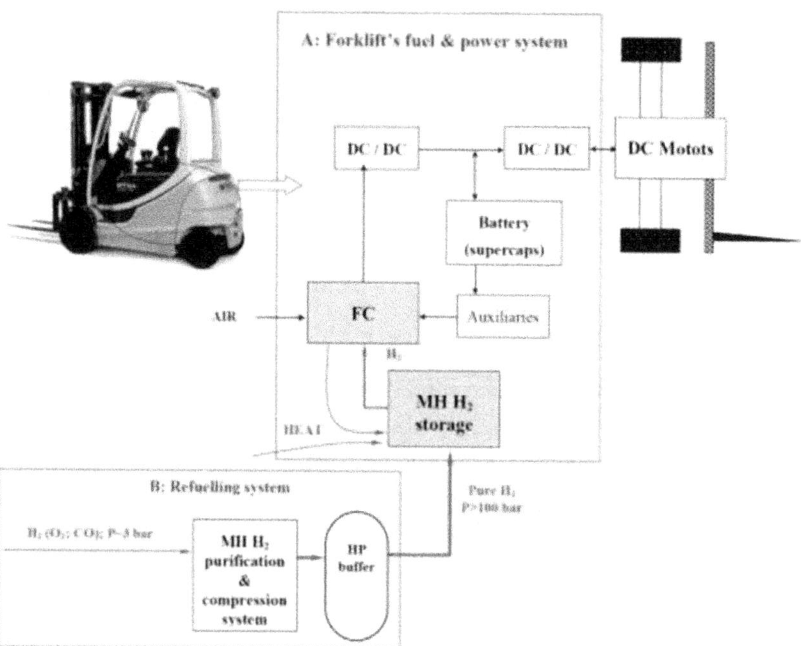

Figure 10. Li based MOFs build on carboxylate links
Reprinted with permission from (Han & Goddard, 2007a) ACS

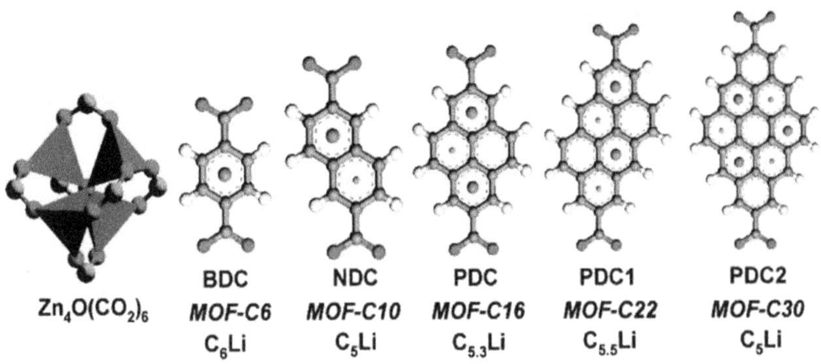

HySA MOFs Research

The HySA R&D program on solid state hydrogen storage has been primarily based on the synthesis of high surface area MOFs and improving their hydrogen storage capacity under ambient temperature and pressure. The three main focus materials are the IRMOF-8, MOF-5 and MOF-177 (Pollet et al., 2015). The post synthetic treatment of these materials primarily focuses on catalysing them to improve their hydrogen uptake for applications in fuel cell electric vehicles. This is achieved by taking into consideration the hydrogen spill-over phenomenon which can improve the hydrogen storage capacity of MOFs at room temperature (Y. Li et al., 2007; Y. Li & Yang, 2006a, 2006b; Shen et al., 2022; Tsao et al., 2009). The platinum based catalysts reportedly enhance the hydrogen spill-over when supported on a high surface area carbon such as activated carbon (Suh et al., 2012). The figure below depicts how the hydrogen spill over takes place on a platinum based catalyst and an MOF.

Figure 11. The hydrogen spill-over phenomenon demonstrated on an MOF
Reprinted with permission from (Suh et al., 2012) ACS

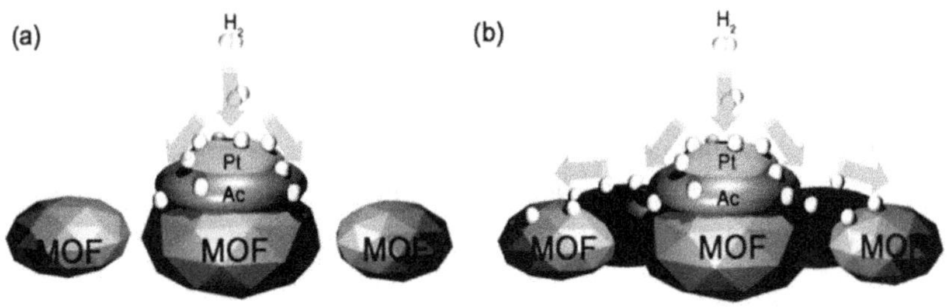

The platinum particles break the bond in the hydrogen molecule resulting in the dissociation of the molecule and the formation of H atoms. The hydrogen then migrate to the substrate material where it can be used or spill over again to the next substrate/receptor. The hydrogen spill over between Pt/C and an MOF is facilitated either by introducing the metal on the MOF (receptor) by chemical doping or by physically mixing the platinum carbon catalyst and the MOF through milling (Suh et al., 2012). In both of these methods, there needs to be bridging done between the receptor and the platinum catalyst to enhance the spill over at standard parameters. The problem identified was then that the hydrogen storage on a bridged material is difficult based on the how the bridge is and the non-uniformity of the Pt particles.

HySA Prototypes

Since the inception of this program in South Africa, a lot of prototypes have been developed, vehicle and battery prototypes precisely. The figure shows some of the vehicle prototypes that have been made under the HySA research including a forklift, an E-scooter, a three wheeler and a golf cart.

These prototypes were part of demonstration of the tank-to-wheel optimization carried out in the center. The golf cart development included the use of an existing electric road-legitimate vehicle which was fitted with a hydrogen low temperature PEMFC as an extender where both metal hydride and high pressure hydrogen storage methods were used. The cart if fitted with a 5Kw E-drive system and a 1.2kW fuel cell to maintain the battery for extended use (Pollet et al., 2015). The E-scooter also uses a hydrogen low temperature PEMFC which is placed in a typical storage box and also has two metal hydride cylinders (Pollet et al., 2015). The addition of these does not compromise the quality. The forklift and three wheeler are also fitted with the same components and function well.

FUTURE RESEARCH DIRECTIONS

Future research direction of the HySA program are to work towards achieving about 25% share of the global fuel cell and hydrogen market through using niche platinum catalysts, systems and components. Current research is focussed on improving, optimizing, upscaling, benchmarking and patenting the work done by HySA and marketing it. A new technology demonstration system is currently under construction at Mintek and more energy is focussed on improving the research and marketing the work on the global scale.

Figure 12. The hydrogen fuelled vehicle prototypes that have been manufactured under the HySA program in South Africa
Reprinted with permission from (Pollet et al., 2015) under Open Access license agreement

CONCLUSION

Going green is the only progressive step everyone is taking globally to try and restore what has been damaged over the course of decades through continued use of fossil fuels. The HySA program has done extremely were in placing itself amongst global competitors in the hydrogen research space. A lot of patents and IP's have been filed and journals have been published highlighting the work done. The development of prototypes such as the hydrogen fuelled forklift, an E-scooter, a three wheeler and a golf cart, hydrogen storage material (MOFs and metal hydrides) that perform to similar to commercial material are some of the highlights. The work still continues with more prototypes being developed and more discoveries being made. All of this, to push towards the common goal of saving the environment, going green and getting a share in the hydrogen global market.

REFERENCES

Andersson, J., & Grönkvist, S. (2019). Large-scale storage of hydrogen. *International Journal of Hydrogen Energy, 44*(23), 11901–11919. doi:10.1016/j.ijhydene.2019.03.063

Asif, U., & Schmidt, K. (2021). Fuel cell electric vehicles (FCEV): Policy advances to enhance commercial success. *Sustainability (Basel), 13*(9), 5149. doi:10.3390u13095149

Bargal, M. H. S., Abdelkareem, M. A. A., Tao, Q., Li, J., Shi, J., & Wang, Y. (2020a). Liquid cooling techniques in proton exchange membrane fuel cell stacks: A detailed survey. *Alexandria Engineering Journal, 59*(2), 635–655. doi:10.1016/j.aej.2020.02.005

Bargal, M. H. S., Abdelkareem, M. A. A., Tao, Q., Li, J., Shi, J., & Wang, Y. (2020b). Liquid cooling techniques in proton exchange membrane fuel cell stacks: A detailed survey. *Alexandria Engineering Journal, 59*(2), 635–655. doi:10.1016/j.aej.2020.02.005

Barrett, S. (2014). Dissecting 'HySA.' *Renewable Energy Focus, 15*(1), 40–44. doi.org/10.1016/S1755-0084(14)70024-9

Bessarabov, D., Human, G., Kruger, A. J., Chiuta, S., Modisha, P. M., du Preez, S. P., Oelofse, S. P., Vincent, I., Van Der Merwe, J., Langmi, H. W., Ren, J., & Musyoka, N. M. (2017). South African hydrogen infrastructure (HySA infrastructure) for fuel cells and energy storage: Overview of a projects portfolio. *International Journal of Hydrogen Energy, 42*(19), 13568–13588. doi:10.1016/j.ijhydene.2016.12.140

Bhatia, S. K., & Myers, A. L. (2006). Optimum conditions for adsorptive storage. *Langmuir, 22*(4), 1688–1700. doi:10.1021/la0523816 PMID:16460092

Blomen, L. J. M. J., & Mugerwa, M. N. (2013). *Fuel cell systems*. Springer Science & Business Media.

Bose, S., Kuila, T., Nguyen, T. X. H., Kim, N. H., Lau, K., & Lee, J. H. (2011). Polymer membranes for high temperature proton exchange membrane fuel cell: Recent advances and challenges. *Progress in Polymer Science, 36*(6), 813–843. doi:10.1016/j.progpolymsci.2011.01.003

Carpetis, C. (1982). Estimation of storage costs for large hydrogen storage facilities. *International Journal of Hydrogen Energy, 7*(2), 191–203. doi:10.1016/0360-3199(82)90146-X

Carrette, L., Friedrich, K. A., & Stimming, U. (2000). Fuel Cells: Principles, Types, Fuels, and Applications. *ChemPhysChem*, *1*(4), 162–193. doi:10.1002/1439-7641(20001215)1:4<162::AID-CPHC162>3.0.CO;2-Z PMID:23696319

Cawthorn, R. G. (2010). The platinum group element deposits of the Bushveld Complex in South Africa. *Platinum Metals Review*, *54*(4), 205–215. doi:10.1595/147106710X520222

Cindrella, L., Kannan, A. M., Lin, J. F., Saminathan, K., Ho, Y., Lin, C. W., & Wertz, J. (2009). Gas diffusion layer for proton exchange membrane fuel cells—A review. *Journal of Power Sources*, *194*(1), 146–160. doi:10.1016/j.jpowsour.2009.04.005

Ciura, A. (2019). *Sustainable development of the public transport sector in EU Countries*.

Cramer, L. A. (2001). The extractive metallurgy of South Africa's platinum ores. *Journal of the Minerals Metals & Materials Society*, *53*(10), 14–18. doi:10.100711837-001-0048-1

Cui, Y., Zhang, J., He, H., & Qian, G. (2018). Photonic functional metal–organic frameworks. *Chemical Society Reviews*, *47*(15), 5740–5785. doi:10.1039/C7CS00879A PMID:30046798

Department-of-Energy-US. (2022). *How Do Fuel Cell Electric Vehicles Work Using Hydrogen?* Energy Efficiency and Renewable Energy.

Ding, M., Flaig, R. W., Jiang, H.-L., & Yaghi, O. M. (2019). Carbon capture and conversion using metal–organic frameworks and MOF-based materials. *Chemical Society Reviews*, *48*(10), 2783–2828. doi:10.1039/C8CS00829A PMID:31032507

DSI. (2020). *Science and Innovation on South Africa's Platinum Valley project*. Department of Science and Innovation, SA Government.

Durbin, D. J., & Malardier-Jugroot, C. (2013). Review of hydrogen storage techniques for on board vehicle applications. *International Journal of Hydrogen Energy*, *38*(34), 14595–14617. doi:10.1016/j.ijhydene.2013.07.058

Eberle, U., Felderhoff, M., & Schueth, F. (2009). Chemical and physical solutions for hydrogen storage. *Angewandte Chemie International Edition*, *48*(36), 6608–6630. doi:10.1002/anie.200806293 PMID:19598190

EDN. (2006). *The promise of more rugged and economical fuel cell stacks*. EDN. https://www.edn.com/the-promise-of-more-rugged-and-economical-fuel-cell-stacks/

Engelbrecht, J. (2012). Potential changes in the physical beneficiation processes that can improve the recovery grade or costs for the platinum group metals. *Proceedings of the Fifth International Platinum Conference: A Catalyst for Change*. Springer.

Evans, J. D., Bon, V., Senkovska, I., Lee, H.-C., & Kaskel, S. (2020). Four-dimensional metal-organic frameworks. *Nature Communications*, *11*(1), 2690. doi:10.103841467-020-16527-8 PMID:32483346

Furukawa, H., Cordova, K. E., O'Keeffe, M., & Yaghi, O. M. (2013). The chemistry and applications of metal-organic frameworks. *Science*, *341*(6149), 1230444. doi:10.1126cience.1230444 PMID:23990564

Gallagher, K. S. (2014). *The globalization of clean energy technology: Lessons from China*. MIT press. doi:10.7551/mitpress/9805.001.0001

Garside, M. (2023). *Production volume of platinum in South Africa from 2010 to 2022*. Statista. https://www.statista.com/statistics/1015227/platinum-production-south-africa/

Gerrard, M. B. (2011). *The law of clean energy: efficiency and renewables.*

Han, S. S., & Goddard, W. A. (2007a). Lithium-Doped Metal-Organic Frameworks for Reversible H2 Storage at Ambient Temperature. *Journal of the American Chemical Society*, *129*(27), 8422–8423. doi:10.1021/ja072599+ PMID:17569539

Han, S. S., & Goddard, W. A. (2007b). Lithium-Doped Metal-Organic Frameworks for Reversible H2 Storage at Ambient Temperature. *Journal of the American Chemical Society*, *129*(27), 8422–8423. doi:10.1021/ja072599+ PMID:17569539

Holdcroft, S. (2014). Fuel cell catalyst layers: A polymer science perspective. *Chemistry of Materials*, *26*(1), 381–393. doi:10.1021/cm401445h

Hwang, H. T., & Varma, A. (2014). Hydrogen storage for fuel cell vehicles. *Current Opinion in Chemical Engineering*, *5*, 42–48. doi:10.1016/j.coche.2014.04.004

HySA_Systems. (2014). *Hydrogen fuelled vehicles*. HySA. http://www.hysasystems.com/index.php/programmes/hydrogen-fuel-cell-vehicles

Irena, I. E. A. (2011). Prospects for the African power sector. *Scenarios and Strategies for Africa Project.*

Ju, H., Meng, H., & Wang, C.-Y. (2005). A single-phase, non-isothermal model for PEM fuel cells. *International Journal of Heat and Mass Transfer*, *48*(7), 1303–1315. doi:10.1016/j.ijheatmasstransfer.2004.10.004

Kandlikar, S. G., & Lu, Z. (2009). Thermal management issues in a PEMFC stack–A brief review of current status. *Applied Thermal Engineering, 29*(7), 1276–1280. doi:10.1016/j.applthermaleng.2008.05.009

Kirubakaran, A., Jain, S., & Nema, R. K. (2009). A review on fuel cell technologies and power electronic interface. *Renewable & Sustainable Energy Reviews, 13*(9), 2430–2440. . doi:10.1016/j.rser.2009.04.004

Latypov, R. M., Heinonen, J. S., & Chistyakova, S. Y. (2022). Magmatic erosion of high-temperature-melting cumulates in the Bushveld Complex by chemical dissolution. *Geosystems and Geoenvironment, 1*(3), 100077.

Li, Q., He, R., Gao, J.-A., Jensen, J. O., & Bjerrum, N. J. (2003). The CO poisoning effect in PEMFCs operational at temperatures up to 200 C. *Journal of the Electrochemical Society, 150*(12), A1599. doi:10.1149/1.1619984

Li, Y., Yang, F. H., & Yang, R. T. (2007). Kinetics and mechanistic model for hydrogen spillover on bridged metal– organic frameworks. *The Journal of Physical Chemistry C, 111*(8), 3405–3411. doi:10.1021/jp065367q

Li, Y., & Yang, R. T. (2006a). Hydrogen storage in metal– organic frameworks by bridged hydrogen spillover. *Journal of the American Chemical Society, 128*(25), 8136–8137. doi:10.1021/ja061681m PMID:16787068

Li, Y., & Yang, R. T. (2006b). Significantly enhanced hydrogen storage in metal–organic frameworks via spillover. *Journal of the American Chemical Society, 128*(3), 726–727. doi:10.1021/ja056831s PMID:16417355

Lochan, R. C., & Head-Gordon, M. (2006). Computational studies of molecular hydrogen binding affinities: The role of dispersion forces, electrostatics, and orbital interactions. *Physical Chemistry Chemical Physics, 8*(12), 1357–1370. doi:10.1039/b515409j PMID:16633617

Luo, X., Wang, J., Dooner, M., & Clarke, J. (2015). Overview of current development in electrical energy storage technologies and the application potential in power system operation. *Applied Energy, 137*, 511–536. doi:10.1016/j.apenergy.2014.09.081

Martin, N. J., & Rice, J. L. (2014). Influencing clean energy laws: An analysis of business stakeholder engagement. *Business Strategy and the Environment, 23*(7), 447–460. doi:10.1002/bse.1795

Mekhilef, S., Saidur, R., & Safari, A. (2012). Comparative study of different fuel cell technologies. *Renewable & Sustainable Energy Reviews, 16*(1), 981–989. doi:10.1016/j.rser.2011.09.020

Mormann, F. (2015). Clean Energy Federalism. *Florida Law Review*, *67*, 1621.

Msimanga, B., & Sebitosi, A. B. (2014). South Africa's non-policy driven options for renewable energy development. *Renewable Energy*, *69*, 420–427. doi:10.1016/j.renene.2014.03.041

Ngqalakwezi, A., & Nkazi, D. B. (2020). Hydrogen Storage: Materials, Kinetics and Thermodynamics. In Advanced Applications of Hydrogen and Engineering Systems in the Automotive Industry. IntechOpen.

Ngqalakwezi, A., Nkazi, D. B., Baloyi, S. J., & Ntho, T. A. (2020). Hydrogen: An Environmental Remediation. In Global Issues and Innovative Solutions in Healthcare, Culture, and the Environment (pp. 168–194). IGI Global.

O'Connor, C., & Alexandrova, T. (2021). The Geological Occurrence, Mineralogy, and Processing by Flotation of Platinum Group Minerals (PGMs) in South Africa and Russia. *Minerals (Basel)*, *11*(1), 54. doi:10.3390/min11010054

O'hayre, R., Cha, S.-W., Colella, W., & Prinz, F. B. (2016). *Fuel cell fundamentals*. John Wiley & Sons. doi:10.1002/9781119191766

Online-Content. (2023). *PGM Markets*. Platinum Group Metals.

Orimo, S., Nakamori, Y., Eliseo, J. R., Züttel, A., & Jensen, C. M. (2007). Complex hydrides for hydrogen storage. *Chemical Reviews*, *107*(10), 4111–4132. doi:10.1021/cr0501846 PMID:17848101

Özcan, O., & Akın, A. N. (2019). Thermodynamic analysis of methanol steam reforming to produce hydrogen for HT-PEMFC: An optimization study. *International Journal of Hydrogen Energy*, *44*(27), 14117–14126. doi:10.1016/j.ijhydene.2018.12.211

Park, S., Lee, J.-W., & Popov, B. N. (2012). A review of gas diffusion layer in PEM fuel cells: Materials and designs. *International Journal of Hydrogen Energy*, *37*(7), 5850–5865. doi:10.1016/j.ijhydene.2011.12.148

Paster, M. D., Ahluwalia, R. K., Berry, G., Elgowainy, A., Lasher, S., McKenney, K., & Gardiner, M. (2011). Hydrogen storage technology options for fuel cell vehicles: Well-to-wheel costs, energy efficiencies, and greenhouse gas emissions. *International Journal of Hydrogen Energy*, *36*(22), 14534–14551. doi:10.1016/j.ijhydene.2011.07.056

Pollet, B. G., Pasupathi, S., Swart, G., Mouton, K., Lototskyy, M., Williams, M., Bujlo, P., Ji, S., Bladergroen, B. J., & Linkov, V. (2014a). Hydrogen South Africa (HySA) Systems Competence Centre: Mission, objectives, technological achievements and breakthroughs. *International Journal of Hydrogen Energy*, *39*(8), 3577–3596. doi:10.1016/j.ijhydene.2013.11.116

Pollet, B. G., Pasupathi, S., Swart, G., Mouton, K., Lototskyy, M., Williams, M., Bujlo, P., Ji, S., Bladergroen, B. J., & Linkov, V. (2014b). Hydrogen South Africa (HySA) Systems Competence Centre: Mission, objectives, technological achievements and breakthroughs. *International Journal of Hydrogen Energy*, *39*(8), 3577–3596. doi:10.1016/j.ijhydene.2013.11.116

Pollet, B. G., Staffell, I., & Adamson, K.-A. (2015). Current energy landscape in the Republic of South Africa. *International Journal of Hydrogen Energy*, *40*(46), 16685–16701. doi:10.1016/j.ijhydene.2015.09.141

Puszkiel, J., Garroni, S., Milanese, C., Gennari, F., Klassen, T., Dornheim, M., & Pistidda, C. (2017). Tetrahydroborates: Development and potential as hydrogen storage medium. *Inorganics*, *5*(4), 74. doi:10.3390/inorganics5040074

Sandrock, G. (1999). A panoramic overview of hydrogen storage alloys from a gas reaction point of view. *Journal of Alloys and Compounds*, *293*, 877–888. doi:10.1016/S0925-8388(99)00384-9

Santhanam, K. S. V., Press, R. J., Miri, M. J., Bailey, A. V., & Takacs, G. A. (2017). *Introduction to hydrogen technology*. John Wiley & Sons.

Scherer, G. G. (1990). Polymer membranes for fuel cells. *Berichte der Bunsengesellschaft fur Physikalische Chemie*, *94*(9), 1008–1014. doi:10.1002/bbpc.19900940926

Schlapbach, L., & Züttel, A. (2001). Hydrogen-storage materials for mobile applications. *Nature*, *414*(6861), 353–358. doi:10.1038/35104634 PMID:11713542

Schouwstra, R. P., Kinloch, E. D., & Lee, C. A. (2000). A short geological review of the Bushveld Complex. *Platinum Metals Review*, *44*(1), 33–39.

Shao, Y., Yin, G., Wang, Z., & Gao, Y. (2007). Proton exchange membrane fuel cell from low temperature to high temperature: Material challenges. *Journal of Power Sources*, *167*(2), 235–242. doi:10.1016/j.jpowsour.2007.02.065

Shen, H., Li, H., Yang, Z., & Li, C. (2022). *Magic of hydrogen spillover: Understanding and application*. Green Energy & Environment.

Sovacool, B. K. (2021). Clean, low-carbon but corrupt? Examining corruption risks and solutions for the renewable energy sector in Mexico, Malaysia, Kenya and South Africa. *Energy Strategy Reviews, 38*, 100723. doi:10.1016/j.esr.2021.100723

Stassen, I., Burtch, N., Talin, A., Falcaro, P., Allendorf, M., & Ameloot, R. (2017). An updated roadmap for the integration of metal–organic frameworks with electronic devices and chemical sensors. *Chemical Society Reviews, 46*(11), 3185–3241. doi:10.1039/C7CS00122C PMID:28452388

Steen, M. L., Jordan, A. C., & Fisher, E. R. (2002). Hydrophilic modification of polymeric membranes by low temperature H2O plasma treatment. *Journal of Membrane Science, 204*(1–2), 341–357. doi:10.1016/S0376-7388(02)00061-3

Suh, M. P., Park, H. J., Prasad, T. K., & Lim, D.-W. (2012). Hydrogen Storage in Metal–Organic Frameworks. *Chemical Reviews, 112*(2), 782–835. doi:10.1021/cr200274s PMID:22191516

Technology, D. of S. and. (2007). Innovation towards a knowledge-based economy: Ten-year plan for South Africa (2008-2018). Department of Science and Technology Pretoria.

Tsao, C.-S., Yu, M.-S., Wang, C.-Y., Liao, P.-Y., Chen, H.-L., Jeng, U.-S., Tzeng, Y.-R., Chung, T.-Y., & Wu, H.-C. (2009). Nanostructure and hydrogen spillover of bridged metal-organic frameworks. *Journal of the American Chemical Society, 131*(4), 1404–1406. doi:10.1021/ja802741b PMID:19140765

Von Colbe, J. B., Ares, J.-R., Barale, J., Baricco, M., Buckley, C., Capurso, G., Gallandat, N., Grant, D. M., Guzik, M. N., & Jacob, I. (2019). Application of hydrides in hydrogen storage and compression: Achievements, outlook and perspectives. *International Journal of Hydrogen Energy, 44*(15), 7780–7808. doi:10.1016/j.ijhydene.2019.01.104

Wang, S., & Jiang, S. P. (2017). Prospects of fuel cell technologies. *National Science Review, 4*(2), 163–166. doi:10.1093/nsr/nww099

Winkler, H. (2005). Renewable energy policy in South Africa: Policy options for renewable electricity. *Energy Policy, 33*(1), 27–38. doi:10.1016/S0301-4215(03)00195-2

Zhang, J. (2008). *PEM fuel cell electrocatalysts and catalyst layers: fundamentals and applications*. Springer Science & Business Media. doi:10.1007/978-1-84800-936-3

ADDITIONAL READINGS

ArendY. (2014). *About.* HySA. https://www.hysasystems.com/index.php/about-hysa

DMR. (2020). *Fuel cells and the future role of south africa through its platinum resources.* DMR. https://www.dmr.gov.za/LinkClick.aspx?fileticket=_RLK8Tx_648%3D&portalid=0

DSI. (2021). *Hydrogen society roadmap for south africa 2021.* DSI. https://www.dst.gov.za/images/South_African_Hydrogen_Society_RoadmapV1.pdf

Langmi, H. W., Ren, J., Musyoka, N. M., North, B. C., Mathe, M. K., & Bessarabov, D. (2015). *Progress in materials-based hydrogen storage at Hysa infrastructure in South Africa.* http://hdl.handle.net/10204/8425

Nekuda Malik, J. A. (2013). South Africa capitalizes on growing hydrogen and fuel-cell market. *MRS Bulletin, 38*(5), 364–365. doi:10.1557/mrs.2013.116

APPENDIX

Acronyms and Definitions

AFCs: Alkaline fuel cell
CHP: Combined heat and power
CSIR: Council of science
DEF: Diethylformamide
DMF: Dimethylformamide
DMFCs: Direct methanol fuel cell
DSI: Department of science and innovation
EVs: Electric vehicles
FCVs: Fuel cell vehicles
GDE: Gas diffusion electrode
HySA: Hydrogen South Africa
IEA: International energy agency
MEA: Membrane electrode assembly
Mintek: Minerals Technology
MOF: Metal organic framework
NWU: North West University
PAFCs: Phosphoric acid fuel cell
PBI: Polybenzimidazoles
PBI-PVDF: Polybenzimidazoles-Polyvinylidene difluroide
PEMFCs: Polymer electrolyte membrane fuel cell
PGM: Platinum group metals
PTFE: Polytetrafluoroethylene
PVDF: Polyvinylidene difluoride
SOFCs: Solid oxide fuel cell
UCT: University of Cape town
UWC: University of the Western Cape

Chapter 5
Advances in Fuel Cell Technologies for Hydrogen Production

Aarti Maurya
Amity University, Noida, India

Amit Kumar Chaurasia
https://orcid.org/0000-0003-1786-3745
Amity University, Noida, India

ABSTRACT

The increasing rate of industrialization and urbanization increases waste and wastewater generation along with global energy demand. That makes the necessity of the current waste treatment and process technologies to be upgraded into the sustainable or green energy generation such as hydrogen production from the waste materials and hydrogen fuel cells-based technologies. Fuel cells are electrochemical devices that use hydrogen as fuel cells for electricity or heat generation. The fuel cells-based technologies are in the infant stage and need to be explored for the vital utilization of hydrogen for the various energy related applications. Stationary application of fuel cells or hydrogen fuel cell vehicles related technologies are hotspot research areas in the 21st century. In this chapter, the authors summarize the recent development of the hydrogen production from waste, current state of the art in hydrogen fuel cells, and hydrogen fuel cell vehicles and their economic and sustainability benefits. It also includes the key challenges of these technologies for commercial applications.

DOI: 10.4018/978-1-6684-6721-3.ch005

INTRODUCTION

Fast urbanization and industrialization utilizing the natural resources rapidly to meet the global energy demand that may generate the huge amount of waste and wastewater that are the global concern in the 21st century. Recent statistics imply that 3 billion people will be added as new energy utilizers in the existing energy consumers by the end of this century. Global primary energy demand (~0.55 quadrillion MJ) is increasing at a rate of 2.11% per year with a predictable total increase of 55% by 2050 as shown in Figure 1 (IEA, 2019).

Figure 1. Forecasts on world population and energy consumption
Source: IEA (2019)

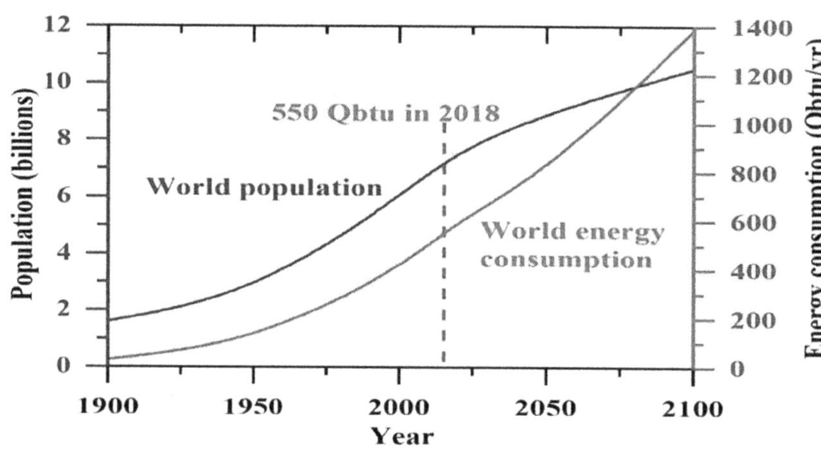

Around 75-85% of global energy consumption by conventional resources consists of petroleum (34%), coal (27%), and other natural gas (24%), which are non-sustainable in nature as well as generate huge amounts of waste and wastewater. A recent literature computed that high-income countries, upper-middle-income countries and low-income countries treat the generated wastewater by 70%, 38% and 8%, respectively (Jones et al., 2021). The increasing rate of pollutants and wastewater generation during the energy generation and utilization have adverse impacts on global climate, human health and ecosystems (Chaurasia, Shankar, et al., 2021a; Rani et al., 2017). Thus, there is an urgent need to generate clean and sustainable energy conversion technologies such as production of hydrogen from waste or wastewater to mitigate the future energy challenges. Also, the hydrogen production or hydrogen utilization technologies should be economical and sustainable.

Hydrogen possesses a high energy density such as 122-142 MJ/Kg that is three times with respect to gasoline (petrol or diesel) (Chaurasia & Mondal, 2022a; Singh Thakur et al., 2022). It also offers sustainable and regenerative stationery applications in the hydrogen fuelled vehicles. Recent literature reported that renewable energy supported by communities that reduce the use of crude oil and coal will drop to 40.5% and 36.7%, respectively by 2030 (Chaurasia et al., 2020, 2022b). As per the statistics reported in 2022, ~70 million tonnes per year of hydrogen produced across the globe in which 90% of hydrogen is produced by steam reforming and electrolysis technologies by refineries and industries. Hydrogen are mainly produced and consumed by refineries ~96% is derived from non-sustainable fuels that consists of ~48% from natural gas, 30% from hydrocarbons based refineries, 18% from coal and about 4% from electrolysis (Lamy & Millet, 2020). Moreover, these processes have several consequences that waste and waste generation and rapid utilization of natural resources. Till date the major share of hydrogen is derived from biological and electrochemical process such as steam reforming, bio-photolysis, photo-fermentation, dark fermentation, bio-electrochemical based system such as microbial electrolysis cells and fuel cells, photolysis and electrolysis (Chaurasia, Siwach, et al., 2021a; Kachroo et al., 2022a). The hydrogen derived from waste or wastewater feedstocks are sustainable in nature and offer several sustainable goals such as value-added products recovery, renewable incentives, and higher energy revenues.

Fuel cells are electrochemical-based technologies that utilize the hydrogen or hydrogen-based fuel in the presence of oxygen and produce electricity and heat along with water as only by-products. Fuel cells are considered as clean power and heat generation from a range of primary energy resources (Chaurasia et al., 2022a; Kachroo et al., 2022b; Shankar et al., 2017). Fuel cells-based technologies are a diverse range of technologies depending upon the fuel sources or applications. However, there are many variants of the fuel cells or fuel cells-based technologies depending on the Fuel Cell type, process and the fuel used. These fuel cells-based technologies have various applications such as stationary power generators, auxiliary power generators, micro power generators, portable and distributed power generators for transportation, military projects and the automotive market (Chaurasia, Siwach, et al., 2021b; Chaurasia & Mondal, 2022b). Fuel cells are used for various energy applications such as portable, backup, transportation, and stationary power applications. There is drastic technological development in the different components of the fuel cells that are translated into the various types of the fuel cells and their applications. These advances in the fuel cells and fuel cells-based energy system have been made feasible due to these technological advances. These fuel cells-based technologies have been extensively tested in automobiles, unmanned machines, space and heating purposes in the industries.

Hydrogen fuel cell and fuel cell based-vehicles are the most potential application of the fuel cells based-technologies that has several benefits such as zero greenhouse gases emission and strong ecological adaptability. However, these technologies have progressively matured and are still in the development stage of the various factors of the fuel cell factors impeding their technological development. The development of hydrogen fuel cell vehicles, which is one of the main application fields of hydrogen in the future, was analyzed and forecasted. Beyond its traditional role as an industrial feedstock, primarily to produce ammonia, methanol, and petroleum refining, hydrogen has emerged as a new energy vector. Across the globe there is aggressive development in the hydrogen infrastructure to meet its sustainability goals and meet rising energy demand. During, Winter Olympics and Paralympics (2022), hydrogen fuel cell vehicles play the major role in the transportation and automobile sector due to their high sustainable adaptability (*Olympics Demonstrate Hydrogen Applications*, n.d.). These applications of the hydrogen fuel cells have been regarded as one of the decisive ways of the development in the transportation sector. The automobile industry's advocacy of green and sustainable hydrogen-based technologies development is the necessity of the development of fuel cells-based technologies development. Hydrogen from water electrolysis or from wastewater through electrolysis or photolysis has the potential to meet the hydrogen production requirement.

INTERNATIONAL HYDROGEN TECHNOLOGY AND HYDROGEN FUEL CELL STRATEGIES

1. Using hydrogen as a secondary energy resources. Higher electrical power generated by renewable sources such as solar and wind is then used to produce H_2 via electrolysis. This H_2, serving as a renewable energy storage route, can be used for (ii) and (iii) below, as well as large-scale industrial operations such as ammonia and steel manufacture.
2. Using H_2 fuel cells as main energy resources for transportation (land, air, and maritime) as well as stationary power sources.
3. Developing the potential conversion technologies for the available natural gas resources that need to be operate wholly or partially with H_2, that are necessity to reduce the CO2 emissions into the atmosphere.

Table 1. Application of hydrogen in transportation industry

Advantages	Disadvantages
High energy density (122-142 MJ/Kg)	Required large storage areas
Most abundant element and only water as by-products	Low existence in free in nature
Generated from numerous primary fuel sources	Low-slung ignition energy (similar to gasoline)
Extensive flammability range, mixture and engines operation	Presently expensive
High diffusivity	-

The improved hydrogen production yield and their transportation for the wider application require more understanding about the hydrogen storage. The hydrogen can be stored in the metal and non-noble-metal-based catalysts in which the non-noble-metal-based catalysts have the potential and futuristic technologies application (ABANADES, 2019). The possible forms of the large hydrogen storage are listed in Figure 2 that enlist their advantages and types.

Use of hydrogen fuel cells:

1. **Utilization in Decentralized Power Plants:** Large-scale power plants, whether nuclear, thermal, or hydropower, transfer emitted electricity to the grid and eventually to the customer. However, due to the varied load of Each user, the voltage is unstable or the power is shut off. Furthermore, typical thermal power generation consumes a huge amount of fossil energy and discharges a large number of hazardous and destructive wastes into the environment.
2. **Utilization in Hydrogen Fuel Cell Vehicles (Table 1):** The use of hydrogen fuel cells in automobiles is now a focus of fuel cell research. Many well-known automobile manufacturers have released hydrogen-powered vehicles. Fuel cells are anticipated to lessen air pollution and modern society's reliance on oil. Because the hydrogen fuel cell vehicle emits only water, it is considerably superior to the oil-powered internal combustion engine model as the ultimate goal of the future car.
3. **Application in Mobile Power:** Portable electronic items such as mobile phones, laptop computers, and digital cameras have grown in popularity over the last ten years, and their functions have become more complex, as has their energy usage. As a result, batteries must have a high energy density and a lengthy use life. However, the present lithium-ion batteries on the market are incapable of meeting people's increasingly demanding needs.

Figure 2. Large scale hydrogen storage options
Source: Andersson and Grönkvist (2019)

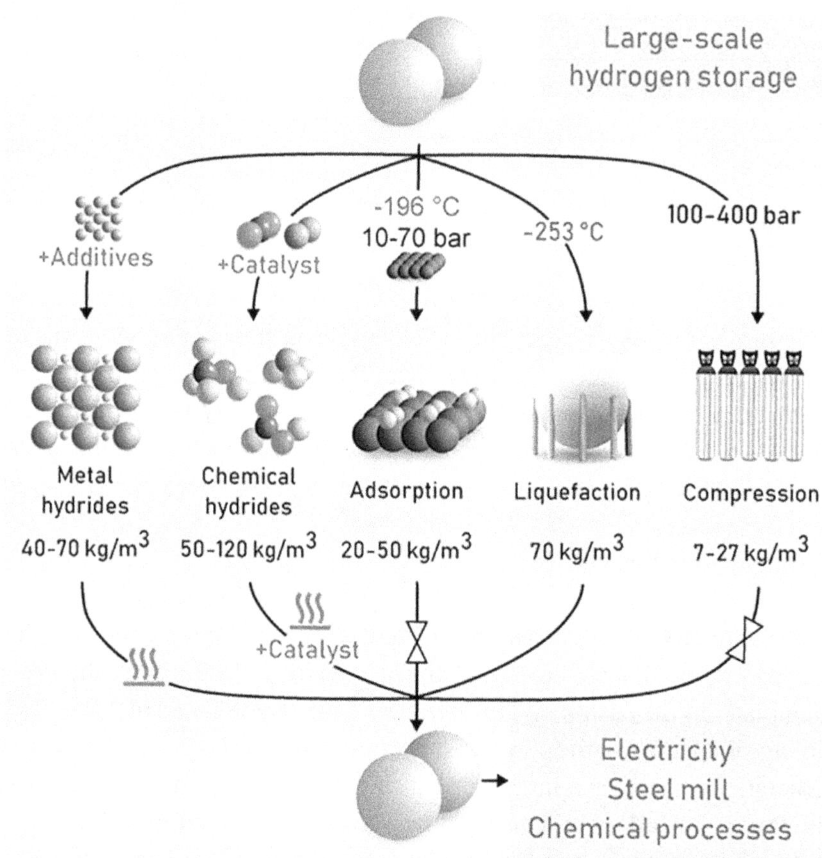

Unleashing Hydrogen Power: Fuel cell technologies are revolutionizing the way we power our lives, providing clean and renewable energy. With hydrogen production, we can tap into a carbon-free energy source that can be used to produce electricity. Fuel cells use an electrochemical reaction to generate electricity from hydrogen. The hydrogen is produced from natural gas or renewable energy sources such as solar, wind, or hydropower. This electricity can then be used to power homes, businesses, and vehicles. Fuel cell technologies are becoming increasingly popular to reduce our reliance on fossil fuels and reduce our carbon footprint. With fuel cells, we can create a more sustainable future for ourselves and for the planet (Chi & Yu, 2018).

The Benefits of Fuel Cells: Fuel cells are a great way to produce clean energy. They are efficient, reliable, and cost-effective. They can be used to power homes, businesses, and vehicles without producing any emissions or noise. Fuel cells can also be used to store energy. This means that the energy can be used when needed, such as during peak demand times or when the sun isn't shining. This helps to reduce electricity costs and improve grid reliability. Fuel cells are also easy to maintain and can last for years. This makes them a great long-term investment in clean energy production.

Fuel Cells for a Sustainable Future: Fuel cells are an important part of a sustainable future. They can provide clean, renewable energy that is reliable and cost-effective. This means that more people can access clean energy, reducing their reliance on fossil fuels. Fuel cells are a great way to power our future. They provide clean, reliable energy that can be used to power homes, businesses, and vehicles. This helps to create a more sustainable future for us and for the planet.

Fuel Cell Vehicle: Fuel cells are an effective and efficient technology for hydrogen production. They can produce large amounts of hydrogen with minimal energy input and produce nearly zero emissions. Additionally, fuel cells are becoming more cost-effective and reliable, making them an ideal choice for hydrogen production. Fuel cells are also becoming more versatile, as they are being used in a variety of applications, such as fuel for vehicles and electricity generation. As the technology continues to improve, fuel cells are becoming an increasingly important tool for hydrogen production. FCV is critical in many countries for relieving environmental stress and reducing fossil fuel consumption depicts some representative automobiles.

The Future of Fuel Cells: Fuel cells are the future of energy production. They are efficient, reliable, and cost-effective. They can be used to power homes, businesses, and vehicles without producing any emissions or noise. Fuel cells are also a great way to store energy. This means that the energy can be used when needed, such as during peak demand times or when the sun isn't shining. This helps to reduce electricity costs and improve grid reliability.

Hydrogen Production With Fuel Cells: Fuel cells are an important part of the hydrogen production process. They use an electrochemical reaction to generate electricity from hydrogen, which can then be used to power homes, businesses, and vehicles. Fuel cells are also a great way to store energy. This means that the energy can be used when needed, such as during peak demand times or when the sun isn't shining. This helps to reduce electricity costs and improve grid reliability. Fuel cells are becoming increasingly popular as a way to reduce our reliance on fossil fuels and reduce our carbon footprint. With fuel cells, we can create a more sustainable future for ourselves and for the planet.

HYDROGEN PRODUCTION FROM THE INDUSTRIAL WASTEWATER

Around 80% of global energy demand is contented by using fossil fuel as a major source that has adverse impacts on the environment along with generating huge amounts of the waste or wastewater. Thus, the biohydrogen production from industrial wastewater is getting immense research interest across the globe. The hydrogen from the wastewater is a sustainable energy source which generates water and energy, it overcomes the negative effects from the usage of fossil fuels. There are certain recent literatures that acclaimed that the economical hydrogen production from the industrial wastewater can be achieved using biological or bio-electrochemical system (MEC). Various recent literature has shown that the industrial wastewater from beverage industry, palm oil effluent, chemical industry, sugar industry, distillery industry wastewater could be potential feedstock for the biohydrogen production using anaerobic digestion, dark, photo or integrated fermentation and MEC technology (Chaurasia & Mondal, 2022). The development of various parameters of the MEC process through technological resources aid for it is scale-up or makes it to commercialization. The H_2 production from industrial wastewater has several attractive incentives that are listed below:

Advantages of hydrogen production from the industrial wastewater are as fallow:

1. Biological living systems such as microbes are capable of degrading almost all types of wastewaters.
2. It reduces waste and wastewater and is capable of meeting energy criteria that are the global concern.
3. It reduces toxic and hazardous pollutants in the aquatic system and the environment.
4. MEC technology has lower operating costs compared to other alternatives technologies.
5. Efficient degradation and removal of organic and inorganic as well as toxic compounds.

Apart from these, there are several properties and applications of the hydrogen as discussed in the previous section that realize the importance of the hydrogen generation from the wastewater. Hydrogen is a concentrated primary energy source that can be converted in different forms of energy with higher energy yield. Possibility to store in various ways consist of liquid hydrogen or as solid hydride as well as ease in the transportation in the wide geographical areas with higher efficiency. Approx. 9.5 kg hydrogen is equivalent to 25 kg gasoline (Badea et al., 2021). Hydrogen can generate from a wide waste feedstock and can be stored for a longer time and easy to transport.

Methods for Simultaneous Removal of Organics and Generation of Hydrogen

Various conventional and advanced techniques are used for the treatment of industrial wastewater including sugar and paper industry wastewater are described in Figure 3.

Figure 3. Various techniques used for the removal of pollutants from the sugar and paper industry wastewater

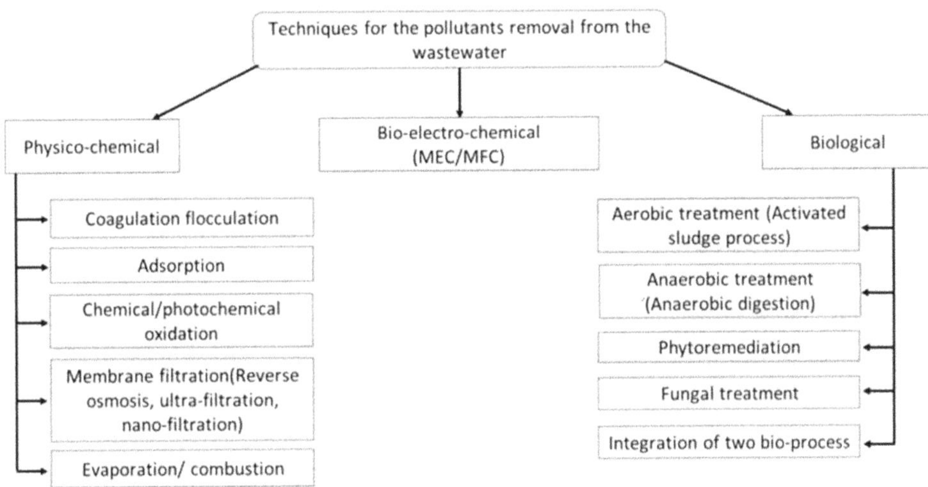

Most of the process is primarily used for the industrial wastewater treatment, while some of them are used for the recovery of value-added products such as microbial electrolysis cells used for the simultaneous hydrogen production and wastewater treatment. Chemical and electro-chemical processes have higher treatment efficiency but they generate secondary pollutants, while biological processes including MEC has competitive treatment efficiency along with energy generation as well as less secondary pollutants. Biological processes have a slower rate of pollutant degradation but advanced biological techniques such as integration of two biological processes (anaerobic digestion/photo fermentation) and MEC or alone MEC have higher treatment efficiency along with hydrogen/electricity generation (Chaurasia et al., 2021). Salient features of some important wastewater treatment techniques used for the removal of organic load, color etc., from wastewater as shown in Figure 2 are compared in Table 2.

Table 2. Salient features of various processes for the treatment of industrial wastewater

Process Property	Physico-Chemical	Biological	Bio-Electrochemical (MEC)
Range of input load	Low and high	Low	Medium
Secondary sludge generation	Higher sludge or residue management is an issue in few cases	Less	Less
Energy production potential	No	High	Medium (scum or residue)
Operating cost	High	low	low
Flexibility in operating temperature	Medium	Less	Medium
Rate of removal	Fast	slow	Fast

Fuel Cell-Based Technologies for the Hydrogen Production From the Wastewater

The industrial wastewater such as sugar and paper industry wastewater contain very high organics load (COD, BOD and TOC) but these wastewaters have low biodegradability index due to presence of several impurities. The presence of impurities, toxic compounds and ions lower the efficiency of the biohydrogen production technologies. Among different hydrogen production technologies as discussed in Table 5 and Table 6, the bio-electrochemical system i.e. MEC can degrade most of the industrial wastewater with higher energy recovery. In addition, depending upon the residual organic load and biodegradability index, the MEC can be a suitable technology for the simultaneous treatment of the industrial wastewater and hydrogen production. It is also notable that the MEC has a higher treatment efficiency than conventional wastewater treatment technologies. As per the discussion in the previous section, Figure 4 and from Table 6, it is clear that the MEC process has several advantages and higher treatment efficiency with hydrogen production. The other wastewater treatment processes mentioned in Table 3 are either time consuming or low efficiency as compared to MEC processes.

The above discussions suggest that most of the wastewater treatment techniques have certain advantages or disadvantages as well as limitations such as biological processes are slower or physico-chemical processes generate secondary pollutants etc. On the other hand, bio-electrochemical processes such as the MEC process offer simultaneous treatment of wastewater and hydrogen production (Chaurasia et al., 2022b; Shankar et al., 2017). The bio-electrochemical system has capability to produce high yield hydrogen from wastewater at ambient temperature with >60%

energy or resources recovery, but rates or yield of hydrogen recovery and generation need to be improved by using efficient cathodes materials or engineered microbes.

Table 3. The comparisons of hydrogen yield between fermentation and Microbial electrolysis cells

Properties	Fermentation	MEC
Hydrogen yield (%)	15–28	24–70
Hydrogen production rate (L/L reactor/d)	0.3–10	0.01–4.2
Complex wastewater treatment	No	Mostly possible
Complete substrate to hydrogen conversion	No	Yes
Production of hydrogen at high yield	No	Yes
Production of hydrogen at high rate	Yes	Yes
Production of hydrogen at ambient conditions	Difficult	Possible
High purity hydrogen production	No	Possible
Treat wastewater for reuse/discharge	No	Possible
Waste bio-refinery application	Long-term	Long-term

Source: Lu and Ren (2016)

FUEL CELLS-BASED BIO-ELECTROCHEMICAL SYSTEM FOR HYDROGEN GENERATION

Microbial electrolysis cells (MECs) are considered as the improved version of MFC, where organic compounds of wastewater are degraded in anodic chambers by microorganisms and H+ ion and free electrons are produced. The free electron is transported from the bulk of the solution to the anode surface and is carried over through an external circuit to cathode, whereas the H+ is transferred from anodic chamber to cathodic chamber through a proton transfer membrane. In MFC, H+ and e- react to form H_2O in presence of a catholyte, whereas in case of MEC, H+ and e- produce H_2 in presence of external voltage at cathode. Both oxidative and fermentative metabolism of microorganisms are observed in MEC (Chaurasia et al., 2022).

Some microbes having different types of metabolic paths and electron transport systems are reported in Table 5. The mechanism of the electron transport in the fuel cells are well understood. There is many proposed mechanisms but none of them has the instrumental or metabolic evident. Still research across the globe are working on this to explore the electron transport.

Table 4. Comparisons of hydrogen generation from wastewater techniques

Process	Advantages	Disadvantages	Reactions	Efficiency (%)	Types of Microorganism
Dark Fermentation	Waste reduction with H_2 production, light and O_2 not required, Technology development required, wide substrate and microorganism can be utilized	Low efficiency and low yield, required high volume of digester, relatively lower H_2 production, as yield increases then thermodynamically unfavorable, produce gas has mixture of gases	$C_6H_{12}O_6 + 6H_2O \rightarrow 12H_2 + 6CO_2$ $\Delta G = -25.83$ kJ/mol $C_6H_{12}O_6 + 2H_2O \rightarrow CH_3CH_2COOH + 2H_2 + 2CO_2$ $C_6H_{12}O_6 + 2H_2O \rightarrow CH_3COOH + 4H_2 + 2CO_2$ $\Delta G^o = -206$ kJ	65-78	*Clostridium* such as *C.buytricum*, *C.thermolacticum*, *C.pasteurianum*
Photo Fermentation	light required, use different waste materials like distillery effluents, waste etc.	Low hydrogen yield, Low efficiency, required light, high volume of reactor needed, sensitive to O_2 efficiency 1–5%	CH3COOH $2H_2O$ + Light energy → $CH_3COOH + 4H_2 + 2CO_2$ $\Delta G^o = +104$ kJ	10-40	*Rhodobacter* such as *Rhodobacter spheroids*, *Rhodobacter capsulatus*, *Rhodovulum sulfidophilum* and *Rhodopseudomonas palustris*
Direct bio-photolysis	Produce H_2 from water and light	Light required throughout the process	$H_2O + 2F \rightarrow H_2 + 1/2O_2$ (where F = 96485 °C)	25-45	Green algae
Indirect biophotolysis	Produce H_2 from water and N_2 from atmosphere	Uptake of hydrogenase enzymes are difficult 30% O_2 present may act as an inhibitor	$H_2O + 2F \rightarrow H_2 + 1/2O_2$	30-40	Cyanobacteria
Sequential bio-process	High process stability, Up to 90% COD removal, High rate of hydrogen production	Irregular hydrogen production Difficulties in design and commercialization	$C_6H_{12}O_6 + 2H_2O \rightarrow CH_3COOH + 4H_2 + 2CO_2$, $2C_3COOH + 4H_2O$ + Light energy → $2CH_3COOH + 8H_2 + 4CO_2$ ΔG^o Overall: $C_6H_{12}O_6 + 6H_2O \rightarrow 12H_2 + 6CO_2$ $\Delta G^o = +3.2$ kJ	40-70	Dark and photo-synthetic microorganism
Anaerobic digestion	Waste reduction with biogas production, light and O_2 not required, Technology development required	Low yield of hydrogen and higher biogas production, Technology development and advancement required	Hydrolysis, acidogenesis, acetogenesis, and methanogenesis, $C_6H_{12}O_6 \rightarrow 3CO_2 + 3CH_4$	50-70	Acetogens Methanogens, Sulfate reducing bacteria, *Methanosarcina*
MEC	Higher treatment efficiency, can degrade almost all types of the waste, resources recovery (hydrogen), can degrade high impurities and complex wastewater, low cost and simple process can be carried out ambient conditions	Required 0.14 V electricity at cathodes, Selection of efficient electrodes (cathodes and anode), mediator required extensive research work	Anode: $CH_3COO^- + 4H_2O \rightarrow 2HCO^{3-} + 9H^+ + 8e^-$ ($E^0 = 0.187$ V) Cathode: $8H^+ + 8e^- \rightarrow 4H_2$ Overall: $CH_3COO^- + 4H_2O$ $2HCO^{3-} + H^+ + 4H_2$ (ΔGr=104.6 kj/mol) (2.3)	60-90	*Escherichia coli*, *Enterobacter aerogenes*, *Clostridium acetobutylicum*, *Clostridium perfringens*

Source: Kumar Gupta et al. (2013) and Rozendal et al. (2006)

Table 5. Some microorganisms with different electron transport systems

Metabolic Type	Transfer Type	Examples of Organisms
Oxidative metabolism	Direct electron transport	*Rhodoferax ferrireducens* *Geobacter sulfurreducens* *Aeromonas hydrophila*
	Indirect electron transport (Mediator-driven)	*Escherichia coli* *Shewanella putrefaciens* *Pseudomonas aeruginosa* *Erwinia dissolvens* *Desulfovibrio desulfuricans*
Fermentative metabolism	Direct electron transport	*Clostridium butyricum*
	Indirect electron transport (Mediator-driven)	*Enterococcus faecium*

The MEC electrode reactions are given in Eq. 1-2 with acetate as substrate.

Anode: $CH_3COO^- + 4H_2O\ 2HCO^{3-} + 9H^+ + 8e^-$ ($E^0 = 0.187$ V) (1)

Cathode: $8H^+ + 8e^-\ 4H_2$ ($E^0 = 0.000$ V) (2)

Overall: $CH_3COO^- + 4H_2O\ 2HCO^{3-} + H^+ + 4H_2$ (ΔGr=104.6 kj/mol) (3)

$$E_{an} = E_{an}^0 = -\frac{RT}{8F} lnln\left(\frac{CH3COO-}{\left(HCO3^{-2}\right)\left(H^+\right)}\right) = 0.187$$

$$-\left(\frac{\left(\frac{8.31J}{molK}\right)(298.15K)}{8\left(\frac{96500C}{mol}\right)} ln\frac{[0.0169]}{[0.005]^2[10^{-7}M]^2}\right) = -0.300V \quad (4)$$

$$E_c = E_c^0 = -\frac{RT}{8F} lnln\left(\frac{H_2}{[H^+]^2}\right) = 0.00 - \left(\frac{\left(\frac{8.31J}{molK}\right)(298.15K)}{2\left(\frac{96500C}{mol}\right)} ln\frac{1}{[10^{-7}M]^2}\right) = -0.414V$$

(5)

The total Nernstian cell potentials (Ecell) necessary for a MEC to produce H_2 at the cathode under these conditions are -0.114 V (-0.414 – (-0.30) = -0.114 V). Alternatively, it can also be calculated as $E_{eq} = -\Delta Gr/nF = (-104.6 \times 10^3)/8 \times 96485$ = -0.114 V and (ΔGr = 104.6 kj/mol). Thus, the minimum cell voltage required in the MEC system for the hydrogen production is -0.114 V. These calculations do not account for the energy or voltage losses due to cathode overpotential, anode overpotential and metabolic loss in the anodic bio-film. From the above reactions (Eq. 15). In case of real industrial wastewater, reactions involved in MEC can be expressed through Eq. 2.6-2.8.

Anode: $n_1C_xH_yO_z + n_2H_2O\ n_3CO_2 + pH^+ + pe^-$ ($E^0 = \Delta V$) (6)

Cathode: $2pH^+ + 2pe^-\ pH_2$ (E^0 = 0.000 V) (7)

$$E_{eq} = -\Delta G_r/pF \tag{8}$$

The above reactions indicate that the hydrogen production in the MEC process for the hydrogen production is influenced by cathode's electro-catalytic activity. There are many challenges such as high electrocatalytic cathodes, mediator, undesired e- sinks etc., and these challenges should be overawed in order for MECs to be applied in large scale system (Randolph and Studer, 2013). When hydrogen is the main target product, then the yield could be considerably decreased as a result of a large number of undesirable e^- sinks of different metabolism. Methanogenesis is the most commonly occurring phenomenon in single chamber MEC systems without separators. This causes the production of CH_4 in place of H_2 after a long time, which was also stated in several works consisting of the MEC first pilot plant using winery wastewater (Chaurasia et al., 2022a; Chaurasia, Shankar, et al., 2021b). Literature showed the contribution of hydrogenotrophic methanogens in the H_2 re-consumption and CH_4 generation, since acetolactic methanogens were usually outcompeted by acetate oxidizing exoelectrogens in low acetate environment for instance MECs (Chaurasia et al., 2022b). Hydrogenotrophic methanogenesis could be a severe issue due to the generation of 1 mol of CH_4 which utilizes 4 mol of H_2 ($4H_2 + CO_2 = CH_4 + 2H_2O$), also these microbes are capable of scavenging hydrogen at nanomolar concentrations. Numerous approaches have been investigated to constrain methanogenesis processes in MECs consisting of physical, chemical, biological methods; chemical inhibitors and mediators may be effective, but these increase the operating costs along with the consequence of new potential contamination. O_2 exposure has considerably limited impact on the growth of methanogens, as other facultative or anaerobic microbes in the films protect these, and large DO might be resulted in the damage on anaerobic exoelectrogens (Kachroo et al., 2022b). The decrease of hydraulic retention time with enhanced flow rate or applied voltage cannot reduce/inhibit methanogens from the reactor. Though the usage of radiation and low temperature attained good performance in limiting methane production at the beginning, they can be difficult due to complex operation and large operating cost. Moreover, the activities of methanogens would recuperate once these operations are stopped (Chaurasia & Mondal, 2022b).

TYPES OF THE FUEL CELLS FOR HYDROGEN GENERATION

Fuel cells, which are a type of bio-electrochemical system, are categorized based on the electrolytes used and the processes involved in the making. Through an electrochemical process, these fuel cells produce various forms of energy, or are used for the generation of heat-based energy or electricity, without the need for the processes like combustion or gasification. There are several fuel cell technologies that are currently either in the prototype development or research phase, out of which, some of the prominent ones differing in terms of their electrolytes used, chemical reactions taking place, catalysts involved, operating temperatures, and the type of fuel used as a feedstock are:

Polymer Electrolyte Membrane Fuel Cells: The high-power density and light weight of polymer electrolyte membrane fuel cells (PEMFCs) make them a popular choice over other types of fuel cells. The name is used interchangeably with proton exchange membrane fuel cells. The solid polymer electrolyte and carbon electrodes with platinum or platinum alloy catalyst make it possible for these fuel cells to operate with just hydrogen, oxygen, and water. Typically, PEMFCs are fueled with pure hydrogen that comes from storage tanks or reformers. These fuel cells operate at lower temperatures, around 80°C (176°F), resulting in faster start-up time and less wear and tear on system components, which means better durability. However, the use of a noble-metal catalyst such as platinum increases system costs. Although PEMFCs are primarily used for transportation purposes, they can be used for stationary applications as well. They are particularly well-suited for vehicle applications like buses, cars, and heavy-duty trucks.

Direct Methanol Fuel Cells: The majority of fuel cells derive their power from hydrogen, which can be directly supplied to the system or generated by reforming hydrogen-rich fuels like methanol, ethanol, or hydrocarbons. But, the operation of the DMFCs (direct methanol fuel cells) is based completely on 100% methanol, which in turn is generally mixed with water before a direct injection to the fuel cell anode. DMFCs do not experience the fuel storage difficulties common to other fuel cell systems, as methanol has a higher energy density than hydrogen (though lower than gasoline or diesel fuel). Furthermore, methanol is a liquid, much like gasoline, making it easier to transport and distribute through existing infrastructure. DMFCs are commonly employed for portable fuel cell applications like cell phones and laptops.

Alkaline Fuel Cells: AFCs were one of the earliest fuel cell technologies created and were extensively employed in the U.S. space program to produce electrical energy and water aboard spacecraft. The use of the solution of potassium hydroxide and water as the electrolyte, which in turns give the liberty of usage for catalysts at anode and cathode from a numerous non-precious metal. Recently, innovative AFCs using a polymer membrane as the electrolyte have emerged. These fuel cells are similar to conventional PEM fuel cells, but use an alkaline membrane instead of an acidic one. One of the primary obstacles for alkaline fuel cells (AFCs) is their vulnerability to carbon dioxide (CO_2) poisoning. Even trace amounts of CO_2 in the air can severely impact the cell's efficiency and lifespan by creating carbonate deposits. While AFCs with liquid electrolytes can be operated in a recirculating mode to regenerate the electrolyte and reduce carbonate formation, this mode introduces issues with shunt currents and other concerns such as wettability, corrosion, and differential pressure handling. Alkaline membrane fuel cells (AMFCs) address these concerns and are less susceptible to CO_2 poisoning than AFCs with liquid electrolytes. Nevertheless, CO_2 can still impact their performance, and they currently lag behind proton exchange membrane fuel cells (PEMFCs) in terms of both performance and durability. AMFCs are being studied for potential use in the range of watts to kilowatts. However, several challenges need to be addressed for their practical application such as tolerance to CO_2, membrane conductivity and durability.

Phosphoric Acid Fuel Cells: The utilization of the phosphoric acid in liquid state as an electrolyte have given the name Phosphoric acid fuel cells (PAFCs), which is encased in a matrix silicon carbide, bonded by Teflon. The electrodes consist of porous carbon containing a platinum catalyst, and the electro-chemical reactions occurring in the cell can be found in the diagram on the right-hand side. PAFCs are often referred to as the "first generation" of modern fuel cells and are considered one of the most well-established cell types, having been used commercially for some time. They are primarily used for stationary power generation, but there have been instances of PAFCs powering large vehicles such as city buses. PAFCs have an advantage over proton exchange membrane (PEM) cells when it comes to tolerating impurities in fossil fuels that have been reformed into hydrogen. PEM cells are susceptible to "poisoning" by carbon monoxide as it binds to the platinum catalyst at the anode, which decreases the fuel cell's efficiency. In contrast, PAFCs have a tolerance to impurities and are over 85% efficient when co-generating electricity and heat. However, they are less efficient when generating electricity alone, with an efficiency range of 37%-42%. This level of efficiency is only slightly better than that of combustion-based power plants, which typically operate at around 33%. Additionally, PAFCs are less powerful and more massive than other fuel cells of similar weight and volume, making them expensive to produce. A significant amount of costly platinum catalysts is necessary to manufacture these fuel cells, which contributes to the higher cost.

Molten Carbonate Fuel Cells: The development of molten carbonate fuel cells (MCFCs) is currently underway, with the aim of using them in natural gas and coal-based power plants for various applications in the electrical utility, industrial, and military sectors. These fuel cells are designed to operate at high temperatures of approximately 650°C (or 1,200°F) and are considered high-temperature fuel cells. The electrolyte used in MCFCs is molten carbonate salt, which is suspended in a matrix of porous, inert ceramic lithium aluminum oxide. One of the benefits of operating at high temperatures is that non-precious metals can be used as catalysts at the anode and cathode, which can help to reduce costs. Significant cost reductions are achievable with MCFCs due to their improved efficiency. When combined with a turbine, these fuel cells can reach efficiencies of up to 65%, far surpassing the 37% to 42% efficiency of phosphoric acid fuel cells. Moreover, when waste heat is captured and utilized, overall fuel efficiencies can exceed 85%. Unlike alkaline, phosphoric acid, and PEM fuel cells, MCFCs do not necessitate an external reformer to convert fuels such as natural gas and biogas into hydrogen. Instead, the fuel cell itself converts methane and other light hydrocarbons to hydrogen through internal reforming, resulting in reduced costs

MCFCs face a significant drawback in terms of durability with their current technology. The harsh operating conditions at high temperatures and the use of corrosive electrolytes accelerate the corrosion and breakdown of components, ultimately reducing the lifespan of the cells. Researchers are actively investigating the use of materials that are resistant to corrosion and the development of fuel cell designs that can potentially double the lifespan from the current 40,000 hours (about 5 years) without sacrificing performance.

Solid Oxide Fuel Cells: Solid oxide fuel cells (SOFCs) utilize a hard, non-porous ceramic compound as an electrolyte, and they are highly efficient at converting fuel to electricity, achieving around 60% efficiency. In co-generation applications that capture and utilize the system's waste heat, the overall fuel use efficiencies can exceed ~84%. These fuel cells operate at very high temperatures, reaching up to 1,830°F. This high-temperature operation eliminates the need for precious-metal catalysts, thus reducing the cost. Additionally, SOFCs can reform fuels internally, allowing for the use of a variety of fuels and reducing the expenses associated with adding a reformer to the system. Solid oxide fuel cells (SOFCs) have distinct advantages over other fuel cell types. For instance, they can tolerate significantly more sulfur than other cell types, and they are not sensitive to carbon monoxide poisoning, which means that they can use various fuels, such as natural gas, biogas, and coal-derived gasses. Additionally, high-temperature operation eliminates the need for precious-metal catalysts, making them less expensive. However, high-temperature operation presents challenges, including slow startup and thermal shielding requirements, which make them unsuitable for transportation. Furthermore, durability is a critical issue due to the harsh conditions at high temperatures. Therefore, developing low-cost materials that can withstand these conditions is essential. One promising approach is to develop lower-temperature SOFCs that have fewer durability problems and are less expensive to operate. Although these lower-temperature SOFCs have not yet achieved the same performance as high-temperature systems, scientists are currently working on developing stack materials that can function at this lower temperature range.

Figure 4. Some common types of fuel cells-based systems
Source: Chaurasia et al. (2022a) and Chaurasia and Mondal (2021)

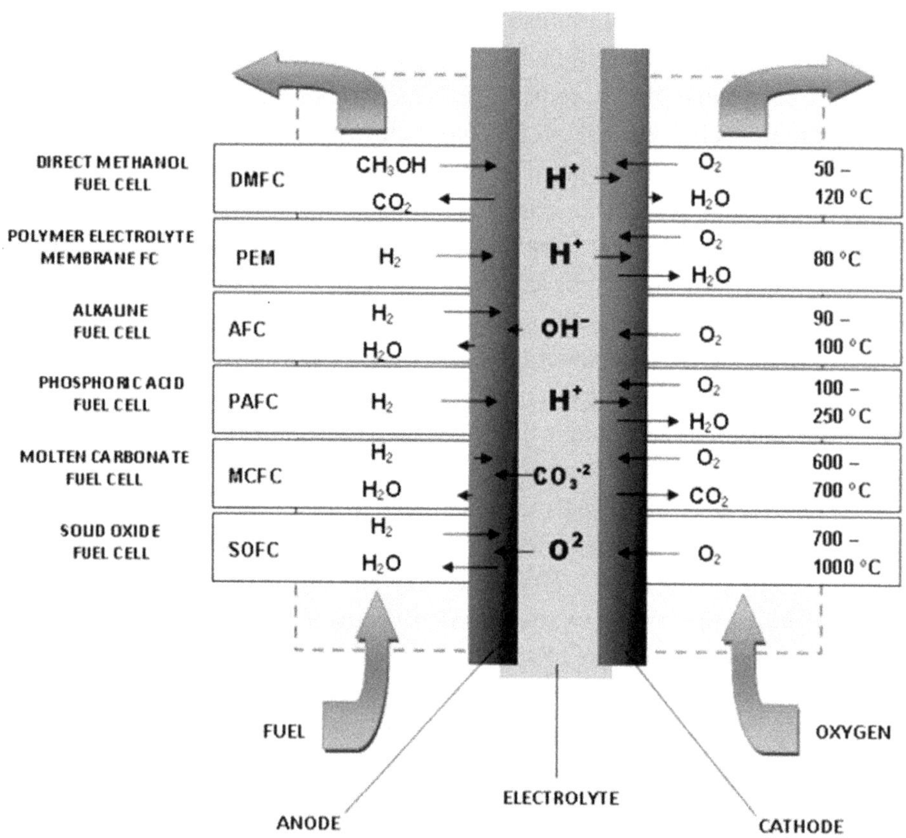

Reversible Fuel Cells: Reversible fuel cells operate similarly to other fuel cells in that they generate electricity and by-products of heat and water from hydrogen and oxygen. However, reversible fuel cell systems have the added capability of using electricity from renewable sources like solar and wind power to carry out electrolysis and produce hydrogen and oxygen fuel. This means that reversible fuel cells not only provide power when needed, but also have the ability to store excess energy in the form of hydrogen when renewable energy technologies produce more energy than is currently needed. This energy storage potential could prove crucial in enabling intermittent renewable energy sources to be more effective and reliable.

Table 6. Some common types of fuel cells-based systems

Type	Electrolyte	T (°C)	Fuel	Advantages	Problems
Proton-Exchange Membrane Fuel Cells	Polymeric membrane	30-100	H_2	Fast start-up, Solid electrolyte	Pure H_2 needed, expensive catalyst
Direct-Methanol Fuel Cells	Polymeric membrane	30-100	CH_3OH	Liquid fuel, no reforming of fuel	Slow-reaction, Fuel cross over from anode to cathode
Alkaline Fuel Cells	KOH (aq)	65-220	KOH	Fast cathodic reaction (Higher current response)	Reactants must be removed
Phosphoric Acid Fuel Cells	H_3PO_4	150-220	H_2	High efficiency with heat cogeneration	Low power and current, expensive materials
Molten Carbonate Fuel Cells	Coronates (Li, Na, K)	600-1000	H_2	High conductivity and current density	Slow start-up, materials issue
Solid Oxide Fuel Cells	(Zr Y) O_2	600-1000	H_2	Low-cost materials, Solid electrolytes	Corrasion of metals, materials issue

Source: Ayol (2022) and Shankar et al. (2017)

ADDITIONAL READINGS

In the area of hydrogen production the photolysis, photocatalysis, electrolyzer and fuel cells-based engines further studied are required for the better understanding of it. For the hydrogen storage and transportation sector, various group are researcher shown that various materials are has high hydrogen storage capacity at the lab-scale experiments. That need to be explored for the higher scale. The hydrogen refueling station in a definite geographical area need to open or installed for the commercial hydrogen fuel cell-based transportation vehicle. The use of nanotechnologies for the stationery application of the hydrogen and hydrogen based-fuel cells are also recommended for the further studied. Hydrogen circular-bioeconomy is also suggested for the further studied.

FUTURE RESEARCH DIRECTIONS

Hydrogen is the fuel of the future and various aspect of the hydrogen production, storage and utilization technologies are in the infant stage that need to be explore at various commercial level. Various research group across the globe are continuous working for it technological development that required the international collaboration and technology sharing that bring many more unexplored aspect of the hydrogen circular-bioeconomy. There is enormous application of the stationery uses of the hydrogen in the various forms such as fuel, heat, electricity and power. But these technologies are costlier and need to be more economical. Economic evaluation of the various hydrogen production process as well as their transportation and utilization should be performed to assess its applicability at the commercial scale. The Development of more effective cathode catalysts by using different metal combinations and evaluation of their performances for the high yield of hydrogen production. Investigation on the impact of cathode geometry, size and shape performance in the various fuel cells applications. Application of pre-treatment of industrial effluents with low biodegradability index to explore its impact on enhanced hydrogen production. Investigation on the MEC process in continuous mode operation to assess its possibility in large scale application. Modelling of the hydrogen fuel cells process for creating design data for installing higher scale hydrogen generation plant. There is various aspect of the hydrogen production, storage and utilization is in continues at research and evaluation phase that open the several application avenues of it.

CONCLUSION

The above discussion enlisted the various ways of hydrogen production from the waste or wastewater along with their potential application. It also noted that hydrogen generation from waste or wastewater can unquestionably be regarded as the progenitor of the decisive clean, efficient and climate-neutral energy systems. It also discussed that hydrogen technologies and hydrogen fuel cells have proven that this becomes the potential future technologies for the energy sector. That is one step forward to the sustainability and green process along with capability to ensure the increasing global energy demand. But it required extensive research work on the various aspects of the fuel cells-based technologies development along with their applications. Hydrogen and fuel cells may serve as a breeding ground for new science and technology, innovations, companies, products, and employment.

REFERENCES

Abanades. (2019). Metal Oxides Applied to Thermochemical Water-Splitting for Hydrogen Production Using Concentrated Solar Energy. *ChemEngineering*, *3*(3), 63. doi:10.3390/chemengineering3030063

Andersson, J., & Grönkvist, S. (2019). Large-scale storage of hydrogen. *International Journal of Hydrogen Energy*, *44*(23), 11901–11919. doi:10.1016/j.ijhydene.2019.03.063

Ayol, A. (2022). *ScienceDirect Yeast industry wastewater treatment with microbial fuel cells : Effect of electrode materials and reactor configurations.* doi:10.1016/j.ijhydene.2022.05.277

Badea, G., Felseghi, R. A., & Aschilean, I. (2021). Hydrogen fuel cell technology for stationary applications. *Hydrogen Fuel Cell Technology for Stationary Applications*, 1–332. doi:10.4018/978-1-7998-4945-2

Chaurasia, A. K., Goyal, H., & Mondal, P. (2020). Hydrogen gas production with Ni, Ni–Co and Ni–Co–P electrodeposits as potential cathode catalysts by microbial electrolysis cells. *International Journal of Hydrogen Energy*, *45*(36), 18250–18265. doi:10.1016/j.ijhydene.2019.07.175

Chaurasia, A. K., Mohapatra, S., Shankar, R., & Thakur, L. S. (2022). Technologies for the clean and renewable energy production for the sustainable environment. *Clean Technologies and Sustainable Development in Civil Engineering*, 141–178. doi:10.4018/978-1-7998-9810-8.ch007

Chaurasia, A. K., & Mondal, P. (2022). Enhancing biohydrogen production from sugar industry wastewater using Ni, Ni–Co and Ni–Co–P electrodeposits as cathodes in microbial electrolysis cells. *Chemosphere, 286*(November). doi:10.1016/j.chemosphere.2021.131728

Chaurasia, A. K., Shankar, R., & Mondal, P. (2021). Effects of nickle, nickle-cobalt and nickle-cobalt-phosphorus nanocatalysts for enhancing biohydrogen production in microbial electrolysis cells using paper industry wastewater. *Journal of Environmental Management*, *298*, 113542. doi:10.1016/j.jenvman.2021.113542 PMID:34426219

Chaurasia, A. K., Siwach, P., Shankar, R., & Mondal, P. (2021). Effect of pre-treatment on mesophilic anaerobic co-digestion of fruit, food and vegetable waste. *Clean Technologies and Environmental Policy, 1*, 1–14. doi:10.100710098-021-02218-5

Chi, J., & Yu, H. (2018). Water electrolysis based on renewable energy for hydrogen production. *Cuihua Xuebao. Chinese Journal of Catalysis, 39*(3), 390–394. doi:10.1016/S1872-2067(17)62949-8

IEA. (2019). *The Future of Hydrogen – Analysis.* IEA.

Jones, E. R., van Vliet, M. T. H., Qadir, M., & Bierkens, M. F. P. (2021). Country-level and gridded estimates of wastewater production, collection, treatment and reuse. *Earth System Science Data, 13*(2), 237–254. doi:10.5194/essd-13-237-2021

Kachroo, H., Chaurasia, A. K., Chaurasia, S. K., & Yadav, V. K. (2022). Sustainable Clean Energy Production from the Bio-electrochemical Process Using Cathode as Nanocatalyst. Handbook of Green and Sustainable Nanotechnology, 1–30. doi:10.1007/978-3-030-69023-6_58-1

Kumar Gupta, S., Kumari, S., Reddy, K., & Bux, F. (2013). Trends in biohydrogen production: Major challenges and state-of-the-art developments. *Environmental Technology, 34*(13–14), 1653–1670. doi:10.1080/09593330.2013.822022 PMID:24350426

Lamy, C., & Millet, P. (2020). A critical review on the definitions used to calculate the energy efficiency coefficients of water electrolysis cells working under near ambient temperature conditions. In *Journal of Power Sources* (Vol. 447). Elsevier B.V. doi:10.1016/j.jpowsour.2019.227350

Lu, L., & Ren, Z. J. (2016). Microbial electrolysis cells for waste biorefinery: A state of the art review. In Bioresource Technology (Vol. 215, pp. 254–264). doi:10.1016/j.biortech.2016.03.034

Olympics Demonstrate Hydrogen Applications. (n.d.). Retrieved March 26, 2023, from https://fuelcellsworks.com/news/olympics-demonstrate-hydrogen-applications/

Rani, M., Shanker, U., & Chaurasia, A. K. (2017). Catalytic potential of laccase immobilized on transition metal oxides nanomaterials: Degradation of alizarin red S dye. *Journal of Environmental Chemical Engineering, 5*(3), 2730–2739. Advance online publication. doi:10.1016/j.jece.2017.05.026

Rozendal, R. A., Hamelers, H. V. M., Euverink, G. J. W., Metz, S. J., & Buisman, C. J. N. (2006). Principle and perspectives of hydrogen production through biocatalyzed electrolysis. *International Journal of Hydrogen Energy, 31*(12), 1632–1640. doi:10.1016/j.ijhydene.2005.12.006

Shankar, R., Pathak, N., Chaurasia, A. K., Mondal, P., & Chand, S. (2017). Energy production through microbial fuel cells. *Sustainable Utilization of Natural Resources*, 354–380. doi:10.1201/9781315153292

Singh Thakur, L., Parmar, H., Kumar Varma, A., Kumar Chaurasia, A., & Mondal, P. (2022). Removal of manganese from synthetic wastewater by Vetiveria zizanioides. *Materials Today: Proceedings*, 8–11. doi:10.1016/j.matpr.2022.08.395

Staffell, I., Scamman, D., Velazquez Abad, A., Balcombe, P., Dodds, P. E., Ekins, P., Shah, N., & Ward, K. R. (2019). The role of hydrogen and fuel cells in the global energy system. *Energy & Environmental Science, 12*(2), 463–491. doi:10.1039/C8EE01157E

Chapter 6
Hydrogen Technology Integration for Energy Support of Electric Vehicle Charging Stations

Maria-Simona Raboaca
National Research and Development Institute for Cryogenic and Isotopic Technologies, ICSI Rm. Valcea, Romania

Traian Candin Mihaltan
Faculty of Building Services Engineering, Technical University of Cluj-Napoca, Romania

ABSTRACT

In a case study, this chapter presents scenarios developed with the aim of analyzing two types of energy management strategies dedicated to charging stations for electric vehicles based on photovoltaic energy, having the power reserve provided with or without hydrogen energy support and its associated technologies (fuel cell and electrolyser). The proposed solution, to solve the problem between the amount of available energy and the number of parking spaces for the electric vehicle charging station, works according to the following principle: the system management is designed to work with one to five electric vehicle charging stations. It can be installed from one to five electric vehicle charging stations with a power of 22 kW. In this research, the state of charge of electric vehicles and the management of distributed energy for electric vehicles were monitored.

DOI: 10.4018/978-1-6684-6721-3.ch006

SCENARIOS USED TO ANALYZE THE TWO ENERGY MANAGEMENT STRATEGIES FOR AN ELECTRIC VEHICLES CHARGING STATION BASED ON PHOTOVOLTAIC ENERGY WITH OR WITHOUT SUPPORT OF THE FUEL CELL AND ELECTROLYZER SYSTEM

With the information from the website (www.parcaripitesti.ro), charging station from our case study is located in Calea Craiovei, Pitesti, Romania and has 10 parking spaces as presented in Figure 1.

Figure 1. Delimitation of parking

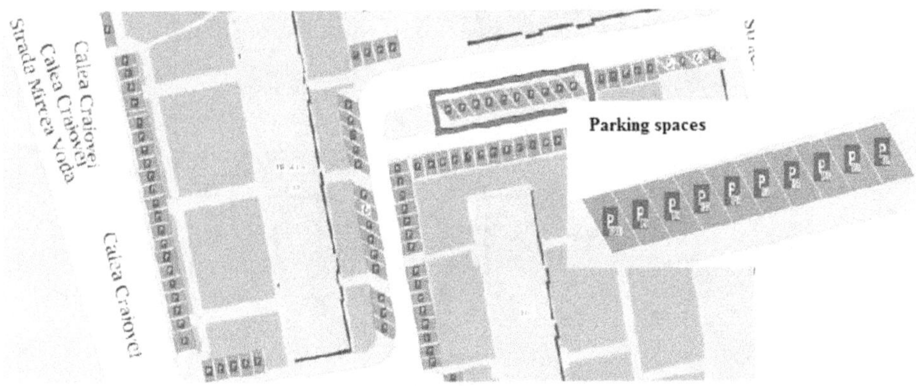

Description of Scenario

The proposed solution to solve the problem between the amount of available energy and the number of parking spaces for charging station electric vehicles, works according to the following principle:

The management of the system is designed to work with one to five EV charging stations.

We can install from one to five EV charging stations, power 22 kW. *In this research were* monitored the status of charging electric vehicles and the management of energy distributed for each type of electric vehicles.

According to equation 13, 38 batteries will be used in the proposed PV energy system, the capacity of battery is 1000 Ah. Energy management for the PV charging station is presented in Table 1.

Table 1. Energy management for the PV charging station

Scenario	Design of charging station	Description
1	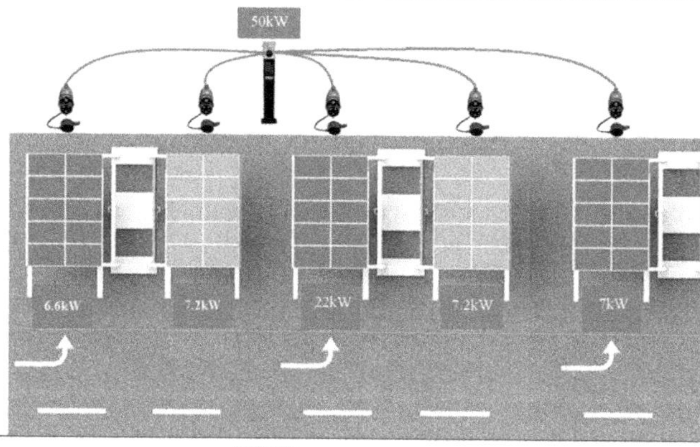 	In this scenario all 5 place of charging station are busy, ***in the same time.***
2	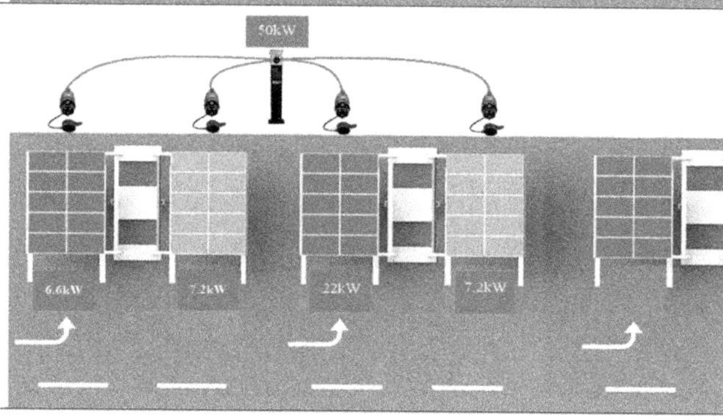 	In this scenario all 5 charging station are busy, but ***not in the same time***.
3	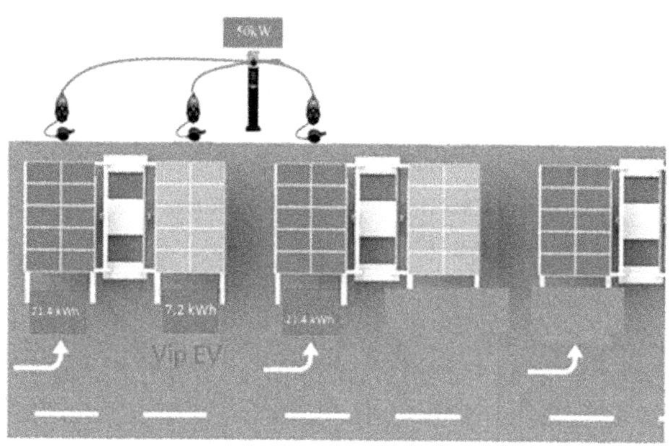 	***VIP electrical vehicle.*** There is the possibility of introducing VIP subscriptions that will permanently receive the maximum power required for charging. With this facility, the user will absorb maximum power needed, and only the remaining available power will be shared for the other stations.

Input Data Electric Vehicles Charging Station Based on Photovoltaic Energy

Case Study 1: Input Data Electric Vehicles Charging Station Based on Photovoltaic Energy Without Support of the Fuel Cell and Electrolyzer. We Propose the Following Configuration.

The input data for an electric vehicles charging station based on photovoltaic energy without support of the fuel cell and electrolyzer we propose the follow configuration:

- Photovoltaic area with Nominal Power by 61 kW;
- Batteries bank;
- 5 places for electric vehicle Charging Station;
- Average Loading time for each electrical vehicles: 5 hours;
- The total energy consumed during a day for all 5-place of EV charging station: 50 kW;
- Supply voltage: 380V c.a.
- Monitoring, command, control and protection elements (programmable automatic, computer, DC-DC and DC/AC converters, measuring transducers and counters, grid modules, etc.).

Figure 2. Total energy produced by PV: Case study 1

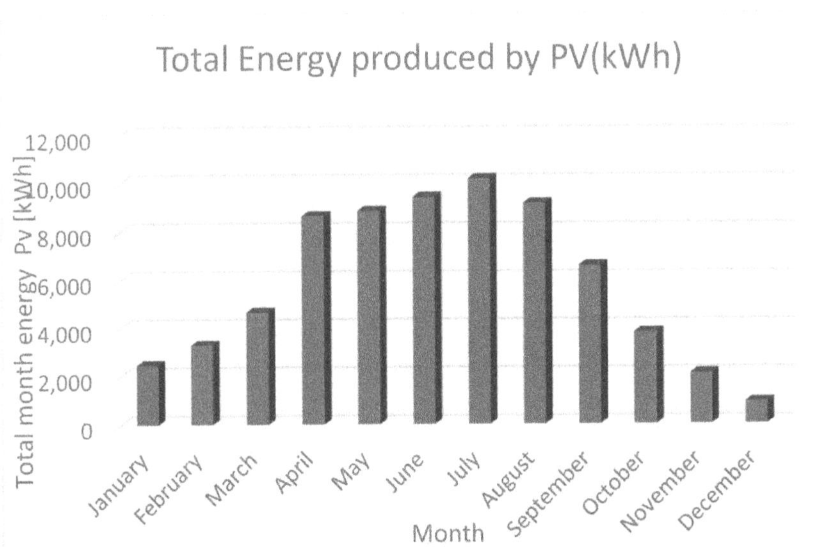

In these conditions, based on solar radiation data, a total energy production is 70,139 kWh estimated for 2020 (see Figure 2 distributed monthly throughout the year according to the graph). The estimation of the amount of energy produced by the photovoltaic park was made using the iHOGA software.

Case Study 2: Input Data Electric Vehicles Charging Station Based on Photovoltaic Energy With Support of the Fuel Cell and Electrolyzer. We Propose the Following Configuration.

The input data for an electric vehicles charging station based on photovoltaic energy without support of the fuel cell and electrolyzer we propose the follow configuration:

- Photovoltaic area with Nominal Power by 61 kW;
- Batteries bank;
- 5 place EV Charging Station;
- PEMFC
- EPEM lectrolyzer
- Hydrogen Tank
- Average Loading time for each electrical vehicles: 5 hours;
- The total energy consumed during a day for all 5 place EV charging station: 50 kW;
- Supply voltage: 380V c.a.
- Monitoring, command, control and protection elements (programmable automatic, computer, DC-DC and DC/AC converters, measuring transducers and counters, grid modules, etc.).

In these conditions, based on solar radiation data, a total energy production is 28,181 kWh estimated for 2020 (see Figure 3) distributed monthly throughout the year according to the graph in Figure 3. The estimation of the amount of energy produced by the photovoltaic park was made using the iHOGA software.

Figure 3. Total energy produced by PV: Case study 2

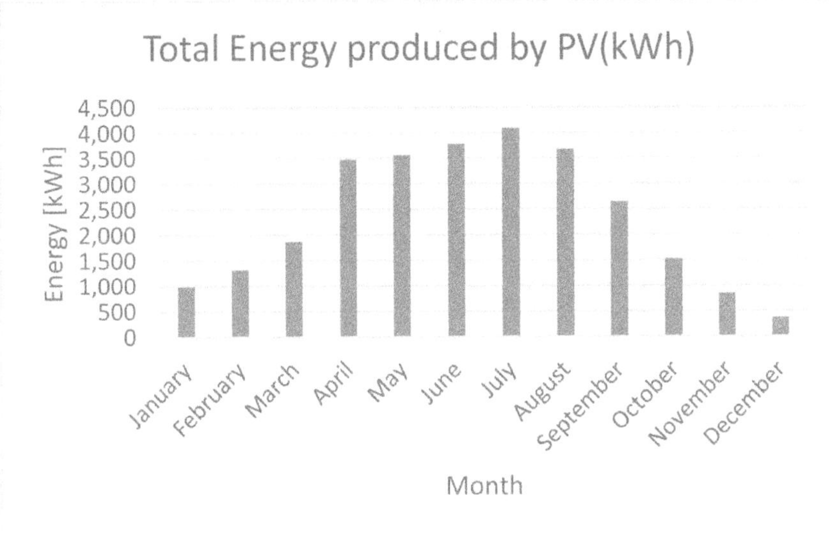

Energy Management Strategies for an Electric Vehicles Charging Station Based on Photovoltaic Energy With and Without Support of the Fuel Cell And Electrolyzer System

Energy Management Strategies for an Electric Vehicles Charging Station Based on Photovoltaic Energy Without Support of the Fuel Cell and Electrolyzer System

Description of Case Study 1 - Scenario 1 for 1 February 2020 presented Figure 4. Five electric vehicles were connected to an EV charging station. The first EV has nominal power of 6 kW and needs 4 hours to charge the full battery of the electric vehicle. The EV was connect to the EV charging station at 11:00 in the morning and finished at 3:00 pm. The second electric vehicle was connected to the EV charging station at 10:00 pm with a nominal power at 7.2 Kw and needs 6 hours to charge the battery bank, so it will finish at 3:00 pm. The third electric vehicle with a nominal power 22Kw was connected at 12:00 pm. and will charge 3 hours until 3:00 pm. At 11:00 pm, electric vehicle 4 was connected on the same day (1 February 2020). That has a nominal power of 7.2kW and needs 5 hours to charge the full battery, so it will be finished at 4:00pm. The last electric vehicle was connected at 9:00 pm with a nominal power at 7 Kw for 7 hours.

Figure 4. Data input: Case Study 1 – Scenario 1 for February 2020

Data	Time	Energy from PV [kWh	Energy Management of Bateries [kWh	Load_Car1 [kWh]	Load_Car2 [kWh]	Load_Car3 [kWh]	Load_Car4 [kWh]	Load_Car5 [kWh]
01.02.20	12:00 AM	0.00	250	0	0	0	0	0
01.02.20	1:00 AM	0.00	250	0	0	0	0	0
01.02.20	2:00 AM	0.00	250	0	0	0	0	0
01.02.20	3:00 AM	0.00	250	0	0	0	0	0
01.02.20	4:00 AM	0.00	250	0	0	0	0	0
01.02.20	5:00 AM	0.00	250	0	0	0	0	0
01.02.20	6:00 AM	0.00	250	0	0	0	0	0
01.02.20	7:00 AM	0.00	250	0	0	0	0	0
01.02.20	8:00 AM	0.23	250	0	0	0	0	0
01.02.20	9:00 AM	1.84	245	0	0	0	0	7
01.02.20	10:00 AM	8.46	239	0	7.2	0	0	7
01.02.20	11:00 AM	17.33	229	6.6	7.2	0	7.2	7
01.02.20	12:00 PM	20.09	199	6.6	7.2	22	7.2	7
01.02.20	1:00 PM	22.56	171	6.6	7.2	22	7.2	7
01.02.20	2:00 PM	19.17	140	6.6	7.2	22	7.2	7
01.02.20	3:00 PM	14.16	133	0	7.2	0	7.2	7
01.02.20	4:00 PM	8.75	142	0	0	0	0	0
01.02.20	5:00 PM	1.78	144	0	0	0	0	0
01.02.20	6:00 PM	0.00	144	0	0	0	0	0
01.02.20	7:00 PM	0.00	144	0	0	0	0	0
01.02.20	8:00 PM	0.00	144	0	0	0	0	0
01.02.20	9:00 PM	0.00	144	0	0	0	0	0
01.02.20	10:00 PM	0.00	144	0	0	0	0	0
01.02.20	11:00 PM	0.00	144	0	0	0	0	0

After analyzing the results, we noticed that between 12:00-15:00 all the locations of EV charging stations were busy at maximum nominal power, subline with green color on the table. At 9:00, EV5 was connected to the EV charging station, so the energy consumed by EVCS between 9:00 to 10:00 was 7 kWh and the energy produced by PV was 1,84 kWh, so 5,16 kWh in this interval was used from batteries. Between 10:00 to 11:00 electric vehicles 5 and 2 are connected to the EVCH and total energy consumed in this interval is 14,2 kWh and the energy produced by PV in the same interval is 8,46 kWh, so 5,74 kWh are used from batteries. Between 12:00 to 3:00, for 3 hours 5 EV are connected to EVCH having a total consume 49,4kWh for each hour, total energy consumed in 3 hour is 148,2 kWh and the energy produced by PV in the same interval is 61,82 kWh, the 86,38 kWh are used by EVCH from batteries. Between 0:00- 9:00 the PV produced 0,23 kWh and between 16:00- 24:00 the PV produced 10,53 kWh energy that was stored in batteries.

Description of Case Study 1 - Scenario 1 for 1 August 2020 are presented Figure 5. It is similar to *Case Study 1 - Scenario 1 for 1 February 2020.*

After analyzing the results, we noticed that between 12:00-15:00 all the locations of EV charging station were busy at maximum nominal power, subline with green color on the table. At 9:00 EV5 was connected to the EV charging station, so the energy consumed by EVCS between 9:00 to 10:00 was 7 kWh and the energy produced by PV was 5,81 kWh, so 1,19 kWh in this interval was used from batteries. Between 10:00 to 11:00 electric vehicles 5 and 2 are connected to the EVCH and total energy consumed in this interval is 14,2 kWh and the energy produced by PV in the same interval is 45,54 kWh, the excess of energy 31,34 kWh was stored in batteries. Between 12:00 to 3:00, for 3 hours 5 EV are connected to EVCH having a total consume 49,4kWh for each hour, total energy consumed in 3 hour is 148,2kWh and the energy produced by PV in the same interval is 120,35kWh, the 27,85kWh are used by EVCH from batteries. Between 0:00- 9:00 the PV produced 37,59 kWh and between 16:00- 24:00 the PV produced 41,61 kWh energy that was stored in batteries.

Figure 5. Data input: Case Study 1 – Scenario 1 for August 2020

Data	Time	Energy from PV [kWh	Energy Management of Bateries [kWh	Load_Car1 [kWh	Load_Car2 [kWh	Load_Car3 [kWh	Load_Car4 [kWh	Load_Car5 [kWh
01.08.20	12:00 AM	0.00	792	0	0	0	0	0
01.08.20	1:00 AM	0.00	792	0	0	0	0	0
01.08.20	2:00 AM	0.00	792	0	0	0	0	0
01.08.20	3:00 AM	0.00	792	0	0	0	0	0
01.08.20	4:00 AM	0.00	792	0	0	0	0	0
01.08.20	5:00 AM	1.27	793	0	0	0	0	0
01.08.20	6:00 AM	8.17	801	0	0	0	0	0
01.08.20	7:00 AM	5.30	807	0	0	0	0	0
01.08.20	8:00 AM	22.85	830	0	0	0	0	0
01.08.20	9:00 AM	5.81	828	0	0	0	0	7
01.08.20	10:00 AM	20.72	835	0	7.2	0	0	7
01.08.20	11:00 AM	45.65	853	6.6	7.2	0	7.2	7
01.08.20	12:00 PM	41.44	844	6.6	7.2	22	7.2	7
01.08.20	1:00 PM	37.24	831	6.6	7.2	22	7.2	7
01.08.20	2:00 PM	41.67	823	6.6	7.2	22	7.2	7
01.08.20	3:00 PM	32.81	834	0	7.2	0	7.2	7
01.08.20	4:00 PM	23.71	858	0	0	0	0	0
01.08.20	5:00 PM	14.16	872	0	0	0	0	0
01.08.20	6:00 PM	3.57	876	0	0	0	0	0
01.08.20	7:00 PM	0.17	876	0	0	0	0	0
01.08.20	8:00 PM	0.00	876	0	0	0	0	0
01.08.20	9:00 PM	0.00	876	0	0	0	0	0
01.08.20	10:00 PM	0.00	876	0	0	0	0	0
01.08.20	11:00 PM	0.00	876	0	0	0	0	0

Figure 6. Energy management: Case Study 1 – Scenario 1

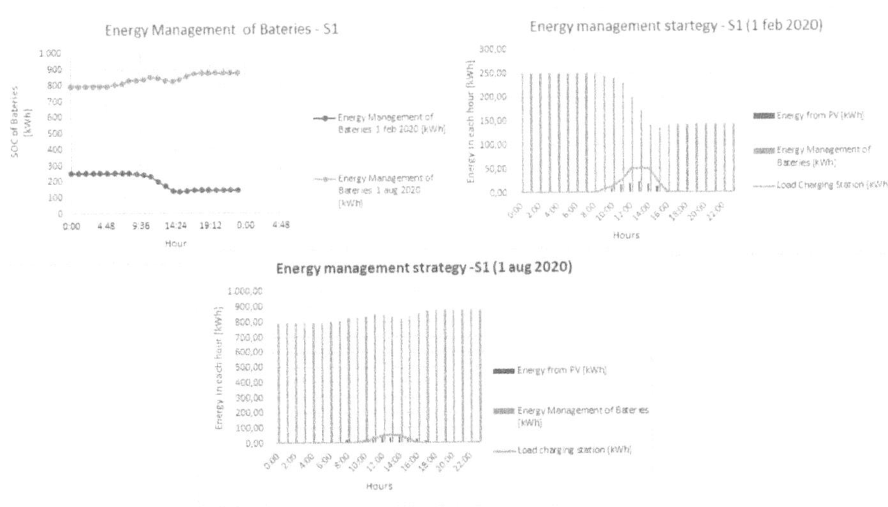

In Case Study 1 - Scenario 1 (Figure 6) is presented the energy management of the batteries for 1 February 2020 and 1 August 2020, in February the trend of batteries is to discharge and in August the trend is to charge.

Description of Case Study 1 - Scenario 2 for 1 February 2020 are presented in Figure 7. Connected were 4 electric vehicles on the EV charging station. The first EV have nominal power of 6 kW and needs 4 hours to charge the full battery of the vehicle, was connected to the EV charging station at 11:00 in the morning and finished at 3:00 pm, electric vehicle 2 was connected to the EV charging station at 10:00 pm with a nominal power at 7.2 kW and need 6 hours to charge the battery bank and finished at 4:00 pm, electric vehicle 3 have a nominal power 22 kW was connected to 12:00 a.m. and charged during 3 hours until 3:00 pm. At 11:00 pm was connected electric vehicle 4 in the same day (1 February 2020) that have a nominal power at 7.2 kW and needs 5 hours to charge the full battery, so it will be finished at 4:00pm.

After analyzing the results, we notice that between 12:00-15:00 4 place of EV charging stations were busy at maxim nominal power, subline with green color on the table. At 10:00 EV2 was connected to the EV charging station, so the energy consumed by EVCS. Between 11:00 to 12:00 electric vehicles 1, 2 and 4 are connected to the EVCH and total energy consumed in this interval is 20,4 kWh and the energy produced by PV in the same interval is 17,33 kWh, so 3,07 kWh are used from batteries. Between 12:00 to 3:00, for 3 hours 4 EV are connected to EVCH having a total consume 42,4kWh for each hour, total energy consumed in 3 hour is 127,2 kWh and the energy produced by PV in the same interval is 61,82 kWh, the 65,38 kWh are used by EVCH from batteries. Between 0:00- 10:00 the PV produced 2,07 kWh and between 16:00- 24:00 the PV produced 10,53 kWh energy that was stored in batteries.

Figure 7. Data input: Case Study 1 – Scenario 2 for 1 February 2020

Data	Time	Energy from PV [kWh	Energy Management of Bateries [kWh	Load_Car1 [kWh]	Load_Car2 [kWh]	Load_Car3 [kWh]	Load_Car4 [kWh]	Load_Car5 [kWh]
01.02.20	12:00 AM	0.00	250	0.00	0.00	0.00	0.00	0.00
01.02.20	1:00 AM	0.00	250	0.00	0.00	0.00	0.00	0.00
01.02.20	2:00 AM	0.00	250	0.00	0.00	0.00	0.00	0.00
01.02.20	3:00 AM	0.00	250	0.00	0.00	0.00	0.00	0.00
01.02.20	4:00 AM	0.00	250	0.00	0.00	0.00	0.00	0.00
01.02.20	5:00 AM	0.00	250	0.00	0.00	0.00	0.00	0.00
01.02.20	6:00 AM	0.00	250	0.00	0.00	0.00	0.00	0.00
01.02.20	7:00 AM	0.00	250	0.00	0.00	0.00	0.00	0.00
01.02.20	8:00 AM	0.23	250	0.00	0.00	0.00	0.00	0.00
01.02.20	9:00 AM	1.84	252	0.00	0.00	0.00	0.00	0.00
01.02.20	10:00 AM	8.46	253	0.00	7.20	0.00	0.00	0.00
01.02.20	11:00 AM	17.33	250	6.60	7.20	0.00	7.20	0.00
01.02.20	12:00 PM	20.09	227	6.60	7.20	22.00	7.20	0.00
01.02.20	1:00 PM	22.56	206	6.60	7.20	22.00	7.20	0.00
01.02.20	2:00 PM	19.17	182	6.60	7.20	22.00	7.20	0.00
01.02.20	3:00 PM	14.16	182	0.00	7.20	0.00	7.20	0.00
01.02.20	4:00 PM	8.75	191	0.00	0.00	0.00	0.00	0.00
01.02.20	5:00 PM	1.78	193	0.00	0.00	0.00	0.00	0.00
01.02.20	6:00 PM	0.00	193	0.00	0.00	0.00	0.00	0.00
01.02.20	7:00 PM	0.00	193	0.00	0.00	0.00	0.00	0.00
01.02.20	8:00 PM	0.00	193	0.00	0.00	0.00	0.00	0.00
01.02.20	9:00 PM	0.00	193	0.00	0.00	0.00	0.00	0.00
01.02.20	10:00 PM	0.00	193	0.00	0.00	0.00	0.00	0.00
01.02.20	11:00 PM	0.00	193	0.00	0.00	0.00	0.00	0.00

Description of Case Study 1 - Scenario 2 for 1 August 2020 are presented Figure 8 is similar with *Case Study 1 - Scenario 2 for 1 February 2020.*

Figure 8. Data input: Case Study 1 – Scenario 1 for 1 August 2020

Data	Time	Energy from PV [kWh	Energy Management of Bateries [kWh	Load_Car1 [kWh]	Load_Car2 [kWh]	Load_Car3 [kWh]	Load_Car4 [kWh]	Load_Car5 [kWh]
01.08.20	12:00 AM	0.00	792.00	0.00	0.00	0.00	0.00	0.00
01.08.20	1:00 AM	0.00	792.00	0.00	0.00	0.00	0.00	0.00
01.08.20	2:00 AM	0.00	792.00	0.00	0.00	0.00	0.00	0.00
01.08.20	3:00 AM	0.00	792.00	0.00	0.00	0.00	0.00	0.00
01.08.20	4:00 AM	0.00	792.00	0.00	0.00	0.00	0.00	0.00
01.08.20	5:00 AM	1.27	793.27	0.00	0.00	0.00	0.00	0.00
01.08.20	6:00 AM	8.17	801.44	0.00	0.00	0.00	0.00	0.00
01.08.20	7:00 AM	5.30	806.74	0.00	0.00	0.00	0.00	0.00
01.08.20	8:00 AM	22.85	829.59	0.00	0.00	0.00	0.00	0.00
01.08.20	9:00 AM	5.81	835.40	0.00	0.00	0.00	0.00	0.00
01.08.20	10:00 AM	20.72	848.92	0.00	7.20	0.00	0.00	0.00
01.08.20	11:00 AM	45.65	873.57	6.60	7.20	0.00	7.20	0.00
01.08.20	12:00 PM	41.44	872.01	6.60	7.20	22.00	7.20	0.00
01.08.20	1:00 PM	37.24	866.25	6.60	7.20	22.00	7.20	0.00
01.08.20	2:00 PM	41.67	864.92	6.60	7.20	22.00	7.20	0.00
01.08.20	3:00 PM	32.81	883.33	0.00	7.20	0.00	7.20	0.00
01.08.20	4:00 PM	23.71	907.05	0.00	0.00	0.00	0.00	0.00
01.08.20	5:00 PM	14.16	921.21	0.00	0.00	0.00	0.00	0.00
01.08.20	6:00 PM	3.57	924.78	0.00	0.00	0.00	0.00	0.00
01.08.20	7:00 PM	0.17	924.95	0.00	0.00	0.00	0.00	0.00
01.08.20	8:00 PM	0.00	924.95	0.00	0.00	0.00	0.00	0.00
01.08.20	9:00 PM	0.00	924.95	0.00	0.00	0.00	0.00	0.00
01.08.20	10:00 PM	0.00	924.95	0.00	0.00	0.00	0.00	0.00
01.08.20	11:00 PM	0.00	924.95	0.00	0.00	0.00	0.00	0.00

After analyzing the results, we notice that between 12:00-15:00 4 place of EV charging stations were busy at maxim nominal power, subline with green color on the table. At 10:00 EV2 was connected to the EV charging station, so the energy consumed by EVCS. Between 11:00 to 12:00 electric vehicles 1, 2 and 4 are connected to the EVCH and total energy consumed in this interval is 20,4 kWh and the energy produced by PV in the same interval is 20,72 kWh, so 0,32 kWh are stored in batteries. Between 12:00 to 3:00, for 3 hours 4 EV are connected to EVCH having a total consume 42,4kWh for each hour, total energy consumed in 3 hour is 127,2 kWh and the energy produced by PV in the same interval is 120,35 kWh, the 6,85 kWh are used by EVCH from batteries. Between 0:00- 10:00 the PV produced 43,04 kWh and between 16:00- 24:00 the PV produced 41,61 kWh energy that was stored in the batteries.

In Case Study 1 - Scenario 2 (Figure 9) is presented the energy management of the batteries for 1 February 2020 and 1 August 2020, in February the trend of batteries is to discharge and in August the trend is to charge.

Description of Case Study 1 - Scenario 3 for 1 February 2020 are presented Figure 10. Connected were 3 electric vehicles. The first vehicle is VIP Vehicle (place 2 of EVCS) that had a reservation with nominal power of 7.2 kW, and needs 6 hours to charge the full battery of the vehicle. It was connected to the charging station at 10:00 in the morning, and will finish at 4:00 pm. Another two electric vehicles (place 3 and 4 of EVCS) were connected to the EV charging station at 12:00 pm with a nominal power at 22 kW each and need 6 hours to charge the battery bank. Because the nominal power of charging station is 50kW, the VIP Vehicles was charged at Nominal Power at 7.2.kw and the two vehicles divided the power so the 2 vehicles in the first 2 hours charged with 21,40 kW and the last hours with 1.80 kW. In the analysis, we can see that between 12:00-15:00, all the charging stations were busy at total nominal power.

Figure 9. Energy management: Case Study 1 – Scenario 2

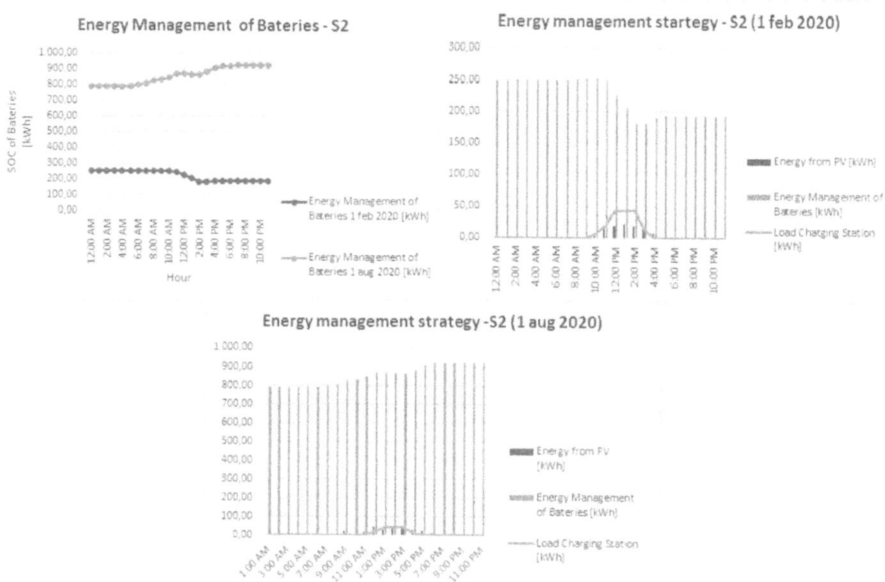

After analyzing the results, we notice that between 12:00-15:00 3 place of EV charging stations were busy at maxim nominal power, subline with green color on the table. At 10:00 to 12:00 EV2 was connected to the EV charging station, so the energy consumed by EVCS total energy consumed in this interval is 14.4 kWh and the energy produced by PV in the same interval is 14,79 kWh, so 0,39 kWh was stored in batteries. Between 12:00 to 3:00, for 3 hours 3 EV are connected to EVCH having a total consume 50 kWh for each hour, total energy consumed in 3 hour is 150 kWh and the energy produced by PV in the same interval is 62,63 kWh, the 87,37 kWh are used by EVCH from batteries. Between 3:00 to 4:00, the energy consumed was 10, 8 kWh and energy produced by PV was 14,16 kWh, the excess of energy 3,8 kWh was stored in batteries. Between 0:00- 10:00 the PV produced

2,07 kWh and between 16:00- 24:00 the PV produced 10,53 kWh energy that was stored in batteries.

Figure 10. Data input: Case Study 1 – Scenario 3 for 1 February 2020

Data	Time	Energy from PV [kWh	Energy Management of Bateries [kWh	Load_Car1 [kWh]	Load_Car2 [kWh]	Load_Car3 [kWh	Load_Car4 [kWh	Load_Car5 [kWh
01.02.20	12:00 AM	0.00	250	0.00	0.00	0.00	0.00	0.00
01.02.20	1:00 AM	0.00	250	0.00	0.00	0.00	0.00	0.00
01.02.20	2:00 AM	0.00	250	0.00	0.00	0.00	0.00	0.00
01.02.20	3:00 AM	0.00	250	0.00	0.00	0.00	0.00	0.00
01.02.20	4:00 AM	0.00	250	0.00	0.00	0.00	0.00	0.00
01.02.20	5:00 AM	0.00	250	0.00	0.00	0.00	0.00	0.00
01.02.20	6:00 AM	0.00	250	0.00	0.00	0.00	0.00	0.00
01.02.20	7:00 AM	0.00	250	0.00	0.00	0.00	0.00	0.00
01.02.20	8:00 AM	0.23	250	0.00	0.00	0.00	0.00	0.00
01.02.20	9:00 AM	1.84	250	0.00	0.00	0.00	0.00	0.00
01.02.20	10:00 AM	8.46	245	0.00	7.20	0.00	0.00	0.00
01.02.20	11:00 AM	17.33	246	0.00	7.20	0.00	0.00	0.00
01.02.20	12:00 PM	20.09	213	0.00	7.20	21.40	21.40	0.00
01.02.20	1:00 PM	22.56	184	0.00	7.20	21.40	21.40	0.00
01.02.20	2:00 PM	19.17	156	0.00	7.20	21.40	21.40	0.00
01.02.20	3:00 PM	14.16	164	0.00	7.20	1.80	1.80	0.00
01.02.20	4:00 PM	8.75	179	0.00	0.00	0.00	0.00	0.00
01.02.20	5:00 PM	1.78	187	0.00	0.00	0.00	0.00	0.00
01.02.20	6:00 PM	0.00	189	0.00	0.00	0.00	0.00	0.00
01.02.20	7:00 PM	0.00	189	0.00	0.00	0.00	0.00	0.00
01.02.20	8:00 PM	0.00	189	0.00	0.00	0.00	0.00	0.00
01.02.20	9:00 PM	0.00	189	0.00	0.00	0.00	0.00	0.00
01.02.20	10:00 PM	0.00	189	0.00	0.00	0.00	0.00	0.00
01.02.20	11:00 PM	0.00	189	0.00	0.00	0.00	0.00	0.00

Description of Case Study 1 - Scenario 3 for 1 August 2020 are presented Figure 11 is similar with *Case Study 1 - Scenario 3 for 1 February 2020.*

Figure 11. Data input: Case Study 1 – Scenario 3 for 1 August 2020

Data	Time	Energy from PV [kWh	Energy Management of Bateries [kWh	Load_Car1 [kWh]	Load_Car2 [kWh]	Load_Car3 [kWh	Load_Car4 [kWh	Load_Car5 [kWh
01.08.20	12:00 AM	0.00	792	0.00	0.00	0.00	0.00	0.00
01.08.20	1:00 AM	0.00	792	0.00	0.00	0.00	0.00	0.00
01.08.20	2:00 AM	0.00	792	0.00	0.00	0.00	0.00	0.00
01.08.20	3:00 AM	0.00	792	0.00	0.00	0.00	0.00	0.00
01.08.20	4:00 AM	0.00	792	0.00	0.00	0.00	0.00	0.00
01.08.20	5:00 AM	1.27	793	0.00	0.00	0.00	0.00	0.00
01.08.20	6:00 AM	8.17	801	0.00	0.00	0.00	0.00	0.00
01.08.20	7:00 AM	5.30	807	0.00	0.00	0.00	0.00	0.00
01.08.20	8:00 AM	22.85	830	0.00	0.00	0.00	0.00	0.00
01.08.20	9:00 AM	5.81	835	0.00	0.00	0.00	0.00	0.00
01.08.20	10:00 AM	20.72	849	0.00	7.20	0.00	0.00	0.00
01.08.20	11:00 AM	45.65	887	0.00	7.20	0.00	0.00	0.00
01.08.20	12:00 PM	41.44	879	0.00	7.20	21.40	21.40	0.00
01.08.20	1:00 PM	37.24	866	0.00	7.20	21.40	21.40	0.00
01.08.20	2:00 PM	41.67	858	0.00	7.20	21.40	21.40	0.00
01.08.20	3:00 PM	32.81	880	0.00	7.20	1.80	1.80	0.00
01.08.20	4:00 PM	23.71	903	0.00	0.00	0.00	0.00	0.00
01.08.20	5:00 PM	14.16	918	0.00	0.00	0.00	0.00	0.00
01.08.20	6:00 PM	3.57	921	0.00	0.00	0.00	0.00	0.00
01.08.20	7:00 PM	0.17	921	0.00	0.00	0.00	0.00	0.00
01.08.20	8:00 PM	0.00	921	0.00	0.00	0.00	0.00	0.00
01.08.20	9:00 PM	0.00	921	0.00	0.00	0.00	0.00	0.00
01.08.20	10:00 PM	0.00	921	0.00	0.00	0.00	0.00	0.00
01.08.20	11:00 PM	0.00	921	0.00	0.00	0.00	0.00	0.00

After analyzing the results, we notice that between 12:00-15:00 3 place of EV charging stations were busy at maxim nominal power, subline with green color on the table. At 10:00 to 12:00 just EV2 was connected to the EV charging station, so the energy consumed by EVCS total energy consumed in this interval is 14.4 kWh and the energy produced by PV in the same interval is 41,44 kWh, so 27,04 kWh was stored in batteries. Between 12:00 to 3:00, for 3 hours, 3 EV are connected to EVCH having a total consumed 50 kWh for each hour. Total energy consumed in 3 hours is 150 kWh and the energy produced by PV in the same interval is 120,35 kWh, the 29,65 kWh are used by EVCH from batteries. Between 3:00 to 4:00, the energy consumed was 10,8 kWh and energy produced by PV was 32,81 kWh, the excess of energy 22,01 kWh was stored in batteries. Between 0:00- 10:00 the PV produced 43,04 kWh and between 16:00- 24:00 the PV produced 41,61 kWh energy was stored in batteries.

Figure 12. Energy management: Case Study 1 – Scenario 3

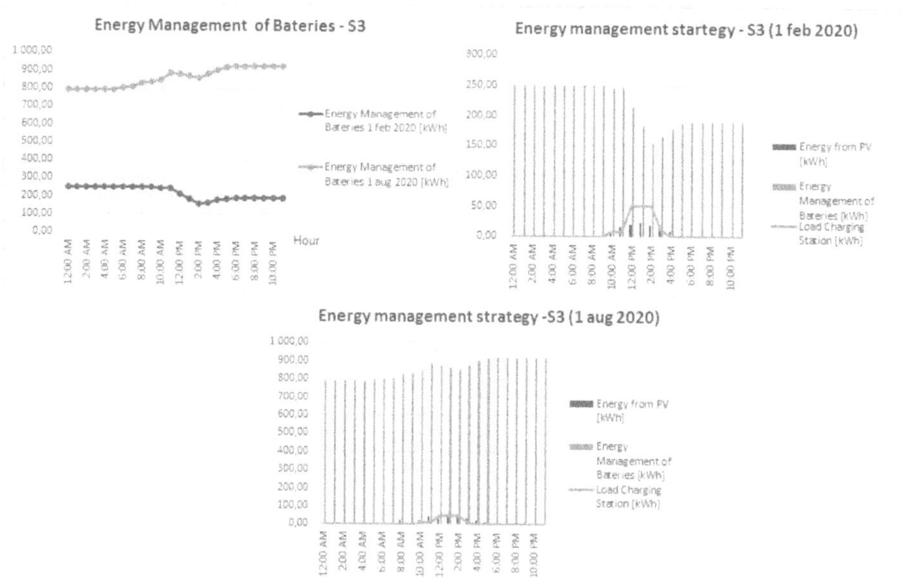

In Case Study 1 - Scenario 3 (Figure 12) is presented the energy management of the batteries for 1 February 2020 and 1 August 2020, in February the trend of batteries is to discharge and in August the trend is to charge

Energy Management Strategies for an Electric Vehicles Charging Station Based on Photovoltaic Energy With Support of the Fuel Cell and Electrolyzer System

Description of Case Study 2 - Scenario 1 for 1 February 2020 are presented figure 13. Was connected 5 electric vehicles on EV charging station, the first EV have nominal power of 6 kW and needs 4 hours to charge the full battery of the electric vehicle, the EV was connect to the EV charging station at 11:00 in the morning and finished at 3:00 pm, the second electric vehicle was connected to the EV charging station at 10:00 pm with a nominal power at 7.2 Kw and need 6 hours to charge the battery bank so it will be finished at 3:00 pm, the third electric vehicle with a nominal power 22Kw was connected to 12:00 a.m. and will charge during 3 hours until 3:00 pm. At 11:00 pm was connected electric vehicle 4 on the same day (1 February 2020) that have a nominal power at 7.2kW and need 5 hours to charge the full battery so it will be finished at 4:00pm. The last electric vehicle was connected at 9:00 pm with a nominal power at 7 Kw for 7 hours.

Figure 13. Data input: Case Study 2 – Scenario 1 for February 2020

Data	Time	Energy from PV [kWh]	Energy Management of Bateries [kWh]	Load_Car1 [kWh]	Load_Car2 [kWh]	Load_Car3 [kW	Load_Car4 [kWh]	Load_Car5 [kWh]	Energy_produced_PEMFC [kW	Energy_consumed_Electrolyzer [kWh]	Tank Hidrogen [kg]
01.02.20	0:00	0.00	125	0.00	0.00	0.00	0.00	0.00	0.80	0.20	5.00
01.02.20	1:00	0.00	125.60	0.00	0.00	0.00	0.00	0.00	0.80	0.20	5.07
01.02.20	2:00	0.00	126.20	0.00	0.00	0.00	0.00	0.00	0.80	0.20	5.14
01.02.20	3:00	0.00	126.80	0.00	0.00	0.00	0.00	0.00	0.80	0.20	5.20
01.02.20	4:00	0.00	127.40	0.00	0.00	0.00	0.00	0.00	0.80	0.20	5.27
01.02.20	5:00	0.00	128.00	0.00	0.00	0.00	0.00	0.00	0.80	0.20	5.34
01.02.20	6:00	0.00	128.60	0.00	0.00	0.00	0.00	0.00	0.80	0.20	5.41
01.02.20	7:00	0.00	129.20	0.00	0.00	0.00	0.00	0.00	0.80	0.20	5.48
01.02.20	8:00	0.09	129.89	0.00	0.00	0.00	0.00	0.00	0.80	0.20	5.54
01.02.20	9:00	0.75	124.25	0.00	0.00	0.00	0.00	7.00	8.00	0.20	4.94
01.02.20	10:00	3.47	121.32	0.00	7.20	0.00	0.00	7.00	8.00	0.20	4.34
01.02.20	11:00	7.10	108.82	6.00	7.20	0.00	7.20	7.00	8.00	0.20	3.74
01.02.20	12:00	8.23	75.45	6.00	7.20	22.00	7.20	7.00	8.00	0.20	3.14
01.02.20	13:00	9.25	43.10	6.00	7.20	22.00	7.20	7.00	8.00	0.20	2.54
01.02.20	14:00	7.86	9.35	6.00	7.20	22.00	7.20	7.00	8.00	0.20	1.94
01.02.20	15:00	5.80	1.56	0.00	7.20	0.00	7.20	7.00	8.00	0.20	1.34
01.02.20	16:00	3.59	12.94	0.00	0.00	0.00	0.00	0.00	0.80	0.20	1.32
01.02.20	17:00	0.73	14.27	0.00	0.00	0.00	0.00	0.00	0.80	0.20	1.30
01.02.20	18:00	0.00	14.87	0.00	0.00	0.00	0.00	0.00	0.80	0.20	1.27
01.02.20	19:00	0.00	15.47	0.00	0.00	0.00	0.00	0.00	0.80	0.20	1.25
01.02.20	20:00	0.00	16.07	0.00	0.00	0.00	0.00	0.00	0.80	0.20	1.22
01.02.20	21:00	0.00	16.67	0.00	0.00	0.00	0.00	0.00	0.80	0.20	1.20
01.02.20	22:00	0.00	17.27	0.00	0.00	0.00	0.00	0.00	0.80	0.20	1.18
01.02.20	23:00	0.00	17.87	0.00	0.00	0.00	0.00	0.00	0.80	0.20	1.15

Condition 1: According to the technical requirements of the Electrolyzer, the manufacture imposes the mandatory condition that the electrolyzer does not fall below 10% of the nominal value.

Condition 2: According to the technical requirements of the PEMFC, the manufacture imposes the mandatory condition that the electrolyzer does not fall below 10% of the nominal value.

The algorithm of energy management: If an electric vehicle request energy from EVCH, the PEFC works at nominal capacity (8kW), else the PEMFC works at

10% from capacity. During the day the electrolyzer works at 10% of its capacity. If the energy needed by EVCS is less than produced by PV and energy produced by PEMFC then the EVCS is supplied from batteries. If the energy needed by EVCS is higher than produced by PV and energy produced by PEMFC, then the energy is stored in batteries.

After analyzing the results, we notice that between 12:00-15:00 all the place of EV charging stations were supplied at maxim nominal power, subline with green color on the table. At 9:00 just EV5 was connected to the EV charging station, so the energy consumed by EVCS between 9:00 to 10:00 was 7 kWh and the energy produced by PV was 0,75 kWh, energy produced by PEMFC was 8 kWh so 1,75 kWh was stored in batteries. Between 10:00 to 11:00 electric vehicles 5 and 2 are connected to the EVCH and total energy consumed in this interval is 14,2 kWh and the energy produced by PV in the same interval is 3,47 kWh, energy produced by PEMFC is 8 kWh so 2,73 kWh are used from batteries. Between 12:00 to 3:00, for 3 hours 5 EV are connected to EVCH having a total consume 49,4kWh for each hour, total energy consumed in 3 hour is 148,2 kWh and the energy produced by PV in the same interval is 25,32 kWh, PEMPC supplied 24 kWh, the 98,88 kWh are used by EVCH from batteries. Between 0:00- 9:00 the PV produced 0,9 kWh and PEMFC 7,2 kWh and between 16:00- 24:00 the PV produced 43,2 kWh and PEMFC 6,4 kWh energy that was stored in batteries.

Description ofCase Study 1 - Scenario 1 for 1 August 2020 are presented Figure 14 is similar with *Case Study 2 - Scenario 1 for 1 February 2020.*

Figure 14. Data input: Case Study 2 – Scenario 1 for August 2020

Data	Time	Energy from PV [kWh]	Energy Management of Bateries [kWh]	Load_Car1 [kWh]	Load_Car2 [kWh]	Load_Car3 [kW	Load_Car4 [kWh]	Load_Car5 [kWh]	Energy_produced_PEMFC [kW	Energy_consumed_Electrolyzer [kWh]	Tank Hidrogen [Kg]
01.08.20	0:00	0.00	125	0.00	0.00	0.00	0.00	0.00	0.8	0.2	5.00
01.08.20	1:00	0.00	125.60	0.00	0.00	0.00	0.00	0.00	0.8	0.2	5.00
01.08.20	2:00	0.00	126.20	0.00	0.00	0.00	0.00	0.00	0.8	0.2	5.00
01.08.20	3:00	0.00	126.80	0.00	0.00	0.00	0.00	0.00	0.8	0.2	5.00
01.08.20	4:00	0.00	127.40	0.00	0.00	0.00	0.00	0.00	0.8	0.2	5.00
01.08.20	5:00	0.52	128.52	0.00	0.00	0.00	0.00	0.00	0.8	0.2	5.00
01.08.20	6:00	3.35	130.67	0.00	0.00	0.00	0.00	0.00	0.8	2	5.14
01.08.20	7:00	2.17	131.64	0.00	0.00	0.00	0.00	0.00	0.8	2	5.29
01.08.20	8:00	9.36	139.80	0.00	0.00	0.00	0.00	0.00	0.8	2	5.43
01.08.20	9:00	2.38	133.99	0.00	0.00	0.00	0.00	7.00	0.8	2	5.58
01.08.20	10:00	8.49	127.08	0.00	7.20	0.00	0.00	7.00	0.8	2	5.72
01.08.20	11:00	18.71	117.18	6.00	7.20	0.00	7.20	7.00	0.8	2	5.86
01.08.20	12:00	16.98	83.57	6.00	7.20	22.00	7.20	7.00	0.8	2	6.01
01.08.20	13:00	15.26	48.23	6.00	7.20	22.00	7.20	7.00	0.8	2	6.15
01.08.20	14:00	17.08	14.71	6.00	7.20	22.00	7.20	7.00	0.8	2	6.30
01.08.20	15:00	13.45	5.56	0.00	7.20	0.00	7.20	7.00	0.8	2	6.44
01.08.20	16:00	9.72	14.07	0.00	0.00	0.00	0.00	0.00	0.8	2	6.58
01.08.20	17:00	5.80	18.68	0.00	0.00	0.00	0.00	0.00	0.8	2	6.73
01.08.20	18:00	1.46	18.94	0.00	0.00	0.00	0.00	0.00	0.8	2	6.87
01.08.20	19:00	0.07	19.61	0.00	0.00	0.00	0.00	0.00	0.8	0.2	6.73
01.08.20	20:00	0.00	20.21	0.00	0.00	0.00	0.00	0.00	0.8	0.2	6.58
01.08.20	21:00	0.00	20.81	0.00	0.00	0.00	0.00	0.00	0.8	0.2	6.44
01.08.20	22:00	0.00	21.41	0.00	0.00	0.00	0.00	0.00	0.8	0.2	6.30
01.08.20	23:00	0.00	22.01	0.00	0.00	0.00	0.00	0.00	0.8	0.2	6.15

Condition 1: According to the technical requirements of the Electrolyzer, the manufacture imposes the mandatory condition that the electrolyzer does not fall below 10% of the nominal value.

Condition 2: According to the technical requirements of the PEMFC, the manufacture imposes the mandatory condition that the electrolyzer does not fall below 10% of the nominal value.

The algorithm of energy management: The electrolyzer works at 100% from capacity started at 6:00a.m. until 19:00, else he works at 10% from capacity. During the day the PEMFC works at 10% of its capacity. If the energy needed by EVCS is less, then produced by PV and energy produced by PEMFC then the EVCS is supplied from batteries. If the energy needed by EVCS is higher than produced by PV and energy produced by PEMFC, then the energy is stored in batteries.

After analyzing the results, we noticed that between 12:00-15:00 all the places of EV charging station were busy at maxim nominal power, subline with green color on the table. At 9:00 just EV5 was connected to the EV charging station, so the energy consumed by EVCS between 9:00 to 10:00 was 7 kWh and the energy produced by PV was 2,38 kWh, and PEMFC produced 0.8 kWh so 3,82 kWh was used from batteries. Between 10:00 to 11:00 electric vehicles 5 and 2 are connected to the EVCH and total energy consumed in this interval is 14,2 kWh and the energy produced by PV is 8,49 kWh and PEMFC 0.8 kWh, so the energy used from batteries is 4,91 kWh. Between 12:00 to 3:00, for 3 hours 5 EV are connected to EVCH having a total consume 49,4kWh for each hour, total energy consumed in 3 hour is 148,2 kWh and the energy produced by PV is 49,32 kWh, and PEMFC 2,4 kWh, so 96,48 kWh are used by EVCH from batteries. Between 0:00- 9:00 the PV produced 15,4 kWh, PEMFC was 7,2 kWh and between 16:00- 24:00 the PV produced 17,05 kWh, PEMFC produced 6,4 kWh energy that was stored in batteries.

Figure 15. Energy produced/energy consumed: Case Study 2 – Scenario 1 for 1 February and 1 August 2020

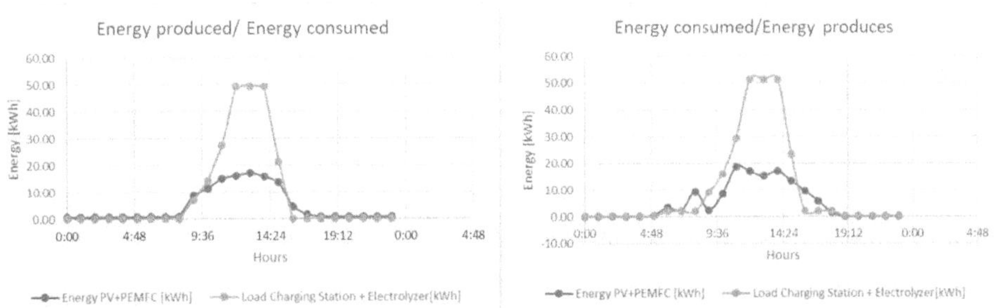

Energy produced and Energy consumed on 1 February 2020 and 1 August 2020 are presented in Figure 15. Energy produced represents the energy produced by PV and energy produce by PEMFC. Energy consumed is composed from energy need by electrolyzer and energy need by EVCS.

Figure 16. Hydrogen Tank – ***Case Study 2 - Scenario 1*** *for 1February and 1 August 2020*

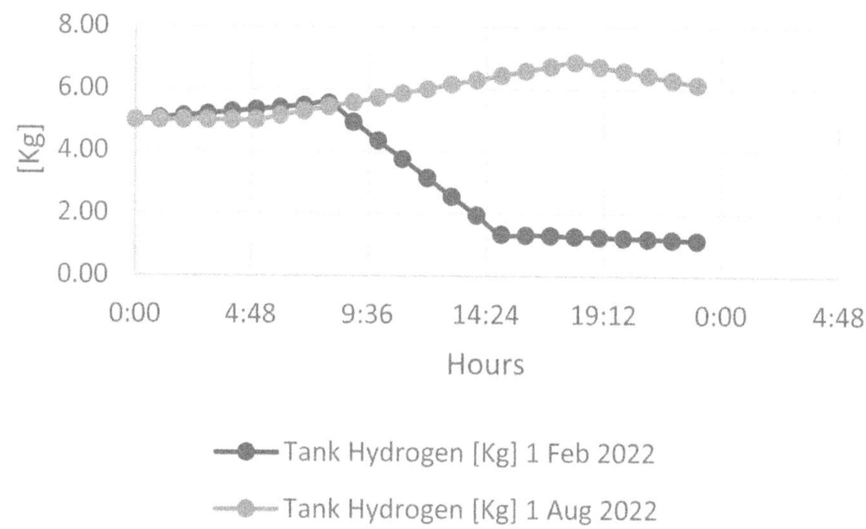

During the winter days the hydrogen from tank decreases and during the summer day the hydrogen from tank increases like in Figure 16.

Description of Case Study 2 - Scenario 2 for 1 February 2020 are presented figure 17. Was connected 4 electric vehicles on EV charging station, the first EV have nominal power of 6 kW and needs 4 hours to charge the full battery of the vehicle, was connected to the EV charging station at 11:00 in the morning and finished at 3:00 pm, electric vehicle 2 was connected to the EV charging station at 10:00 pm with a nominal power at 7.2 kW and need 6 hours to charge the battery bank and finished at 4:00 pm, electric vehicle 3 have a nominal power 22 kW was connected to 12:00 a.m. and charged during 3 hours until 3:00 pm. At 11:00 pm was connected electric vehicle 4 on the same day (1 February 2020) that have a nominal power at 7.2 kW and need 5 hours to charge the full battery so it will be finished at 4:00pm.

Figure 17. Data input: Case Study 2 – Scenario 1 for February 2020

Data	Time	Energy from PV [kWh]	Energy Management of Bateries [kWh]	Load_Car1 [kWh]	Load_Car2 [kWh]	Load_Car3 [kW	Load_Car4 [kWh]	Load_Car5 [kWh]	Energy_produced_PEMFC [kW	Energy_consumed_Electrolyzer [kWh]	Tank Hidrogen [Kg]
01.02.20	0:00	0.00	125.00	0.00	0.00	0.00	0.00	0.00	0.80	0.20	5.00
01.02.20	1:00	0.00	125.80	0.00	0.00	0.00	0.00	0.00	0.80	0.20	5.07
01.02.20	2:00	0.00	126.60	0.00	0.00	0.00	0.00	0.00	0.80	0.20	5.14
01.02.20	3:00	0.00	127.40	0.00	0.00	0.00	0.00	0.00	0.80	0.20	5.20
01.02.20	4:00	0.00	128.20	0.00	0.00	0.00	0.00	0.00	0.80	0.20	5.27
01.02.20	5:00	0.00	129.00	0.00	0.00	0.00	0.00	0.00	0.80	0.20	5.34
01.02.20	6:00	0.00	129.80	0.00	0.00	0.00	0.00	0.00	0.80	0.20	5.41
01.02.20	7:00	0.00	130.60	0.00	0.00	0.00	0.00	0.00	0.80	0.20	5.48
01.02.20	8:00	0.09	131.49	0.00	0.00	0.00	0.00	0.00	0.80	0.20	5.54
01.02.20	9:00	0.75	133.05	0.00	0.00	0.00	0.00	0.00	8.00	0.20	4.94
01.02.20	10:00	3.47	137.32	0.00	7.20	0.00	0.00	0.00	8.00	0.20	4.34
01.02.20	11:00	7.10	132.02	6.00	7.20	0.00	7.20	0.00	8.00	0.20	3.74
01.02.20	12:00	8.23	105.85	6.00	7.20	22.00	7.20	0.00	8.00	0.20	3.14
01.02.20	13:00	9.25	80.70	6.00	7.20	22.00	7.20	0.00	8.00	0.20	2.54
01.02.20	14:00	7.86	54.15	6.00	7.20	22.00	7.20	0.00	8.00	0.20	1.94
01.02.20	15:00	5.80	53.56	0.00	7.20	0.00	7.20	0.00	8.00	0.20	1.34
01.02.20	16:00	3.59	65.14	0.00	0.00	0.00	0.00	0.00	0.80	0.20	1.32
01.02.20	17:00	0.73	66.67	0.00	0.00	0.00	0.00	0.00	0.80	0.20	1.30
01.02.20	18:00	0.00	67.47	0.00	0.00	0.00	0.00	0.00	0.80	0.20	1.27
01.02.20	19:00	0.00	68.27	0.00	0.00	0.00	0.00	0.00	0.80	0.20	1.25
01.02.20	20:00	0.00	69.07	0.00	0.00	0.00	0.00	0.00	0.80	0.20	1.22
01.02.20	21:00	0.00	69.87	0.00	0.00	0.00	0.00	0.00	0.80	0.20	1.20
01.02.20	22:00	0.00	70.67	0.00	0.00	0.00	0.00	0.00	0.80	0.20	1.18
01.02.20	23:00	0.00	71.47	0.00	0.00	0.00	0.00	0.00	0.80	0.20	1.15

Condition 1: According to the technical requirements of the Electrolyzer, the manufacture imposes the mandatory condition that the electrolyzer does not fall below 10% of the nominal value.

Condition 2: According to the technical requirements of the PEMFC, the manufacture imposes the mandatory condition that the electrolyzer does not fall below 10% of the nominal value.

The algorithm of energy management: If an electric vehicle request energy from EVCH, the PEFC works at nominal capacity (8kW), else the PEMFC works at 10% from capacity. During the day the electrolyzer works at 10% of its capacity. If the energy needed by EVCS is less than produced by PV and energy produced by PEMFC then the EVCS is supplied from batteries. If the energy need by EVCS is high, then it is produced by PV and energy produce by PEMFC then the energy is stored in batteries.

After analyzing the results, we notice that between 12:00-15:00 4 place of EV charging stations were busy at maxim nominal power, subline with green color on the table. At 10:00 EV2 was connected to the EV charging station, so the energy consumed by EVCS. Between 11:00 to 12:00 electric vehicles 1, 2 and 4 are connected to the EVCH and total energy consumed in this interval is 20,4 kWh and the energy produced by PV in the same interval is 7,10 kWh, PEMFC 8 kWh so 5,3 kWh are used from batteries. Between 12:00 to 3:00, for 3 hours 4 EV are connected to EVCH having a total consume 42,4 kWh for each hour, total energy consumed in 3 hour is 127,2 kWh and the energy produced by PV in the same interval is 25,34 kWh, PEMFC 24 kWh the 77,86 kWh are used by EVCH from batteries. Between 0:00-10:00 the PV produced 0,84 kWh, PEMFC 8 kWh and between 16:00- 24:00 the PV produced 4,32 kWh energy, and by PEMFC 6,4 kWh that was stored in batteries.

Description of Case Study 2 - Scenario 2 for 1 August 2020 are presented Figure 18 is similar with *Case Study 1 - Scenario 2 for 1 February 2020.*

Figure 18. Data input: Case Study 2 – Scenario 1 for August 2020

Data	Time	Energy from PV [kWh]	Energy Management of Bateries [kWh]	Load_Car1 [kWh]	Load_Car2 [kWh]	Load_Car3 [kW	Load_Car4 [kWh]	Load_Car5 [kWh]	Energy_produced_PEMFC [kW	Energy_consumed_Electrolyzer [kWh]	Tank Hidrogen [kg]
01.08.20	0:00	0.00	125.00	0.00	0.00	0.00	0.00	0.00	0.80	0.20	5.00
01.08.20	1:00	0.00	125.60	0.00	0.00	0.00	0.00	0.00	0.80	0.20	5.00
01.08.20	2:00	0.00	126.20	0.00	0.00	0.00	0.00	0.00	0.80	0.20	5.00
01.08.20	3:00	0.00	126.80	0.00	0.00	0.00	0.00	0.00	0.80	0.20	5.00
01.08.20	4:00	0.00	127.40	0.00	0.00	0.00	0.00	0.00	0.80	0.20	5.00
01.08.20	5:00	0.52	128.52	0.00	0.00	0.00	0.00	0.00	0.80	0.20	5.00
01.08.20	6:00	3.35	130.67	0.00	0.00	0.00	0.00	0.00	0.80	2.00	5.14
01.08.20	7:00	2.17	131.64	0.00	0.00	0.00	0.00	0.00	0.80	2.00	5.29
01.08.20	8:00	9.36	139.80	0.00	0.00	0.00	0.00	0.00	0.80	2.00	5.43
01.08.20	9:00	2.38	140.99	0.00	0.00	0.00	0.00	0.00	0.80	2.00	5.58
01.08.20	10:00	8.49	141.08	0.00	7.20	0.00	0.00	0.00	0.80	2.00	5.72
01.08.20	11:00	18.71	138.18	6.00	7.20	0.00	7.20	0.00	0.80	2.00	5.86
01.08.20	12:00	16.98	111.57	6.00	7.20	22.00	7.20	0.00	0.80	2.00	6.01
01.08.20	13:00	15.26	83.23	6.00	7.20	22.00	7.20	0.00	0.80	2.00	6.15
01.08.20	14:00	17.08	56.71	6.00	7.20	22.00	7.20	0.00	0.80	2.00	6.30
01.08.20	15:00	13.45	54.56	0.00	7.20	0.00	7.20	0.00	0.80	2.00	6.44
01.08.20	16:00	9.72	63.07	0.00	0.00	0.00	0.00	0.00	0.80	2.00	6.58
01.08.20	17:00	5.80	67.68	0.00	0.00	0.00	0.00	0.00	0.80	2.00	6.73
01.08.20	18:00	1.46	67.94	0.00	0.00	0.00	0.00	0.00	0.80	2.00	6.87
01.08.20	19:00	0.07	68.61	0.00	0.00	0.00	0.00	0.00	0.80	0.20	6.73
01.08.20	20:00	0.00	69.21	0.00	0.00	0.00	0.00	0.00	0.80	0.20	6.58
01.08.20	21:00	0.00	69.81	0.00	0.00	0.00	0.00	0.00	0.80	0.20	6.44
01.08.20	22:00	0.00	70.41	0.00	0.00	0.00	0.00	0.00	0.80	0.20	6.30
01.08.20	23:00	0.00	71.01	0.00	0.00	0.00	0.00	0.00	0.80	0.20	6.15

Condition 1: According to the technical requirements of the Electrolyzer, the manufacture imposes the mandatory condition that the electrolyzer does not fall below 10% of the nominal value.

Condition 2: According to the technical requirements of the PEMFC, the manufacture imposes the mandatory condition that the electrolyzer does not fall below 10% of the nominal value.

The algorithm of energy management: The electrolyzer works at 100% from capacity started at 6:00a.m. until 19:00, else he works at 10% from capacity. During the day the PEMFC works at 10% of its capacity. If the energy needed by EVCS is less than produced by PV and energy produced by PEMFC then the EVCS is supplied from batteries. If the energy needed by EVCS is high, then it is produced by PV and energy produced by PEMFC then the energy is stored in batteries.

After analyzing the results, we notice that between 12:00-15:00 4 place of EV charging stations were busy at maxim nominal power, subline with green color on the table. At 10:00 EV2 was connected to the EV charging station, so the energy consumed by EVCS. Between 11:00 to 12:00 electric vehicles 1, 2 and 4 are connected to the EVCH and total energy consumed in this interval is 20,4 kWh and the energy produced by PV in the same interval is 18,71 kWh and PEMFC 0,8 kWh, so 0,89 kWh are used from batteries. Between 12:00 to 3:00, for 3 hours 4 EV are connected to EVCH having a total consume 42,4kWh for each hour, total energy consumed in 3 hour is 127,2 kWh and the energy produced by PV is 49,32 kWh, PEMFC 2,4 kWh so the 75,48 kWh are used by EVCH from batteries. Between 0:00- 10:00 the PV produced 17,78 kWh, PEMFC 8 kWh, and between 16:00- 24:00 the PV produced 17,05 kWh energy, PEMFC 6,4 kWh that was stored in batteries.

Figure 19. Energy produced/energy consumed: Case Study 2 – Scenario 1 for 1 February and 1 August 2020

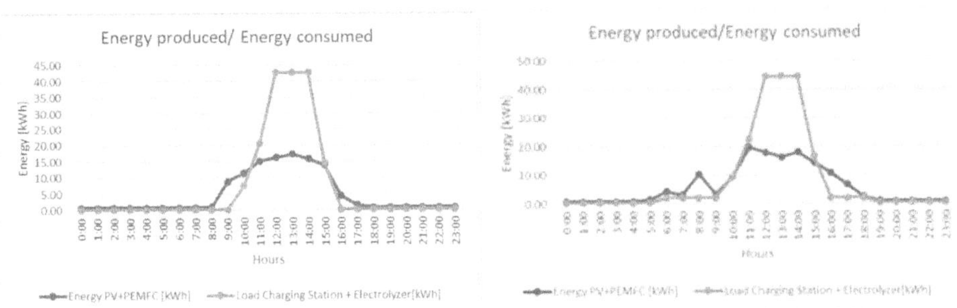

Energy produced and Energy consumed on 1 February 2020 and 1 August 2020 are presented in Figure 19. Energy produced represents the energy produced by PV and energy produced by PEMFC. Energy consumed is composed of energy needed by electrolyzer and energy needed by EVCS.

Description of Case Study 2 - Scenario 3 for 1 February 2020 are presented figure 20. Was connected 3 electric vehicles, first vehicle is VIP Vehicle (place 2 of EVCS) that have a reservation with nominal power of 7.2 kW and needs 6 hours to charge the full battery of the vehicle, will connect to the charging station at 10:00 in the morning and will finish at 4:00 pm, another two electric vehicles (place 3 and 4 of EVCS) were connected to the EV charging station at 12:00 pm with a nominal power at 22 kW each and need 6 hours to charge the battery bank, but because the nominal power of charging station is 50kW, the VIP Vehicles was charged at Nominal Power at 7.2.kw and the two vehicles divided the power so the 2 vehicles in the first 2 hours charged with 21,40 kW and the last hours with 1.80 kW. In the analyze we can see that between 12:00-15:00 all the charging station was busy at total nominal power.

Condition 1: According to the technical requirements of the Electrolyzer, the manufacture imposes the mandatory condition that the electrolyzer does not fall below 10% of the nominal value.

Condition 2: According to the technical requirements of the PEMFC, the manufacture imposes the mandatory condition that the electrolyzer does not fall below 10% of the nominal value.

Figure 20. Data input: Case Study 2 – Scenario 3 for 1 February 2020

Data	Time	Energy from PV [kWh]	Energy Management of Bateries [kWh]	Load_Car1 [kWh]	Load_Car2 [kWh]	Load_Car3 [kW	Load_Car4 [kWh]	Load_Car5 [kWh]	Energy_produced_PEMFC [kW	Energy_consumed_Electrolyzer [kWh]	Tank Hidrogen [Nm3]
01.02.20	0:00	0.00	125.00	0.00	0.00	0.00	0.00	0.00	0.80	0.20	5.00
01.02.20	1:00	0.00	125.60	0.00	0.00	0.00	0.00	0.00	0.80	0.20	5.07
01.02.20	2:00	0.00	126.20	0.00	0.00	0.00	0.00	0.00	0.80	0.20	5.14
01.02.20	3:00	0.00	126.80	0.00	0.00	0.00	0.00	0.00	0.80	0.20	5.20
01.02.20	4:00	0.00	127.40	0.00	0.00	0.00	0.00	0.00	0.80	0.20	5.27
01.02.20	5:00	0.00	128.00	0.00	0.00	0.00	0.00	0.00	0.80	0.20	5.34
01.02.20	6:00	0.00	128.60	0.00	0.00	0.00	0.00	0.00	0.80	0.20	5.41
01.02.20	7:00	0.00	129.20	0.00	0.00	0.00	0.00	0.00	0.80	0.20	5.48
01.02.20	8:00	0.09	129.89	0.00	0.00	0.00	0.00	0.00	0.80	0.20	5.54
01.02.20	9:00	0.75	131.25	0.00	0.00	0.00	0.00	0.00	8.00	0.20	4.94
01.02.20	10:00	3.47	135.32	0.00	7.20	0.00	0.00	0.00	8.00	0.20	4.34
01.02.20	11:00	7.10	143.02	0.00	7.20	0.00	0.00	0.00	8.00	0.20	3.74
01.02.20	12:00	8.23	109.05	0.00	7.20	21.40	21.40	0.00	8.00	0.20	3.14
01.02.20	13:00	9.25	76.10	0.00	7.20	21.40	21.40	0.00	8.00	0.20	2.54
01.02.20	14:00	7.86	41.75	0.00	7.20	21.40	21.40	0.00	8.00	0.20	1.94
01.02.20	15:00	5.80	44.56	0.00	7.20	1.80	1.80	0.00	8.00	0.20	1.34
01.02.20	16:00	3.59	55.94	0.00	0.00	0.00	0.00	0.00	0.80	0.20	1.32
01.02.20	17:00	0.73	57.27	0.00	0.00	0.00	0.00	0.00	0.80	0.20	1.30
01.02.20	18:00	0.00	57.87	0.00	0.00	0.00	0.00	0.00	0.80	0.20	1.27
01.02.20	19:00	0.00	58.47	0.00	0.00	0.00	0.00	0.00	0.80	0.20	1.25
01.02.20	20:00	0.00	59.07	0.00	0.00	0.00	0.00	0.00	0.80	0.20	1.22
01.02.20	21:00	0.00	59.67	0.00	0.00	0.00	0.00	0.00	0.80	0.20	1.20
01.02.20	22:00	0.00	60.27	0.00	0.00	0.00	0.00	0.00	0.80	0.20	1.18
01.02.20	23:00	0.00	60.87	0.00	0.00	0.00	0.00	0.00	0.80	0.20	1.15

The algorithm of energy management: If an electric vehicle request energy from EVCH, the PEFC works at nominal capacity (8kW), else the PEMFC works at 10% from capacity. During the day the electrolyzer works at 10% of its capacity. If the energy needed by EVCS is less than produced by PV and energy produced by PEMFC then the EVCS is supplied from batteries. If the energy needed by EVCS is high, then it is produced by PV and energy produced by PEMFC then the energy is stored in batteries.

After analyzing the results, we notice that between 12:00-15:00 3 place of EV charging stations were busy at maxim nominal power, subline with green color on the table. At 10:00 to 12:00 just EV2 was connected the EV charging station, so the energy consumed by EVCS total energy consumed in this interval is 14,4 kWh and the energy produced by PV in the same interval is 10,57 kWh, PEMFC 1,6 kWh, so 2,23 kWh are used from batteries. Between 12:00 to 3:00, for 3 hours 3 EV are connected to EVCH having a total consume 50 kWh for each hour, total energy consumed in 3 hour is 150 kWh and the energy produced by PV in the same interval is 25,34 kWh, PEMFC 2,4 kWh so the 122,26 kWh are used by EVCH from batteries. Between 3:00 to 4:00, the energy consumed was 10, 8 kWh and energy produced by PV was 5,8 kWh, PEMFC 8 kWh, the excess of energy 3 kWh was stored in batteries. Between 0:00- 10:00 the PV produced, 84 kWh, PEMFC 16 kWh and between 16:00- 24:00 the PV produced 4,32 kWh energy, PEMFC 6,4 kWh that was stored in batteries.

Description of Case Study 1 - Scenario 3 for 1 August 2020 are presented Figure 21 is similar with *Case Study 1 - Scenario 3 for 1 February 2020.*

Condition 1: According to the technical requirements of the Electrolyzer, the manufacture imposes the mandatory condition that the electrolyzer does not fall below 10% of the nominal value.

Condition 2: According to the technical requirements of the PEMFC, the manufacture imposes the mandatory condition that the electrolyzer does not fall below 10% of the nominal value.

Figure 21. Data input: Case Study 2 – Scenario 1 for August 2020

Data	Time	Energy from PV [kWh]	Energy Management of Bateries [kWh]	Load_Car1 [kWh]	Load_Car2 [kWh]	Load_Car3 [kW	Load_Car4 [kWh]	Load_Car5 [kWh]	Energy_produced_PEMFC [kW	Energy_consumed_Electrolyzer [kWh]	Tank Hidrogen [Nm3]
01.08.20	0:00	0.00	125.00	0.00	0.00	0.00	0.00	0.00	0.80	0.20	5.00
01.08.20	1:00	0.00	125.60	0.00	0.00	0.00	0.00	0.00	0.80	0.20	5.00
01.08.20	2:00	0.00	126.20	0.00	0.00	0.00	0.00	0.00	0.80	0.20	5.00
01.08.20	3:00	0.00	126.80	0.00	0.00	0.00	0.00	0.00	0.80	0.20	5.00
01.08.20	4:00	0.00	127.40	0.00	0.00	0.00	0.00	0.00	0.80	0.20	5.00
01.08.20	5:00	0.52	128.52	0.00	0.00	0.00	0.00	0.00	0.80	0.20	5.00
01.08.20	6:00	3.35	130.67	0.00	0.00	0.00	0.00	0.00	0.80	2.00	5.14
01.08.20	7:00	2.17	131.64	0.00	0.00	0.00	0.00	0.00	0.80	2.00	5.29
01.08.20	8:00	9.36	139.80	0.00	0.00	0.00	0.00	0.00	0.80	2.00	5.43
01.08.20	9:00	2.38	140.99	0.00	0.00	0.00	0.00	0.00	0.80	2.00	5.58
01.08.20	10:00	8.49	141.08	0.00	7.20	0.00	0.00	0.00	0.80	2.00	5.72
01.08.20	11:00	18.71	151.38	0.00	7.20	0.00	0.00	0.00	0.80	2.00	5.86
01.08.20	12:00	16.98	117.17	0.00	7.20	21.40	21.40	0.00	0.80	2.00	6.01
01.08.20	13:00	15.26	81.23	0.00	7.20	21.40	21.40	0.00	0.80	2.00	6.15
01.08.20	14:00	17.08	47.11	0.00	7.20	21.40	21.40	0.00	0.80	2.00	6.30
01.08.20	15:00	13.45	48.56	0.00	7.20	1.80	1.80	0.00	0.80	2.00	6.44
01.08.20	16:00	9.72	57.07	0.00	0.00	0.00	0.00	0.00	0.80	2.00	6.58
01.08.20	17:00	5.80	61.68	0.00	0.00	0.00	0.00	0.00	0.80	2.00	6.73
01.08.20	18:00	1.46	61.94	0.00	0.00	0.00	0.00	0.00	0.80	2.00	6.87
01.08.20	19:00	0.07	62.61	0.00	0.00	0.00	0.00	0.00	0.80	0.20	6.73
01.08.20	20:00	0.00	63.21	0.00	0.00	0.00	0.00	0.00	0.80	0.20	6.58
01.08.20	21:00	0.00	63.81	0.00	0.00	0.00	0.00	0.00	0.80	0.20	6.44
01.08.20	22:00	0.00	64.41	0.00	0.00	0.00	0.00	0.00	0.80	0.20	6.30
01.08.20	23:00	0.00	65.01	0.00	0.00	0.00	0.00	0.00	0.80	0.20	6.15

The algorithm of energy management: The electrolyzer works at 100% from capacity started at 6:00a.m. until 19:00, else he works at 10% from capacity. During the day the PEMFC work at 10% of capacity. If the energy needed by EVCS is less than produced by PV and energy produced by PEMFC then the EVCS is supplied from batteries. If the energy need by EVCS is high, then it is produced by PV and energy produce by PEMFC then the energy is stored in batteries.

After analyzing the results, we notice that between 12:00-15:00 3 place of EV charging stations were busy at maxim nominal power, subline with green color on the table. At 10:00 to 12:00 just EV2 was connected the EV charging station, so the energy consumed by EVCS total energy consumed in this interval is 14.4 kWh and the energy produced by PV in the same interval is 27,2 kWh, PEMFC 1,6 kWh so 16 kWh was stored in batteries. Between 12:00 to 3:00, for 3 hours 3 EV are connected to EVCH having a total consume 50 kWh for each hour, total energy consumed in 3 hour is 150 kWh and the energy produced by PV is 49,32 kWh, PEMFC 2,4 kWh, the 98,28 kWh are used by EVCH from batteries. Between 3:00 to 4:00, the energy consumed was 10,8 kWh and energy produced by PV was 13,45 kWh, PEMFC 0.8 kWh, the excess of energy 3,45 kWh was stored in batteries. Between 0:00- 10:00 the PV produced 17,78 kWh, PEMFC 8 kWh and between 16:00- 24:00 the PV produced 17,05 kWh energy PEMFC 6,4 kWh that was stored in batteries.

Figure 22. Energy produced/energy consumed: Case Study 2 – Scenario 1 for 1 February and 1 August 2020

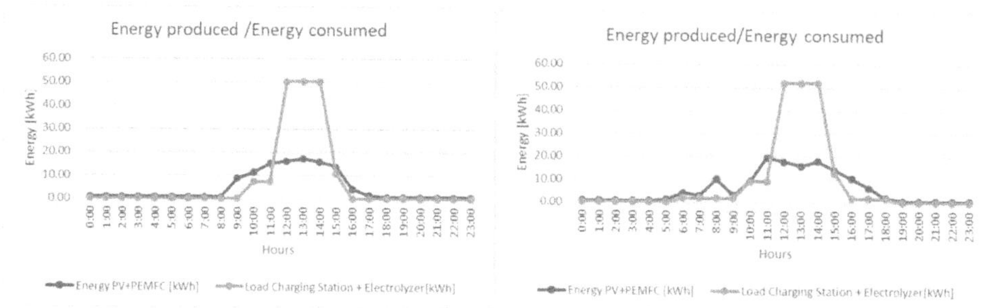

Energy produced and Energy consumed on 1 February 2020 and 1 August 2020 are presented in Figure 22. Energy produced represents the energy produced by PV and energy produced by PEMFC. Energy consumed is composed of energy needed by electrolyzer and energy needed by EVCS.

Chapter 7
Hydrogen Storage Challenge in the Hydrogen-Based Civilization

Ankita Jaiswal
Indian Institute of Technology, Dhanbad, India

Sridhar Sahu
Indian Institute of Technology, Dhanbad, India

ABSTRACT

Industrial revolution and sustainable development have intensified the need for clean resources of energy. Among all the possible alternative sources of green energy, hydrogen energy is the most promising one because of its abundancy, maximum energy density per unit mass, and clean combustion. The energy content per unit mass of hydrogen is estimated to be the highest among all known chemical fuels. Therefore, it has enough potential to meet the rising energy demand if some major barriers in its production, storage, and effective commercial or vehicular use are resolved efficiently. One of the most significant challenges in the development of a global hydrogen-based economy is the hydrogen storage problem, which essentially means minimizing the enormous volume of hydrogen gas to achieve optimum gravimetric and volumetric density synergistically. This chapter briefly discusses the significant challenges, especially the hydrogen storage problem in the path of the development of a hydrogen-based economy and the possible approaches to meet the challenges.

DOI: 10.4018/978-1-6684-6721-3.ch007

INTRODUCTION

The past two centuries have been marked by technical breakthroughs, rapid globalization, and the industrial revolution, which have enhanced our living standards and reliance on technology (De Vries, 1994). These global advancements, population growth, and the need for sustainable development have significantly increased our energy demand, which has now reached approximately 15 Terawatt (Smalley, 2005). In this era of technology and digitalization, the per capita demand for energy has increased exponentially, thereby intensifying our need for additional energy resources. Till today, fossil fuels like coal, petroleum, diesel, etc. are the prime sources of energy and approximately 85% of the global energy (Abbott, 2009) needs are fulfilled by them, as depicted in Figure1(a). The formation rate of fossil fuels is slower than their consumption rate, resulting in their rapid depletion; therefore, there is a high probability that they will run out within the next few decades. In addition, day by day increase in energy demand, wastage and improper use of energy, less exploration and minimal use of renewable sources of energy has brought about a serious shortage of energy often called as energy crisis (Dudley, 2018). Energy crisis basically means that our global energy demand exceeds the amount of prime energy reserve and this ratio is getting higher and higher with the passage of time because of the above outlined reasons. Therefore, we must urgently look for some renewable energy sources that can effectively replace fossil fuels and satisfy our increasing energy needs. In addition, the green-house gases such as carbon monoxide, carbon dioxide, and methane produced by the combustion of fossil fuels trap solar heat in the earth's atmosphere, resulting in global warming and environmental temperature imbalance (Coyle & Simmons, 2014).

Figure 1. (a) Percentage contribution of different energy sources used to meet global energy demand; (b) global average temperature rise with year (datasets obtained from NASA, NOAA, Berkeley Earth, and meteorological offices of the U.K. and Japan)
Source: (a) Reprinted from Dudley (2018), (b) reprinted from NCEI.Monitoring.info@noaa.gov

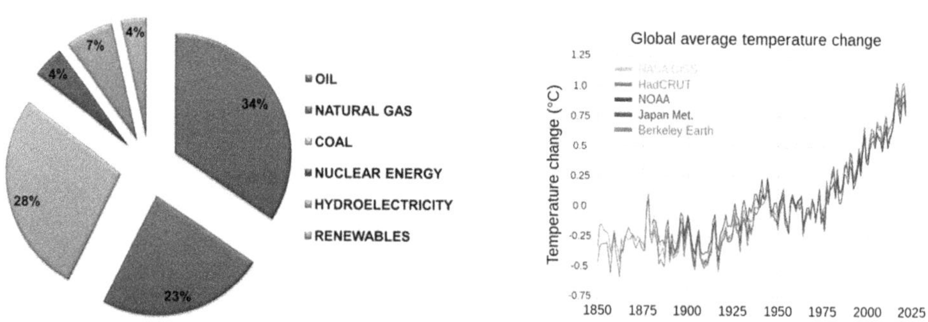

The Intergovernmental Panel on Climate Change (IPCC) believes that human activities are directly responsible for the unusual rapid rise in global warming (Letcher, 2019). Further, scientific studies predicts that the surface temperatures of the earth are rising at the rate of 0.2°C per decade (I.C Change, 2013). The global rise of temperature since 1850 is depicted in Figure 1(b). It is found that the present level of warming is 1.5°C higher than that of the pre-industrial era, and many of the resulting negative effects are already evident (NCEI.Monitoring.info@noaa.gov, n.d.-a, Wikipedia contributors, 2022). Increased temperature in cold regions has contributed to permafrost melting, deglaciation, and sea ice loss. Many species are being forced to the brink of relocation or extinction due to the rapid pace of environmental change and climate disturbances in mountainous regions, coastal areas, and the polar regions (Cahill et al., 2013). The effects of climate change also include low agricultural yield (Hernandez-Ochoa et al., 2018) food and water scarcity (Rosenzweig & Parry, 1994; Gosling & Arnell, 2013), increased flooding (Dankers & Feyen, 2008), extreme heat, an increase in disease, and economic loss (Lhotka et al., 2017). The World Health Organization (WHO) identifies climate change as the greatest health threat of the 21st century (McMichael et al., 2003). Further impacts of global warming will amplify these effects, potentially leading to tipping points such as the melting of the Greenland ice sheet and other larger glacial masses. According to scientific studies, the average global temperature rises and its negative consequences may become more apparent in the coming decades (Lu et al., 2020). Further, the scientific community and the environmentalists also warn that if this trend continues, the frequency of natural disasters like floods, heat waves, glacial melting, droughts, and so on will increase, making earth uninhabitable. Hence, though excessive use of fossil fuels can satisfy our energy needs, but it also contributes to an increase in air pollution, global warming, and climate changes, which is currently a source of global environmental concern (Kam et al., 2021).

BACKGROUND

Green energy is the key to reduce our reliance on fossil fuels and mitigating climate change caused by pollution and global warming (Aithal & Acharya, 2016). It is estimated that about 85% of the global energy requirement is met by fossil fuels, thereby leading to their overconsumption. There is growing momentum to minimize the emissions of greenhouse gases in order to mitigate the progression of climate change. For instance, France enacted law no. 2015-992, mandating a 40% reduction in greenhouse gas emissions by the year 2030 as compared to 1990 (France Hydrogène, n.d.). Although fossil fuel combustion is not the only major source of greenhouse gas emissions, it is by far the most significant. As per United States

Environmental Protection Agency (USEPA), about 76% of the total emissions are contributed by fossil fuel combustion (*Inventory of U.S. Greenhouse Gas Emissions and Sinks: 1990–2015, 2017*). Thus, to address the global energy crisis and mitigate global warming, it is very urgent to minimize the usage of conventional fossil fuels. Therefore, we should seek out and efficiently implement renewable energy sources such as bio-energy, solar power, hydro-energy, hydrogen energy, and so on for the next generation energy resources. It is already predicted that renewable energy would become the dominant form of electricity generation to achieve carbon neutrality by 2050, rising to 85% or more in some scenarios. According to 2015 UN Paris Agreement on climate change, Japan pledged to reduce its carbon emissions by 26% by 2030 and by 80% by 2050 (*Prime Minister Suga's Attendance at the Leaders Summit on Climate*, n.d.).

Hydrogen energy is known as one of the most promising and compatible alternatives to the fossil fuels due to its high energy density per unit mass (120kJ/g), which is roughly three times that of existing fossil fuel gasoline, as shown in Figure 2(a). Because of its high energy density, hydrogen is considered to be one of the most promising alternative energy carriers. Moreover, hydrogen is highly abundant and clean energy carrier which produces only water as by product when it undergoes combustion (Züttel et al., 2010). The combustion reaction between hydrogen and oxygen can be expressed by the following reaction $2H_2+O_2=2H_2O+\Delta H$ where $\Delta H=120\text{-}140$ kJ/g of energy. Thus, it can be regarded as green and eco-friendly energy carrier which makes it more promising source of future fuel as compared to other renewable sources of energy. These criteria make hydrogen the most feasible and promising future fuel for transportation and industrial purposes. Japan plans to run 800,000 hydrogen-powered fuel cell cars on the road by 2030 (*Japan's Hydrogen Industrial Strategy*, n.d.). For transportation, refuelling must be quick, safety priority, and the size and weight of the storage system must be reasonably low. Nowadays hydrogen fuel cells offer a tremendous amount of potential for the production of electrical power vehicles. In the near future, hydrogen may eventually replace electricity as an important "energy carrier," that can be used for the purpose of storing, transporting, and delivering energy to customers. Renewable energy sources, such as the sun, don't have the ability to produce energy indefinitely. However, hydrogen can store renewable energy until it is needed. Therefore, we must develop the technology for clean alternative energy carriers, such as hydrogen today, in order to facilitate the transition to an environmental friendly, sustainable, and green economy in the future.

CHALLENGES IN THE HYDROGEN-BASED CIVILIZATION

Even though hydrogen has a lot of advantages over fossil fuels and other renewable energy resources, it has not yet been able to completely replace fossil fuels in transportation and mobile applications. This is because cost-effective production and storage of hydrogen presents a lot of technical challenges and safety concerns (Tseng et al., 2005). Despite being a better alternative, commercialising the hydrogen economy necessitates resolving the following issues:

- Efficient production of hydrogen and its distribution
- Effective storage of hydrogen for industrial and on-board applications
- Design of fuel cells with adequate durability for long-term use
- Minimum energy loss during energy conversion
- Financial and safety constraints on the implementation of hydrogen-powered vehicles

Among all of the issues discussed above, pure hydrogen production and its effective storage for on-board applications are the most important ones that must be carefully addressed in order to establish a hydrogen-based global economy.

Pure Hydrogen Production

First of all, hydrogen is not an energy source rather it is an energy carrier unlike other renewable energy resources. Although hydrogen is available in abundant quantities in the nature, but it is mostly in the form of hydrides, water and organic compounds (Ishaq et al., 2022). Therefore, pure hydrogen needs to be extracted from these compounds via different techniques like electrolysis, water splitting, dehydrogenation, etc. to use for energy generation and transportation purposes. This indicates that we must pay for the energy, which creates a tough economic quandary because we have grown accustomed to consuming energy for free since the beginning of the industrial revolution. As a result, one of the major challenges in the field of hydrogen-based economy today is pure hydrogen production from water, hydrides and other organic compounds.

The Hydrogen Storage Problem

Hydrogen is the lightest gas, most reactive and has a very low critical temperature (33K), which indicates that hydrogen remains in gaseous form at ambient temperature and pressure. It can be well described using Van der waal's equation for real gas-

$$\left(P+\frac{n^2a}{V^2}\right)(V-nb)=nRT$$

where V, P, T, n are the volume, pressure, temperature in Kelvin scale and number of moles of the H_2 gas respectively. R is the universal gas constant whereas a and b are the Van der Waal's constant.

This means that even a small amount of hydrogen will tend to occupy the maximum volume available, resulting in a very low volume density for the H_2 gas (0.089 kg/m^3). At room temperature and atmospheric pressure, 1 kg of H_2 gas is estimated to occupy a volume of approximately 11 m^3 (Zuttel, 2004). This simply means that if a fuel cell car kept at atmospheric pressure is allowed to store1 kg of hydrogen required to travel 100 km, the size of the fuel tank would have to be 11m in length. Therefore, the second significant challenge associated with hydrogen as an energy carrier is its storage, which implies reducing the enormous volume of hydrogen gas to enhance its volume density. Further, compared to gasoline, the energy content of hydrogen by weight is approximately three times higher, but the energy content by volume is about four times lower. A comparison of mass energy density and volumetric energy density of hydrogen with other known chemical fuels are depicted in Figure 2(a) and 2(b) respectively. Both Figure 2(a) and 2(b) are reprinted from ref. (Tzimas et al., 2003). Among all the fuels known, hydrogen possesses the highest energy density per mass (120kJ/g), but it has the lowest energy density per unit volume (8MJ/L) at normal atmospheric pressure due to its low volumetric density at ambient temperature and pressure, which makes it difficult to store hydrogen within a given volume with high volumetric density. Despite being the most promising green energy source, hydrogen cannot be stored effectively like other conventional fuels because of its very light wight and low density. As a result, unusually large amount of space is required to store a given amount of hydrogen gas. In order to store an optimal amount of energy, the size of the hydrogen storage tank would have to be impractically large for practical vehicle applications; thus, "hydrogen storage" appears to be the most significant challenge in the path to a future hydrogen-based economy. Thus, it is important and necessary to develop materials that can store hydrogen with optimum volumetric density and gravimetric density in a safe, secure, economical, and cheap manner without affecting the vehicle's cargo space. Therefore, modern fuel cell vehicles use compressed hydrogen gas, wherein 5 kg of hydrogen gas is compressed under 700 bar pressure and stored in a carbon fibre reinforced storage tank. But even at 700 bar the storage criterions don't satisfy the current transportation and industrial targets. Hence, high density hydrogen storage is a difficult challenge in stationary, portable, and transportation applications. Currently available H_2 storage options necessitate a large volume for gaseous hydrogen storage. Further, it also requires the development of advanced and modern storage technologies that can store hydrogen with highest possible energy density. We need to enhance H_2 storage techniques and develop a new storage mechanism. In other words, the extremely low volume density of gaseous hydrogen at room temperature is a significant barrier to the development of a "hydrogen economy" (www.sc. doe.gov/bes/hydrogen.pdf, 2004). Thus, for the development of efficient hydrogen storage material for practical applications, volumetric and gravimetric density of hydrogen becomes a very vital parameter. Therefore, the establishment of cost-effective hydrogen-based civilizations is technically very much dependent on the efficient storage of hydrogen with optimum volumetric and gravimetric density.

Figure 2. Variation of (a) mass energy density of different fuels and (b) volume energy density of different fuels
Source: Reprinted from Tzimas et al. (2003). Copyright © JRC Publications Repository.

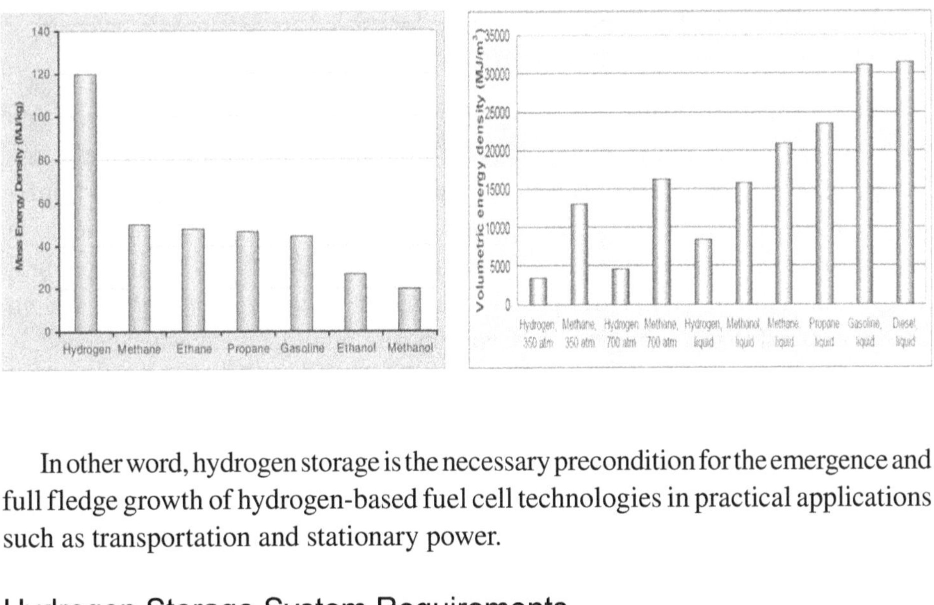

In other word, hydrogen storage is the necessary precondition for the emergence and full fledge growth of hydrogen-based fuel cell technologies in practical applications such as transportation and stationary power.

Hydrogen Storage System Requirements

Before comparing different ways to store hydrogen, we can figure out which storage medium is best for mobility by measuring the properties of each system. At ambient temperature and pressure, pure hydrogen has a very high gravimetric energy density of 120 MJ/kg (100 wt%) but a poor volumetric energy density of 0.01 MJ/L. For mobile applications, high volumetric and gravimetric densities are significant requirements. Currently, petrol and diesel are the most used fuels, so they are being used to benchmark hydrogen carrier as a "future fuel" as presented in Table 1. In Table 1, we presented a comparative study of the physico-chemical properties of hydrogen, natural gas, gasoline and diesel at ambient thermodynamic conditions (*Fossil and Alternative Fuels - Energy Content*, n.d.; France Hydrogène, 2013, Rocha & Sheen, 2019).

Table 1. Physico-chemical properties of gaseous hydrogen, natural gas, gasoline and diesel at ambient thermodynamic conditions

Property	Hydrogen	Natural Gas	Gasoline (Petrol)	Diesel
Energy per kg (MJ/kg)	142	53.6	46.4	45.4
Energy per liter (MJ/L)	0.01	9	34.2	34.6
Density at NTP (kg/m^3)	0.089	0.68	770	850
Flashpoint (C)	-231	-188	-45	62
Concentration range in air for flammability (vol %)	4-75	5-15	1-7.6	0.6-5.5

From the tabulated data (Ref. Table 1), we can clearly visualize that hydrogen has very low density, low energy content per unit volume and extremely low flashpoint which makes hydrogen to lag behind other fossil fuels. Thus, in order to make hydrogen as energy carrier for the "future fuel", we need to improve these physico-chemical properties especially the density and the energy content per unit volume. In addition, the kinetic speed i.e the rate at which a given storage system should begin releasing hydrogen and then stop with minimal delay when needed, is another performance parameter that should be taken into account when characterising a fuel. The rates should correspond to the transportation applications. Thus, for ideal hydrogen storage requirements, we must focus on hydrogen storage techniques with overall low cost, minimum risk and high efficiency at ambient temperature and pressure conditions. Based on these conditions, we can state that temperature dependent and pressure dependent techniques may not fulfil the criterions of ideal hydrogen storage requirements because it will necessitate the addition of a heat management system, pressure management system, which increases costs, complexity, potentially mass and safety considerations. The desired hydrogen storage materials should be highly efficient. This indicates if hydrogen is used as energy carrier, then it should be utilized to generate maximum usable energy out of it efficiently and with minimum energy wastage during conversion or other purposes. Ideal storage methods, especially those intended for widespread use, must also comply with strict safety standards. Toxic properties, flammability etc., make a storage material highly risky and undesirable. Therefore, it is best to refrain from using materials whose extraction requires a lot of resources or from creating designs that make recycling challenging or impossible. To ensure market penetration, all of the aforementioned qualities should be within reach.

US-DoE Targets for Ideal Hydrogen Storage Systems

To enhance the technical performance for hydrogen storage systems onboard light-duty vehicles, following criteria have been proposed which are summarized in Table 2. These goals were set by the U.S. DRIVE Partnership, a collaboration between the United States-Department of Energy (US-DoE), the United States Council for Automotive Research (USCAR), utility & energy companies (*Hydrogen and Fuel Cell Technologies*, n.d.). The targets thus set by US-DoE for hydrogen storage systems provide a remarkable indication of their anticipated potential performance. Because molecular hydrogen has a relatively low volumetric energy density when compared to other chemical fuels used in transportation, the primary goal of the US-DoE is to design and develop storage materials that can meet on-board hydrogen storage targets of high volumetric and gravimetric density set for light duty vehicles. According to US-DoE goals for 2025, efficient hydrogen storage systems should be able to achieve a minimum gravimetric density of 5.5 wt% and a minimum volumetric density of 40 g/L. In addition, they must be able to operate at nearby room temperature under a maximum delivery pressure of 5-12 bar.

DIFFERENT HYDROGEN STORAGE METHODS

Because of its low critical temperature and very low density at ambient conditions, hydrogen storage turns out to be one of the major challenges standing in the path of the development of a hydrogen-based civilization. Storage of hydrogen can be broadly divided into two major classes –

System-Based Hydrogen Storage

It refers to the conventional ways of storing hydrogen in which hydrogen is physically confined in storage systems. This method basically includes storing hydrogen in systems either in gaseous or liquid form (Rivard et al., 2019). These types of storage methods can again be divided into two the following two subcategories-

Table 2. Onboard Hydrogen Storage targets for light duty vehicles (updated May 2017)

Storage Parameter	Units	2020	2025	Ultimate
System Gravimetric Capacity:	kWh/kg	1.5	1.8	2.2
Usable, specific-energy from H_2 (net useful energy/max system mass)	(kg H_2/kg system)	(0.045)	(0.055)	(0.065)
System Volumetric Capacity:	kWh/L	1.0	1.3	1.7
Usable energy density from H_2 (net useful energy/max system volume)	(kg H_2/L system)	(0.030)	(0.040)	(0.050)
Storage System Cost:	$/kWh net	10	9	8
	($/kg H_2)	333	300	266
• Fuel cost	$/gge at pump	4	4	4
Durability/Operability:				
• Operating ambient temperature	°C	-40/60 (sun)	-40/60 (sun)	-40/60 (sun)
• Min/max delivery temperature	°C	-40/85	-40/85	-40/85
• Operational cycle life (1/4 tank to full)	Cycles	1500	1500	1500
• Min delivery pressure from storage system	bar (abs)	5	5	5
• Max delivery pressure from storage system	bar (abs)	12	12	12
• Onboard Efficiency	%	90	90	90
• "Well" to Powerplant Efficiency	%	60	60	60
Charging / Discharging Rates:				
• System fill time	min	3-5	3-5	3-5
• Minimum full flow rate (e.g., 1.6 g/s target for 80kW rated fuel cell power)	(g/s)/kW	0.02	0.02	0.02
• Average flow rate	(g/s)/kW	0.004	0.004	0.004
• Start time to full flow (20°C)	s	5	5	5
• Start time to full flow (-20°C)	s	15	15	15
• Transient response at operating temperature 10%–90% and 90%–0% (based on full flow rate)	s	0.75	0.75	0.75
Environmental Health & Safety: • Permeation & leakage • Toxicity • Safety				

Taken from https://www.energy.gov/eere/fuelcells/doe-technical-targets-onboard-hydrogen-storage-light-duty-vehicles

Storage of Hydrogen in Gaseous Form in High Pressure (700-800 Bar) Tanks

Hydrogen is a gas at normal thermodynamic conditions with very low density of 0.089kg/m^3. So, even a small quantity of H_2 gas tends to occupy the maximum possible volume available. Initially, high-pressure gas cylinders with a maximum pressure of 20 MPa were considered as the most prevalent type of storage system. However, with the development of lightweight composite storage cylinders capable of withstanding pressures of about 80 MPa, hydrogen can be stored with a volumetric density of 36 kg/m^3, which is roughly half of its density in liquid form at its normal boiling point (Durbin & Malardier-Jugroot, 2013). This implies that energy density of hydrogen gas can be increased by storing it under high pressure.

However, the gravimetric density of hydrogen follows a decreasing trend with increasing pressure because of the increasing thickness of the pressure cylinder walls, and the gravimetric density is at its maximum at zero overpressure (Zuttel, 2004). Therefore, the increase in volumetric storage density in pressurized gas systems is sacrificed by the decrease in gravimetric storage density. Further, compression of the gas under such a high pressure releases heat which causes heating of the gas inside the tank. In order to prevent the storage tank from overheating during a compression refill, gas is typically cooled beforehand. This requires an additional 1.8–3.6% of energy (Gye et al., 2019; Gardiner & Satyapal, 2009). Thus, to achieve an optimum gravimetric density from a small volume of gas, we need to compress it under a high pressure of about 70-80 MPa which adds up safety considerations and extra cost in transportation. Besides this, dealing with such high pressure in transportation can be associated with safety issues and security concerns. Inspite of all challenges, the technology of storing hydrogen as a gas in tanks has somewhat reached a mature stage and is now being utilized for the purpose of transportation at this time. Nissan Company has recently developed high pressure (70 MPa) hydrogen tanks in order to achieve the desired driving range (Aoyama et al., 2008). These high-pressure tanks are designed from a new lightweight material, such as carbon fiber composite. Modern compressed-gas H_2 storage systems can store only 3–4 wt% and less than 25 g/L at 350 -700 bar pressures at room temperature. Therefore, today's fuel cell vehicles employ carbon fibre reinforced tanks as hydrogen storage systems, with 5 kg of hydrogen gas compressed under 70 MPa of pressure. The High Pressure Gas Safety Institute of Japan has certified this newly developed high pressure hydrogen storage tank as a 70 MPa high pressure hydrogen storage vessel (Otsubo et al., 2007). In comparison to the previous high pressure hydrogen storage systems, which compressed and stored hydrogen at 35 MPa, approximately 30% more hydrogen can be stored in the same tank layout space. Despite the fact that this technology can store hydrogen in small volumes, it is not suitable for wide applications in industries and transportations.

There are major drawbacks of this storage methods. Even at a pressure of 69 MPa bar, the energy content is significantly less than the same volume of gasoline. High cost, safety concerns, and a sudden change in pressure are limitations of high pressure tanks, and only 20% of the available energy can be utilized with this alternative. Significant amount of energy is lost during the conversion process. The safety of pressurized cylinders is a major concern, particularly in densely populated areas. The technically straightforward and well-proven high-pressure storage method has some significant drawbacks, including the relatively low hydrogen density and the extremely high gas pressures in the systems.

Storing Hydrogen in Liquid Form at Cryogenic Temperature (20K)

The volumetric density of hydrogen can be increased by another conventional way: liquification. The fundamental requirement for storing hydrogen in liquid form is to cool it to cryogenic temperature of the order of about 20K, which is considered as the boiling point of hydrogen at normal atmospheric pressure (Jena, 2011). In the cryogenic hydrogen storage technique, the H_2 gas is liquefied by cooling it to 21.2 K. At 21.2 K, the density of liquid hydrogen is 70.8kg/m^3. Liquid hydrogen has a threefold greater energy density than gasoline. Since less space is required to store liquid hydrogen, smaller and lighter containers can be used. However, because the critical temperature of hydrogen is only 33 K, only open systems are capable of storing liquid hydrogen, as there is no liquid phase above the critical temperature. But, achieving such a low temperature requires advanced cooling systems and temperature maintenance arrangements. Even a little fluctuation of temperature may lead to an easy "boil off" of liquid hydrogen into a gaseous form. Moreover, the critical temperature of gaseous hydrogen is about 33K which is very low and can be easily achieved by thermal fluctuations of H_2 stored at cryogenic temperature. Even a small temperature gain can increase the internal pressure inside the cryogenic tank. Typically, a liquid hydrogen tank is not capable of withstanding internal pressure because it is designed to store liquid at cryogenic temperatures. The pressure inside the tank rises as heat transfers from the environment to the liquified hydrogen. Since the tank is not designed to withstand high pressure, a certain amount of the the stored hydrogen is allowed to get escape via a connecting relief valve, also known as "boil-off". As thermal insulation is imperfect, unused hydrogen kept in a warm environment will gradually deplete over time, resulting in an unusual energy loss. Therefore, in order to store liquid hydrogen, the cryogenic storage tank must be adequately insulated to minimize heat transfer. As a result, thermal insulation of the cryogenic storage system is required to minimize the hydrogen evaporation and leakage in order to store liquefied hydrogen efficiently. In addition, current liquid-hydrogen systems consume about more than 30% of the hydrogen's energy for liquefaction and hence this method too cannot meet the DOE's long-term volumetric capacity goals. Thus, storing hydrogen in liquid form is not very convenient and cost-effective and needs extra cryogenic temperature management system. It is estimated that approximately 30% of the total energy is required for the liquefaction of hydrogen, which indicates the difficulty as well as energy wastage in liquid storage. Furthermore, from a safety perspective, liquid hydrogen's high flammability makes cryogenic storage an undesirable alternative. To overcome these drawbacks, fresh approaches are required.

Material-Based Hydrogen Storage

The H_2 storage requirements necessary for the on-board hydrogen storage cannot be efficiently met by hydrogen liquefaction or storage in high pressure tanks. Further, storing hydrogen in traditional ways is not very convenient in terms of volumetric as well as gravimetric density. Therefore, the scientific community is actively looking for new approaches of hydrogen storage which can store H_2 with optimum volumetric and gravimetric density. Since the density of hydrogen can be increased through adsorption into highly porous solid state materials, material-based storage of H_2 molecules in a host material is a practical strategy to satisfy onboard storage needs and has shown potential advantages for the storage of hydrogen (McWhorter et al., 2011). Material-based hydrogen storage is basically based on interaction of hydrogen with matter. The most popular solid state storage materials include carbon nanostructures (Züttel et al., 2002; Hirscher et al., 2002), metal hydrides (Schneemann et al., 2018; Schüth et al., 2004), complex metal hydrides (Khalil, 2013; Ley et al., 2015), metal clusters (Trivedi & Bandyopadhyay, 2015; Jaiswal et al., 2022), metal organic frameworks (MOFs) (Yang et al., 2012; Li et al., 2009), and other nanostructure materials (Jaiswal et al., 2022, Lan et al., 2009). Hydrogen can be stored both reversibly and irreversibly in these materials. In material-based storage systems hydrogen is stored via the adsorption phenomena.

Based on the strength of interaction (ref. Figure 3), adsorption can be divided into the following three main classes (Jena, 2011):

- **Chemisorption:** In this type of adsorption the H-H bond of the H_2 molecules is broken, and each H in atomic form enters into the void spaces or makes the bond with an absorbent complex. As the atomic hydrogen is chemically bonded to the system, the adsorption energy remains high and is found to be of the order of a 2-4 eV. Because of strong chemical binding between the adsorbent complex and the atomic hydrogen, the adsorbed hydrogens cannot be reversed easily and desorption takes place at very high temperature. Thus, the storage materials in which the hydrogens are adsorbed via chemisorption are not favourable for ideal hydrogen storage materials.

- **Physisorption:** In this type of adsorption, hydrogen is adsorbed weakly over the complex in molecular fashion. The H-H bond length of the H_2 molecule remains comparable to that of the bond length of the isolated H_2 molecule. In general, the binding energy of the physiosorbed H_2 molecules lie in the range of few meV range. Because of low binding energy of the adsorbed H_2 molecules, desorption occurs at very low temperature.
- **Quasi-Binding:** In this case of adsorption, the H_2 molecules are adsorbed in quasi-molecular fashion and hence it is termed as quasi-binding. It means that the adsorption of H_2 molecules occur with slight elongated bond length. But the H-H bond is not broken. The binding strength of H_2 molecules adsorbed via this mechanism lies in between that of physisorption and chemisorption. Typically, the range of binding energy can be estimated in the range of 0.1-0.8 eV which is ideal for hydrogen storage under ambient temperature and pressure. There are two different reasons for quasi-molecular binding: the first one is charge donation and back donation, and the other is charge polarization. Kubas demonstrated that in the case of transition metal atoms, quasi-molecular bonding is caused by charge donation from the filled bonding orbital of the adsorbed H_2 molecule to the unfilled d orbitals of the transition metal atoms and back-donation of charge from the filled d orbitals of the transition metal atoms to the unfilled antibonding orbital of the H_2 molecule (Kubas, 2001). Niu et al., on the other hand, stated that the electric field generated by a positively charged metal ion can polarise the H_2 molecule, allowing it to bind to the metal cation in quasi-molecular form (Niu et al., 1992).

Based on different types of hydrogen adsorption, material-based systems can be classified into three major categories -sorbent materials, complex hydrides and nanostructured materials (Jena, 2011).

- **Sorbent Materials:** It includes materials like graphene, fullerene, metal-organic frameworks (MOFs), covalent-organic frameworks (COFs) etc. in which hydrogen adsorption occurs via physisorption. Metal-organic frameworks (MOFs) are a type of material that works well for hydrogen storage at low temperatures around 77 K (Hirscher & Panella, 2007). Rosi et al. have reported that MOF-5 can achieve a gravimetric density of about 4.5 wt % at 78 K and 1.0 wt % at room temperature and 20 bar pressure (Rosi et al., 2003). A thermal management system is necessary when adsorption is triggered via temperature variation. Therefore, even at cryogenic temperatures,

MOFs cannot completely satisfy US-DOE criteria. The use of expensive and energy-intensive refrigeration techniques results in low efficiency at cryogenic temperatures. Furthermore, proper tank insulation and thermal management systems are required to maintain the cryogenic temperature. Despite the fact that the pressure is not excessively high, it still poses the risks because cryogenic tanks cannot withstand pressure fluctuations.

Figure 3. Different types of adsorption: (a) physisorption, (b) chemisorption, (c) quasi-binding
Source: Reprinted with permission from Jena (2011). Copyright © 2011, American Chemical Society.

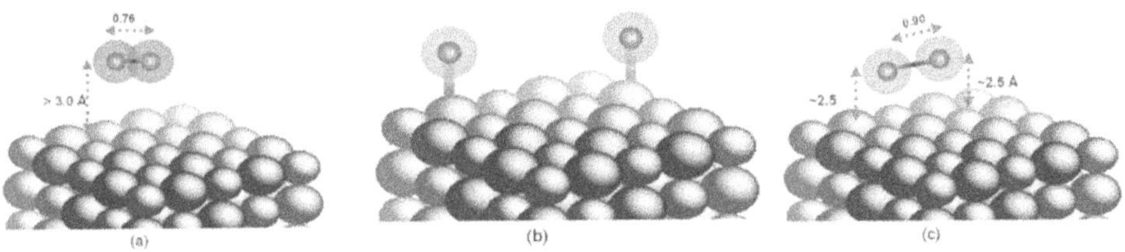

- **Complex Hydrides:** It mainly indicates to the class of light metal hydrides and chemical hydride type of materials in which hydrogen adsorption occurs via chemical bond formation i.e. via chemisorption. Materials consisting of group I and II salts of alanates, amides, and borohydrides ([AlH_4]-, [NH_2]-, and [BH_4]-) are currently referred to as "complex hydrides". In the majority of instances, chemical hydrides are produced by chemical reaction which cannot be reversed easily for on board vehicular and industrial applications. In case of amides, alanates, and borohydrides, hydrogen is in general bonded to the central atoms in "complex" anions via covalent bond. Even though these materials can achieve a hydrogen density, they were initially overlooked as potential hydrogen storage systems because of their lack of ambient thermodynamical working condition, poor kinetics, and reversibility (Sandi, 2004). Thus, complex hydrides were not considered as potential candidates for use as rechargeable hydrogen carriers until recently. However, this situation was totally altered by the pioneering work of Bogdanovic´ and Schwickardi who demonstrated that, upon doping with selected titanium compounds, the dehydriding of anionic aluminum hydrides could be kinetically enhanced and made reversible under moderate conditions in the solid state (Bogdanović & Schwickardi, 1997). This breakthrough discovery has prompted a worldwide effort to develop doped alanates as practical hydrogen storage materials, which has expanded rapidly to include amides and borohydrides.

- **Nanostructured Materials:** These are materials in which adsorption of H_2 molecules occur in quasi-molecular fashion via quasi-molecular binding. In general, functionalized sorbent materials, nanoclusters, nanoparticles etc. comes under this category. At the nanoscale, the physico-chemical properties of materials change drastically, so research nowadays focuses on the nanostructures as potential hydrogen storage materials. Extensive research onto nanostructured material research has clearly revealed that reduced size, low coordination, and low dimensionality can result in properties that differ significantly from the corresponding bulk materials (Jena, 2011). It has also been investigated that reducing the size of complex hydrides to the nanoscale regime can enhance the thermodynamics and kinetics of hydrogen storage. Pure carbon nanostructures adsorb H_2 molecules with very low binding energy which is not feasible for practical hydrogen storage conditions. However, metal functionalization over pure nanostructures creates open metal sites which are capable of holding multiple H_2 molecules with optimum binding energy. Significant theoretical work has also been done on pure and doped carbon nanostructures (Hirscher et al., 2002), BN-based nanostructures (Kumar et al., 2015), SiC based nanostructures (nanoclusters, nanotubes, fullerenes, graphene, graphyne etc.) (Banerjee et al., 2012; Tabtimsai et al., 2018), which demonstrate their potential as reversible hydrogen storage materials under ambient environmental conditions. Zhang et al. have demonstrated that transition metal doped C_{60} fullerene and carbon nanotubes can adsorb H_2 molecules up to 9 wt% reversibly and with favourable kinetics (Zhao et al., 2005). However, due to their high cohesive energy, transition metals decorated on nanostructures tend to cluster, reducing their hydrogen storage capacity (Krasnov et al., 2007). In this situation, functionalization with alkali and alkaline earth metals can prevent metal aggregation due to their low cohesive energy compared to transition metal atoms, making them viable choice for metal dopants. Thus, nanostructured materials when decorated by alkali or alkaline earth metals not only enhance gravimetric density because of their low atomic weight but also increase their hydrogen adsorption capacity. Our investigation demonstrates that Li decorated Si doped C_{20} fullerene and benzene analogues of Si-Li/Na nanoclusters can be regarded as potential candidates for hydrogen storage systems (Jaiswal et al., 2022), as shown in Figure 4.

Figure 4. Optimized geometries of maximum hydrogen adsorbed: (a) $Si_{10}Li_{10}$, (b) $Si_{10}Na_{10}$, (c) $Si_2C_{18}Li_6$, and (d) $Si_4C_{16}Li_6$ complexes
Source: Reprinted from Jaiswal et al. (2022)

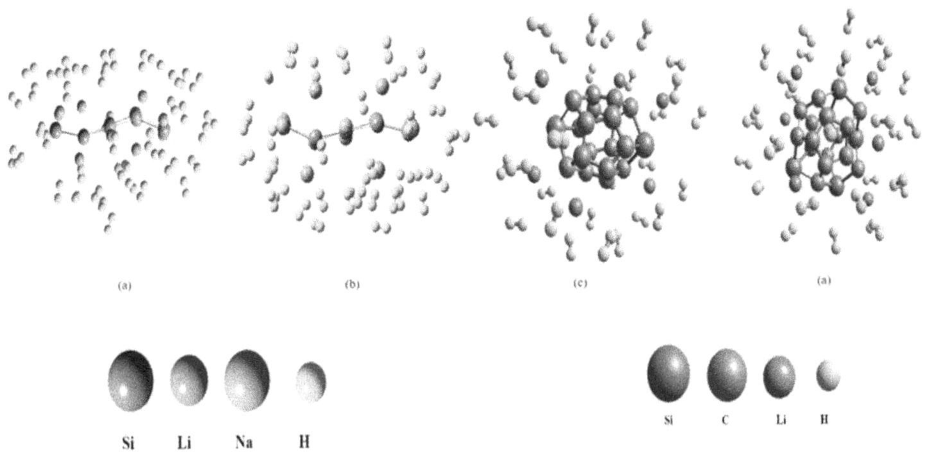

FUTURE RESEARCH DIRECTIONS

The major two problems discussed above in the field of green hydrogen economy are still not effectively solved. A lot of research is going on to tackle these issues and to advance the hydrogen-based civilization. But no results have yet fully satisfied the industrial and transportation needs for the advancement of green hydrogen economy in terms of efficiency, cost effectiveness, reliability and above all safety measures. This urges to develop new methods of hydrogen generation, its storage mechanisms, its distribution and effective practical applications. In addition, efficient fuel cell design, rapid refuelling process, the installation of hydrogen refuelling stations, and low-cost transportation will be crucial future research directions in order to realize the green hydrogen economy. Thus, development of full fledge green hydrogen economy is still loaded with significant challenges and thus opens a lot of future research scopes.

SUMMARY AND CONCLUSION

Our review mainly focusses on the hydrogen storages; especially, the challenges and probable solutions while effectively storing hydrogen at affordable and compatible environment. Hydrogen is one of the most promising green energy carriers with maximum energy density among all the fossil fuels known. Despite having maximum energy density and clean combustion, hydrogen economy has not yet developed with full fledge. One of the major challenges behind this lag is the hydrogen storage problem. Various hydrogen storage methods have been employed till date, but none of them completely fulfils the US-DoE criterions. Hydrogen stored in high pressure tanks are associated with drawbacks such as low gravimetric density, poor efficiency and accidental risk. Although hydrogen stored under high pressure in cylinders does not meet the US-DOE requirements but still it is the current choice for industrial and vehicular applications. Further, storage at cryogenic temperature may improve the gravimetric and volumetric density up to some extent but still it is not optimum for the industrial applications and is limited by severe drawbacks like easy boil off of hydrogen even at little thermal fluctuations and flammability risk. In such scenario, hydrogen storage in material-based systems has proven to be quite efficient with minimum risk but are very hard to realize in practical applications. Thus, development of hydrogen based global economy solely depends on the efficient hydrogen storge mechanism.

ACKNOWLEDGMENT

We thank Indian Institute of Technology (Indian School of Mines) Dhanbad, India for providing research facilities. We are also grateful to American Chemical Society publishers, JRC Publications Repository and United States-Department of Energy for granting us permission to reprint their figures and datasets. We also acknowledge NASA, NOAA, Berkeley Earth, and other meteorological offices, and *BP Statistical Review* for making their data public and open access.

REFERENCES

Abbott, D. (2009). Keeping the energy debate clean: How do we supply the world's energy needs? *Proceedings of the IEEE*, *98*(1), 42–66. doi:10.1109/JPROC.2009.2035162

AFHYPAC. (2013). *Les données de base physico-chimiques sur l'hydrogène. Mémento de l'Hydrogène*. Available online:http://www.afhypac.org/documents/toutsavoir/fiche_1.2_donnees_physicochimiques_rev. mars_2013.pdf

AFHYPAC & FNCCR. (2018). *Déployer les stations hydrogène dans votre territoire*. Available online: http://www.afhypac.org/documents/divers/GUIDE-STATION-HYDROGENE-WEB.pdf

Aithal, P. S., & Acharya, S. (2016). Impact of Green Energy on Global Warming-A Changing Scenario. *International Journal of Scientific Research and Modern Education.*

Aoyama, T., Iiyama, A., Shinohara, K., Kamegaya, S., Yamamoto, S., & Ban, Y. (2009). Status of FCV development at Nissan and future issues. *SAE International Journal of Engines*, *1*(1), 314–323. doi:10.4271/2008-01-0423

Banerjee, S., Nigam, S., Pillai, C. G. S., & Majumder, C. (2012). Hydrogen storage on Ti decorated SiC nanostructures: A first principles study. *International Journal of Hydrogen Energy*, *37*(4), 3733–3740. doi:10.1016/j.ijhydene.2011.05.078

Basic research needs for the hydrogen economy. (2004). www.sc.doe.gov/bes/hydrogen.pdf

Bogdanović, B., & Schwickardi, M. (1997). Ti-doped alkali metal aluminium hydrides as potential novel reversible hydrogen storage materials. *Journal of Alloys and Compounds*, *253*, 1–9. doi:10.1016/S0925-8388(96)03049-6

Cahill, A. E., Aiello-Lammens, M. E., Fisher-Reid, M. C., Hua, X., Karanewsky, C. J., Yeong Ryu, H., . . . Wiens, J. J. (2013). How does climate change cause extinction? *Proceedings of the Royal Society B: Biological Sciences, 280*(1750), 20121890. 10.1098/rspb.2012.1890

Change, I. C. (2013). The physical science basis. *Contribution of working group I to the fifth assessment report of the intergovernmental panel on climate change, 1535.*

Coyle, E. D., & Simmons, R. A. (2014). *Understanding the global energy crisis.* Purdue University Press. doi:10.26530/OAPEN_469619

Dankers, R., & Feyen, L. (2008). Climate change impact on flood hazard in Europe: An assessment based on high-resolution climate simulations. *Journal of Geophysical Research*, *113*(D19), D19105. doi:10.1029/2007JD009719

de Carvalho Rocha, W. F., & Sheen, D. A. (2019). Determination of physicochemical properties of petroleum derivatives and biodiesel using GC/MS and chemometric methods with uncertainty estimation. *Fuel*, *243*, 413–422. doi:10.1016/j.fuel.2018.12.126

De Vries, J. (1994). The industrial revolution and the industrious revolution. *The Journal of Economic History*, *54*(2), 249–270. doi:10.1017/S0022050700014467

Dudley, B. (2018). BP statistical review of world energy. *BP Statistical Review, 6*(2018).

Durbin, D. J., & Malardier-Jugroot, C. (2013). Review of hydrogen storage techniques for on board vehicle applications. *International Journal of Hydrogen Energy*, *38*(34), 14595–14617. doi:10.1016/j.ijhydene.2013.07.058

Engineering ToolBox. (2008). *Fossil and Alternative Fuels - Energy Content*. Available at: https://www.engineeringtoolbox.com/fossil-fuels-energy-content-d_1298.html

Gardiner, M., & Satyapal, S. (2009). *Energy Requirements for Hydrogen Gas Compression and Liquefaction as Related to Vehicle Storage Needs*. DOE Hydrogen and Fuel Cells Program Record. Available online:https://www.hydrogen.energy.gov/pdfs/9013_energy_requirements_for_hydrogen_gas_compression.pdf

Gosling, S. N., & Arnell, N. W. (2016). A global assessment of the impact of climate change on water scarcity. *Climatic Change*, *134*(3), 371–385. doi:10.100710584-013-0853-x

Gye, H. R., Seo, S. K., Bach, Q. V., Ha, D., & Lee, C. J. (2019). Quantitative risk assessment of an urban hydrogen refueling station. *International Journal of Hydrogen Energy*, *44*(2), 1288–1298. doi:10.1016/j.ijhydene.2018.11.035

Hernandez-Ochoa, I. M., Asseng, S., Kassie, B. T., Xiong, W., Robertson, R., Pequeno, D. N. L., & Hoogenboom, G. (2018). Climate change impact on Mexico wheat production. *Agricultural and Forest Meteorology*, *263*, 373–387. doi:10.1016/j.agrformet.2018.09.008

Hirscher, M., Becher, M., Haluska, M., Quintel, A., Skakalova, V., Choi, Y. M., Dettlaff-Weglikowska, U., Roth, S., Stepanek, I., Bernier, P., Leonhardt, A., & Fink, J. (2002). Hydrogen storage in carbon nanostructures. *Journal of Alloys and Compounds*, *330*, 654–658. doi:10.1016/S0925-8388(01)01643-7

Hirscher, M., & Panella, B. (2007). Hydrogen storage in metal–organic frameworks. *Scripta Materialia*, *56*(10), 809–812. doi:10.1016/j.scriptamat.2007.01.005

Ishaq, H., Dincer, I., & Crawford, C. (2022). A review on hydrogen production and utilization: Challenges and opportunities. *International Journal of Hydrogen Energy*, *47*(62), 26238–26264. doi:10.1016/j.ijhydene.2021.11.149

Jaiswal, A., Chakraborty, B., & Sahu, S. (2022). High capacity reversible hydrogen storage in Si substituted and Li decorated C20 fullerene: Acumen from density functional theory simulations. *International Journal of Energy Research*, *46*(14), 19521–19537. doi:10.1002/er.8524

Jaiswal, A., Sahoo, R. K., Ray, S. S., & Sahu, S. (2022). Alkali metals decorated silicon clusters (SinMn, n= 6, 10; M= Li, Na) as potential hydrogen storage materials: A DFT study. *International Journal of Hydrogen Energy*, *47*(3), 1775–1789. doi:10.1016/j.ijhydene.2021.10.228

Japan's Hydrogen Industrial Strategy. (n.d.). Center for Strategic and International Studies. https://www.csis.org/analysis/japans-hydrogen-industrial-strategy

Jena, P. (2011). Materials for hydrogen storage: Past, present, and future. *The Journal of Physical Chemistry Letters*, *2*(3), 206–211. doi:10.1021/jz1015372

Kam, P. M., Aznar-Siguan, G., Schewe, J., Milano, L., Ginnetti, J., Willner, S., McCaughey, J. W., & Bresch, D. N. (2021). Global warming and population change both heighten future risk of human displacement due to river floods. *Environmental Research Letters*, *16*(4), 044026. doi:10.1088/1748-9326/abd26c

Khalil, Y. F. (2013). Experimental determination of dust cloud deflagration parameters of selected hydrogen storage materials: Complex metal hydrides, chemical hydrides, and adsorbents. *Journal of Loss Prevention in the Process Industries*, *26*(1), 96–103. doi:10.1016/j.jlp.2012.09.010

Krasnov, P. O., Ding, F., Singh, A. K., & Yakobson, B. I. (2007). Clustering of Sc on SWNT and reduction of hydrogen uptake: Ab-initio all-electron calculations. *The Journal of Physical Chemistry C*, *111*(49), 17977–17980. doi:10.1021/jp077264t

Kubas, G. J. (2001). Metal–dihydrogen and σ-bond coordination: The consummate extension of the Dewar–Chatt–Duncanson model for metal–olefin π bonding. *Journal of Organometallic Chemistry*, *635*(1-2), 37–68. doi:10.1016/S0022-328X(01)01066-X

Kumar, E. M., Sinthika, S., & Thapa, R. (2015). First principles guide to tune h-BN nanostructures as superior light-element-based hydrogen storage materials: Role of the bond exchange spillover mechanism. *Journal of Materials Chemistry. A, Materials for Energy and Sustainability*, *3*(1), 304–313. doi:10.1039/C4TA04706K

Lan, J., Cao, D., & Wang, W. (2009). Li12Si60H60 fullerene composite: A promising hydrogen storage medium. *ACS Nano*, *3*(10), 3294–3300. doi:10.1021/nn900842j PMID:19761195

Letcher, T. M. (2019). Why do we have global warming? In *Managing global warming* (pp. 3–15). Academic Press. doi:10.1016/B978-0-12-814104-5.00001-6

Ley, M. B., Meggouh, M., Moury, R., Peinecke, K., & Felderhoff, M. (2015). Development of hydrogen storage tank systems based on complex metal hydrides. *Materials (Basel), 8*(9), 5891–5921. doi:10.3390/ma8095280 PMID:28793541

Lhotka, O., Kyselý, J., & Farda, A. (2018). Climate change scenarios of heat waves in Central Europe and their uncertainties. *Theoretical and Applied Climatology, 131*(3), 1043–1054. doi:10.100700704-016-2031-3

Li, J., Cheng, S., Zhao, Q., Long, P., & Dong, J. (2009). Synthesis and hydrogen-storage behavior of metal–organic framework MOF-5. *International Journal of Hydrogen Energy, 34*(3), 1377-1382.

Lu, H., Guan, Y., He, L., Adhikari, H., Pellikka, P., Heiskanen, J., & Maeda, E. (2020). Patch aggregation trends of the global climate landscape under future global warming scenario. *International Journal of Climatology, 40*(5), 2674–2685. doi:10.1002/joc.6358

McMichael, A. J., Campbell-Lendrum, D. H., Corvalán, C. F., Ebi, K. L., Githeko, A., Scheraga, J. D., & Woodward, A. (2003). *Climate change and human health: risks and responses*. World Health Organization.

McWhorter, S., Read, C., Ordaz, G., & Stetson, N. (2011). Materials-based hydrogen storage: Attributes for near-term, early market PEM fuel cells. *Current Opinion in Solid State and Materials Science, 15*(2), 29–38. doi:10.1016/j.cossms.2011.02.001

Ministry of Foreign Affairs of Japan Prime minister Suga's Attendance at the Leaders Summit on Climate. (2021). https://www.mofa.go.jp/ic/ch/page6e_000236.html

Niu, J., Rao, B. K., & Jena, P. (1992). Binding of hydrogen molecules by a transition-metal ion. *Physical Review Letters, 68*(15), 2277–2280. doi:10.1103/PhysRevLett.68.2277 PMID:10045354

Noaa, N. (2019). *State of the climate: Global climate report for annual 2018*. www. ncdc. noaa. gov/sotc/global/201813

Otsubo, H., Mizuno, M., Negishi, Y., & Ogami, N. (2007). *High-pressure hydrogen tank for FCHV* (No. 2007-01-2009). SAE Technical Paper.

Rivard, E., Trudeau, M., & Zaghib, K. (2019). Hydrogen storage for mobility: A review. *Materials (Basel), 12*(12), 1973. doi:10.3390/ma12121973 PMID:31248099

Rosenzweig, C., & Parry, M. L. (1994). Potential impact of climate change on world food supply. *Nature, 367*(6459), 133–138. doi:10.1038/367133a0

Rosi, N. L., Eckert, J., Eddaoudi, M., Vodak, D. T., Kim, J., O'Keeffe, M., & Yaghi, O. M. (2003). Hydrogen storage in microporous metal-organic frameworks. *Science*, *300*(5622), 1127–1129. doi:10.1126cience.1083440 PMID:12750515

Sandí, G. (2004). Hydrogen storage and its limitations. *The Electrochemical Society Interface*, *13*(3), 40–44. doi:10.1149/2.F06043IF

Schneemann, A., White, J. L., Kang, S., Jeong, S., Wan, L. F., Cho, E. S., Heo, T. W., Prendergast, D., Urban, J. J., Wood, B. C., Allendorf, M. D., & Stavila, V. (2018). Nanostructured metal hydrides for hydrogen storage. *Chemical Reviews*, *118*(22), 10775–10839. doi:10.1021/acs.chemrev.8b00313 PMID:30277071

Schüth, F., Bogdanović, B., & Felderhoff, M. (2004). Light metal hydrides and complex hydrides for hydrogen storage. *Chemical Communications*, (20), 2249–2258. doi:10.1039/B406522K PMID:15489969

Scientific consensus on climate change. (2022, October 15). In *Wikipedia*. https://en.wikipedia.org/wiki/Scientific_consensus_on_climate_change

Smalley, R. E. (2005). Future global energy prosperity: The terawatt challenge. *MRS Bulletin*, *30*(6), 412–417. doi:10.1557/mrs2005.124

Tabtimsai, C., Ruangpornvisuti, V., Tontapha, S., & Wanno, B. (2018). A DFT investigation on group 8B transition metal-doped silicon carbide nanotubes for hydrogen storage application. *Applied Surface Science*, *439*, 494–505. doi:10.1016/j.apsusc.2017.12.255

Trivedi & Bandyopadhyay. (2015). Hydrogen storage in small size MgnCo clusters: A density functional study. *International Journal of Hydrogen Energy, 40*(37), 12727-12735.

Tseng, P., Lee, J., & Friley, P. (2005). A hydrogen economy: Opportunities and challenges. *Energy*, *30*(14), 2703–2720. doi:10.1016/j.energy.2004.07.015

Tzimas, E., Filiou, C., Peteves, S., & Veyret, J. (2003). *Hydrogen Storage: State-of-the-Art and Future Perspective*. EUR 20995 EN. cat. No. LD-NA-20995-EN-C.

United States Environmental Protection Agency. (2017). *Inventory of U.S. Greenhouse Gas Emissions and Sinks: 1990–2015*. Available online: https://www.epa.gov/sites/production/files/2017-02/documents/2017_ complete_report.pdf

U.S. Department of Energy's Energy Efficiency and Renewable Energy Website. (n.d.). https://www1.eere.energy.gov/hydrogenandfuelcells/storage/current_technology.html

Yang, S. J., Kim, T., Im, J. H., Kim, Y. S., Lee, K., Jung, H., & Park, C. R. (2012). MOF-derived hierarchically porous carbon with exceptional porosity and hydrogen storage capacity. *Chemistry of Materials, 24*(3), 464–470. doi:10.1021/cm202554j

Zhao, Y., Kim, Y. H., Dillon, A. C., Heben, M. J., & Zhang, S. B. (2005). Hydrogen storage in novel organometallic buckyballs. *Physical Review Letters, 94*(15), 155504. doi:10.1103/PhysRevLett.94.155504 PMID:15904160

Züttel, A. (2004). Hydrogen storage methods. *Naturwissenschaften, 91*(4), 157–172. doi:10.100700114-004-0516-x PMID:15085273

Züttel, A., Remhof, A., Borgschulte, A., & Friedrichs, O. (2010). Hydrogen: the future energy carrier. *Philosophical Transactions of the Royal Society A: Mathematical, Physical and Engineering Sciences, 368*(1923), 3329-3342.

Züttel, A., Sudan, P., Mauron, P., Kiyobayashi, T., Emmenegger, C., & Schlapbach, L. (2002). Hydrogen storage in carbon nanostructures. *International Journal of Hydrogen Energy, 27*(2), 203–212. doi:10.1016/S0360-3199(01)00108-2

ADDITIONAL READING

Ball, M., & Wietschel, M. (Eds.). (2009). *The hydrogen economy: opportunities and challenges*. doi:10.1017/CBO9780511635359

Busby, R. L. (2005). *Hydrogen and fuel cells*. Pennwell Corporation.

Crabtree, G. W., Dresselhaus, M. S., & Buchanan, M. V. (2004). The hydrogen economy. *Physics Today, 57*(12), 39–44. doi:10.1063/1.1878333

Hirose, K. (2010). *Handbook of hydrogen storage: new materials for future energy storage*. John Wiley & Sons.

National Research Council. (2004). *The hydrogen economy: opportunities, costs, barriers, and R&D needs*. National Academies Press.

Pahwa, P. K., & Pahwa, G. K. (2014). *Hydrogen economy*. The Energy and Resources Institute (TERI).

Rand, D. A. J., & Dell, R. M. (2007). *Hydrogen energy: Challenges and prospects*. Academic Press.

Sherif, S. A., Goswami, D. Y., Stefanakos, E. L., & Steinfeld, A. (Eds.). (2014). *Handbook of hydrogen energy*. CRC Press. doi:10.1201/b17226

KEY TERMS AND DEFINITIONS

Carbon-Free: Free of carbon containing compounds.

Critical Temperature: The highest temperature at which the substance can exist as a liquid.

Energy Density: Amount of energy stored in a given system per unit volume.

Gravimetric Density: The available energy per unit mass of a substance.

Green Energy: Energy generated from natural resources.

Hydrogen Storage: Storage of hydrogen.

Nanocluster: A collection of few atoms.

Renewable: Able to be renewed by natural processes.

Volumetric Density: Mass per unit volume.

Chapter 8
Hydrogen Storage Technologies and Related Heat and Mass Transfer Studies

Rahul Uday Urunkar
https://orcid.org/0000-0001-8377-9475
Department of Mechanical Engineering, Rajarambapu Institute of Technology, Shivaji University, Kolhapur, India

Sharad Dattatray Patil
Department of Mechanical Engineering, Rajarambapu Institute of Technology, Shivaji University, Kolhapur, India

ABSTRACT

The energy demands of the future are ever increasing, and hydrogen as an ideal energy carrier can fulfil these demands. The production, purification, delivery, storage, and application are the significant measures of the hydrogen-based economy. The utmost challenge to utilize hydrogen as a fuel lies in the improvement of storage techniques. Hydrogen storage technologies comprise of high-pressure compression, cryogenic liquefaction, and absorption in solid state such as metal hydrides and complex hydrides. As compared with other techniques, hydrogen storage in solid form seems to be one of the utmost likely solutions. However, it involves extremely coupled transport processes such as chemical kinetics, heat, and mass transfer. Complex hydrides are capable substitute aspirants for solid state hydrogen storage because of many advantages, but many of such hydrides suffer from poor kinetics as well as great thermodynamic stability. Significant heat transfer techniques and issues associated with hydrogen storage methods are discussed, with emphasis on metal hydride.

DOI: 10.4018/978-1-6684-6721-3.ch008

Nomenclature:

ΔH: Enthalpy of formation [J/mol H_2]
ΔS: Entropy of formation [J/mol H_2-K]
c: Hydrogen concentration
C_p: Specific heat [J/kg-K]
E_a: Activation energy [J/mol H_2]
k: Thermal conductivity [W/m-K]
h: Heat transfer coefficient [W/m^{2}K]
m: Mass of material [kg]
M: Molecular mass [kg/kmol]
P: Pressure [bar]
$R_{u:}$ Ideal gas constant [J/mol-K]
T: Temperature [J/mol-K]
V: Velocity [m/s]

Subscripts:

e: Effective
eq: Equilibrium
inj: Injection
H_2: Hydrogen gas
t: Time
a: Alloy

INTRODUCTION

The storage and transfer of energy is very important and interesting area in our everyday lifecycle. Rapidly growing population and development in industrial sectors demands for more and more energy (Mahfuz et al., 2014). The major issue to be tackled is the conflict among demand as well as supply of the energy. The depletion of non-renewable energy sources, such as fossil fuels, is another important matter. To cultivate systems of energy which will accomplish energy requirements from sustainable as well as renewable sources is one of the paramount defies at the present time. Such energy systems should be practical, economic, consistent as well as bearable, with little or no impact on the environment system.

To address all these issues hydrogen (H_2) can be a better option as an unconventional fuel (Gökçek & Kale, 2018; Patil & Gopal, 2013). In recent years, there is increasing interest in technologies based on hydrogen fuel. The production, purification, delivery, storage as well as application are the significant measures of the H_2 based economy. For many years, storage of hydrogen has been in the main focus of research. Currently, there are number of research happenings in evaluating the need, influence, opportunities as well as challenges of enactment of H_2 based economy. The utmost challenge to utilize hydrogen as a fuel lies in the improvement of the techniques used to store hydrogen. At this time a number of studies have been done and numerous techniques to store hydrogen have been suggested but very few of them meet the recent U.S. Department of Energy (DOE) targets. Each storage technique has its own limitations. This area of research is still in developing phase and continuous investigation to develop existing mechanisms is being thoroughly undertaken. The hydrogen storage techniques considered in this chapter include compressed hydrogen, liquid hydrogen and solid-state hydrogen storage (metallic and complex hydrides). Each of this method involves extremely coupled transport processes such as chemical kinetics, heat transfer as well as mass transfer. Thorough knowledge of such processes as well as their comparative influence on execution of system is essential to facilitate the design and optimization of hydrogen storage systems.

STORAGE TYPES

Consistent and harmless storage as well as distribution of hydrogen are the important stages to the extensive usage of hydrogen. The hydrogen based green economy necessitates stationary as well as mobile storage systems for hydrogen to succeed. Stationary hydrogen storage systems are mostly for stationary generation of power. Mobile hydrogen storage systems are mostly for usage in a vehicle. But, these systems of hydrogen storage have their weird necessities and defies (Prabhukhot et al., 2016). The different storage methods for hydrogen are revealed in Figure 1. Three primary methods used to store hydrogen include high pressure gas, cryogenic liquid and solid state hydride. In order to enhance gravimetric densities, many new materials such as metal organic frameworks, various forms of carbon, complex hydrides etc. are also being recommended by many researchers.

Figure 1. Hydrogen storage methods

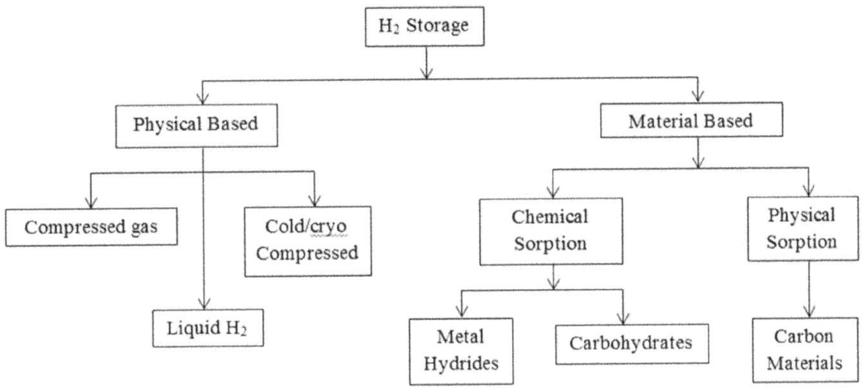

DOE TARGET

Table 1 shows targets for on-board hydrogen storage set by the United States Department of Energy (US-DOE). These technical performance targets provide the guiding principles for the onboard storage of hydrogen study and put up a long run perspective for storage of hydrogen. However, meeting all these targets simultaneously has been acknowledged as a challenging task.

Table 1. DOE targets for on-board hydrogen storage

Storage Variables	Unit	2020	2025	Ultimate
Gravimetric capacity (Material-based)	kWh/kg	1.50	1.80	2.20
Gravimetric capacity (System-based)	kg-H_2/kg system	0.045	0.055	0.065
Volumetric capacity (Material based)	kWh/L	1	1.3	1.7
Volumetric capacity (System based)	kgH_2/L system	0.03	0.04	0.05
Cost of storage system	$/kWh net	10	9	8
Cost of fuel	$/kg$H_2$	333	300	266
Atmospheric temperature	^{0}C	-40 / 60 (sun)	-40 / 60 (sun)	-40 / 60 (sun)
Delivery temperature (Min / Max)	^{0}C	-40 / 85	-40 / 85	-40 / 85
Cycle life (1/4th tank to full)	cycles	1500.0	1500.0	1500.0
Min. delivery pressure from tank	bar	5.0	5.0	5.0
Max. delivery pressure from tank	bar	12.0	12.0	12.0
Charging / Discharging rates	min	3-5	3-5	3-5

Source: Zhang and Wu (2017)

COMPRESSED HYDROGEN STORAGE

Hydrogen stored in compressed form, as a gas, is one of the commercially existing technology and easiest way to store hydrogen. In order to be competitive with systems based on fossil fuel, it is desirable to store normally 4 to 7 kg of hydrogen on-board. Due to lower density of hydrogen, for on-board usage with adequate quantity it is required to compress to higher pressures. For the fuel cell application the compressed hydrogen pressure is in the range of 350 bar to 700 bar. At normal temperature, the density of H_2 corresponding to 350 bar is about 23 kg/m^3 and at 700 bar it is about 37 kg/m^3 as shown in Figure 2. Even though hydrogen is compressed up to 700 bar its density is still low, which is a severe problem (Al-zareer et al., 2018).

Figure 2. Mass density of hydrogen under various conditions
Source: Al-zareer et al. (2018)

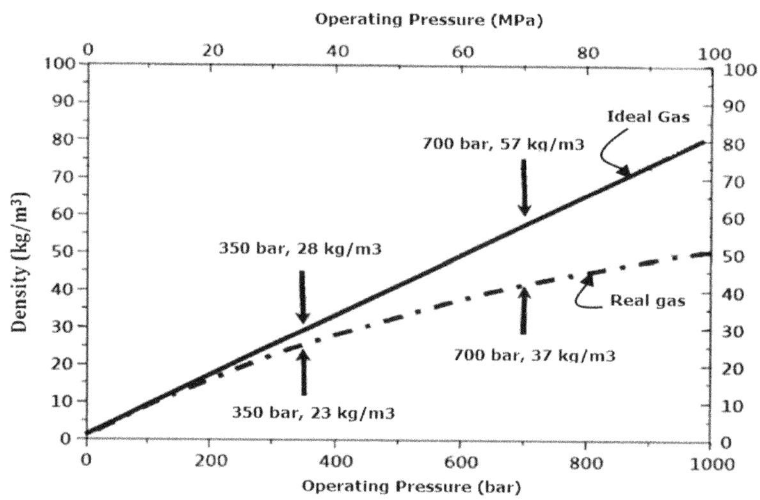

At normal temperature hydrogen acts identical to an ideal gas and it fulfills the ideal gas law. The hydrogen compression process requires a huge quantity of energy. Thermodynamically, reversible isothermal compression requires the minimum work (Cengel & Boles, 2002) which can be calculated as follows

$$w = mRT_1 \ln\left(\frac{P_2}{P_1}\right) \tag{1}$$

$= 1 \times 4.1570 \times 300 \times \ln(700/1)$

$= 8170.00 \text{kJ/kg}$

Theoretically, pressurizing affects around 11 to 13% of the hydrogen energy adversely (Pesonen & Alakunnas, 2017). The isothermal compression process of hydrogen gas from 1 bar to 800 bar requires 2.210 kWh/kg energy, however actual process requires much more energy. Strictly speaking, there is a considerable rise in temperature of hydrogen gas throughout compression process even after incorporating multi-stage intercooling. As a result, the compression of hydrogen is well defined as polytropic process (Cengel & Boles, 2002). The energy required for such a process is given by

$$w = \frac{n}{n-1} RT_1 \left[\left(\frac{P_2}{P_1} \right)^{(n-1)/n} - 1 \right] \quad (2)$$

Equations (1) and (2) indicated that compression work can be decreased proportionally by reducing the inlet hydrogen temperature. Intercooling techniques have been designed for the same.

Compressors for Hydrogen

Two different kinds of compressors are extensively used to enhancement pressure of hydrogen upto minimum 150 bar; namely diaphragm compressors and reciprocating piston compressors. Diaphragm compressors are mainly utilised for minor scale uses, whereas reciprocating piston compressors may be utilised for major scale uses (such as pipelines) as well as minor scale uses (such as fueling stations) (Tzimas et al., 2003). The energy efficiency of both types of compressor can be improved by increasing heat transfer.

For diaphragm type compressors the compression ratio for every stage is restricted to around 7 and hence minimum 4 compression stages should be required to achieve 700 bar pressure. Even after considering the perfect intercooling between the compression stages, the required minimum work is still 24% larger than that needed for isothermal process. Therefore, substantial prospective and concern present in attaining savings of energy by virtue of improved heat exchange throughout the compression process (Tzimas et al., 2003).

Compressed Hydrogen Storage Vessel

The highly pressurized hydrogen can be stored in different vessels based on its application such as bulk transportation, vehicular and stationary (Zheng et al., 2011). The key requirements for materials used for hydrogen storage vessel are small density, great tensile strength, small diffusivity and non-reactivity with hydrogen. Extreme permissible pressure in the tank is limited by the tensile strength of material which is turn puts limit for the volumetric density as well as mass of stored hydrogen. It should be noted that enhancing operating pressure of the tank requires safety features which in turn results in thicker walls as well as increased system weight than hydrogen capacity. Following are some important materials utilized for highly pressurized hydrogen storage vessels (Hibbeler, 2018):

Steel alloys: Sy = 703 MPa, ρ= 8160 Kg/m^3
Titanium alloys: Sy = 924 MPa, ρ= 4430 Kg/m^3
Carbon composites: Sy = 2070 MPa, ρ= 1900 Kg/m^3

Followings are different types of pressure vessels used for storing high pressure hydrogen (Barthelemy et al., 2016):

- **Type I:** These pressure vessels are fully metallic type. Such type of pressure vessels are usually made up from Al or steel. It is possible to store hydrogen up to 500 bar with these type of pressure vessels. This type is the utmost conventional, less expensive, and also the heaviest.
- **Type II:** These pressure vessels are made up of steel which overwrapped with glass fiber composite. The manufacturing cost of the type II vessel is around 50% higher than that of type I. However, it is about 30 to 40% lighter than type I.
- **Type III:** These pressure vessels are made up of full wrapping of composite material along with metal liner. The liner is utilized for the purpose of sealing whereas the composite structure is utilized to take the structural load. Up to working pressure of 450 bar such type of pressure vessels are found to be reliable, however, still facing difficulties for clearing the ageing tests at 700 bar. The manufacturing cost of the type III vessel is almost twice than that of type II.
- **Type IV:** Fully composite. Generally liner is made from a polymer like High Density Polyethylene (HDPE) (Figure 3). For carrying the structural load, carbon glass composites or carbon fiber are utilized. Type IV vessel is the lightest among others but the cost is too high. Such pressure vessels are found to be reliable up to 1000 bar working pressure.

Figure 3. Type IV compressed gaseous hydrogen vessel
Source: Barthelemy et al. (2016)

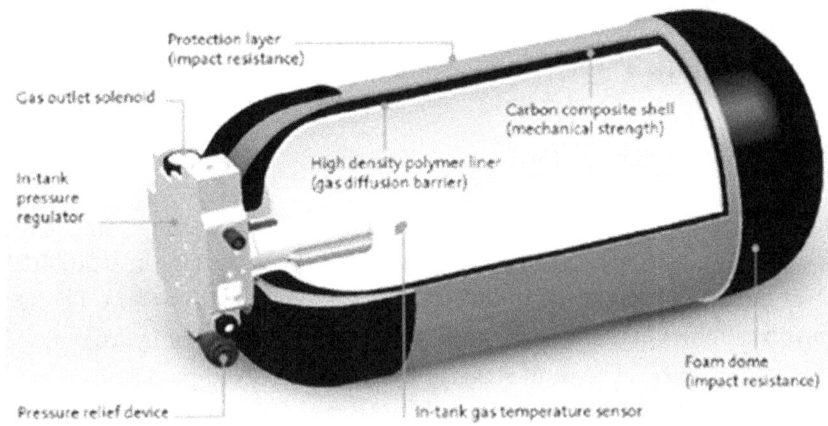

Merits-Demerits

The major advantages of storing hydrogen as high pressure gas are the ease of the technology and fast uptake-release rate. As a result several prototype fuel-cell vehicles implemented compressed hydrogen in it. However, this storage method has significant drawbacks. Although the pressure can be increased, the volumetric and gravimetric densities of compressed hydrogen do not increase proportionally, making it less effective compared to other storage methods. This results in bulkiness and difficulty integrating it into vehicles. Additionally, the extremely lightweight nature of hydrogen increases the risk of leakage from containing vessels under high pressure. Safety concerns surrounding pressurized cylinders are also significant, particularly in densely populated regions.

LIQUID HYDROGEN STORAGE

Another commercially available technology for storage of hydrogen is in liquid form at cryogenic condition. Hydrogen is normally in a gaseous state at ambient conditions but can be liquefied by cooling it down to 20 K at atmospheric pressure, which is lower than its critical point temperature of 33 K at 12.9 bar (College of the Desert, 2001). Storing hydrogen in a liquid state offers a significant advantage, as it can greatly increase its volumetric density at comparatively lesser tank pressures. This allows for more energy to be stored per unit volume. At normal boiling point of 20 K, liquid hydrogen (LH_2) has a density of about 71 kg/m^3, which is approximately 2 times the density of compressed gaseous hydrogen at 700 bar & 288 K. Operational pressures for the liquid storage of hydrogen range from 1 bar to 3.5 bar (Wolf, 2002). Compared to compressed hydrogen, the energy density of liquid hydrogen can be as high as 8400 MJ/m^3, while compressed hydrogen has an energy density of 4400 MJ/m^3 (Sandi, 2004).

Process of Liquefaction

The liquefaction of hydrogen includes three distinct heat exchange stages from a thermodynamic perspective. The first stage aims to remove sensible heat (4000 kJ/kg) and lower the temperature from 300 K to 20 K (Jacobsen et al., 1997). The second stage involves removing latent heat (450 kJ/kg) to condense the hydrogen at 20 K (Jacobsen et al., 1997). The third stage is referred to as ortho-para conversion, that necessitates the removal of 703 kJ/kg of heat (Baker & Shaner, 1978). The overall process of liquefaction necessitates the removal of a total of 5153 kJ/kg of heat, which is provided by a liquefaction cycle.

The Joule-Thompson cycle is considered to be the most basic cycle for hydrogen liquefaction (Timmerhaus & Flynn, 1989). To initiate this cycle, first hydrogen in gaseous state is compressed to a large pressure (e.g., P=100 bar) and later cooled through a heat exchanger earlier going through a throttling valve. As the gas undergoes isenthalpic Joule-Thomson expansion through the valve, some liquid is produced. The cooled gas is then detached from liquid and directed back to the compressor, after passing through another heat exchanger. Since hydrogen has an inversion temperature of almost 200 K, which is the lower temperature than at which throttling yields a cooling effect, expansion at ambient temperature will not result in a cooling effect. As a result, liquid nitrogen (T=77 K) is frequently utilized to pre-cool hydrogen prior to expansion. The theoretical work required to liquefy hydrogen gas at ambient temperature is approximately 3.23 kWh/kg, however the actual work required is around 15.2 kWh/kg, which equates to nearly 50% of the lower heating value of hydrogen.

Heat exchangers are a critical component in the production of liquid hydrogen, as they are used to transfer heat during the liquefaction process. Cryogenic liquefiers require heat exchanger effectiveness of at least 85% in order to produce liquid due to the low operating temperatures (Barron & Nellis, 1999). There are several ways to improve the effectiveness of heat exchangers, such as having high heat transfer coefficients and large surface area to volume ratios, maintaining low temperature differences among inlet and outlet streams. However, due to the deviation of hydrogen's specific heat throughout heat exchange, a situation may arise where the lowest temperature difference among the heat transfer streams present in the middle of the flow path, as depicted in Figure 4 (Timmerhaus & Flynn, 1989). This situation should be reflected when trying to enhance efficiency of heat exchanger by reducing temperature differences among inlet and outlet streams. However, two challenges must be addressed while using this method: the efficiency of the liquefaction process and the boil-off of the liquid.

Figure 4. Thermal profile in a counter-flow hydrogen heat exchanger
Source: Timmerhaus and Flynn (1989)

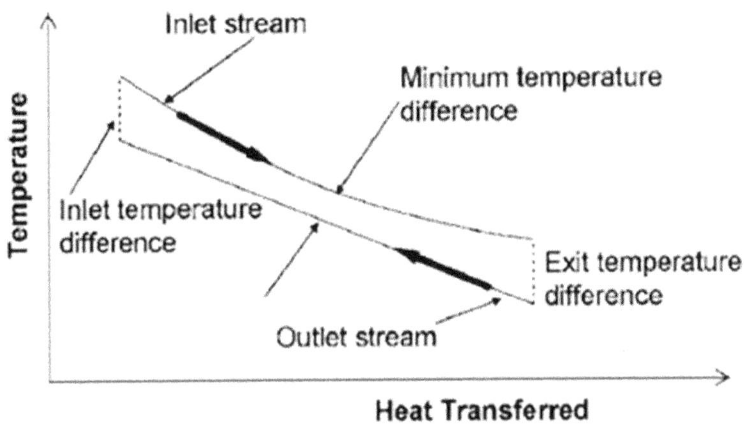

Liquid Hydrogen Storage Tank and Related Heat-Mass Transfer Issues

The storage of liquid hydrogen requires special double-walled vessel with good thermal insulation systems. Figure 5 illustrates the liquid hydrogen tank (Krainz et al., 2004). Heat can be transferred to the hydrogen tank from its surroundings via three different modes, namely conduction, convection, as well as radiation. To prevent heat transmission via conduction, tank walls should be constructed using materials having less thermal conductivities. However, it is important to note that the thermal conductivities as well as other properties of materials may change significantly at cryogenic temperatures. To reduce convection, the gap between the inner and outer vessels is usually evacuated to large vacuum, typically around 1.3×10^{-8} bar (Barron & Nellis, 1999). This technique is also known as vacuum superinsulation and involves covering the layers surrounding the vacuum with specialized insulating materials.

To minimize radiation heat transfer, low-emissivity radiation shields are placed between the hot and cold surfaces. The heat transmission rate due to radiation is directly proportionate to the emissivity of the shields as well as inversely proportionate to their number. The use of multi-layered insulation (MLI) on the inner tank face is an effective way to reduce radiation heat transfer. MLI is made up of 30 to 50 less-emissivity radiation shields, such as aluminum foil with an emissivity of 0.05, detached by less-conductivity spacers like fiberglass paper or glass wool. This arrangement effectively prevents both radiation and thermal conduction between the layers. The use of MLI helps to minimize heat conductivity and reduce heat transfer due to radiation. The thickness of insulation and the boil-off rate have an inverse relationship, with small increments in thermal resistance leading to a significant decrease in boil-off. This relationship is highly non-linear (Colozza, 2002).

The substitute insulation to MLI is microsphere insulation that shows an isotropic thermal conductivity. A comparison was made between vacuum superinsulation with multilayer insulation and microsphere insulation in pressure vessels. It was observed that tanks with vacuum zones and multilayer insulation lose less hydrogen than those with microsphere insulation. To decrease heat exchange, contacting areas with the environment should be minimized. Spheres have the minimum ratio of surface area to volume, which allows for less heat exchange from the surroundings. Tank size, as well as tank shape, is a parameter that affects heat gains. Cylindrical liquid hydrogen tanks are favoured than spherical tanks owing to ease of built-up and cost, and ease of fitting into vehicles, especially for enlarged tank size to deliver adequate fuel capacity as well as large storage efficiency. As shown in Table 2, there is an inverse relationship among tank size and evaporation rate, which means that the boil-off effect decreases as the tank size increases.

Figure 5. Liquid hydrogen tank system
Source: Krainz et al. (2004)

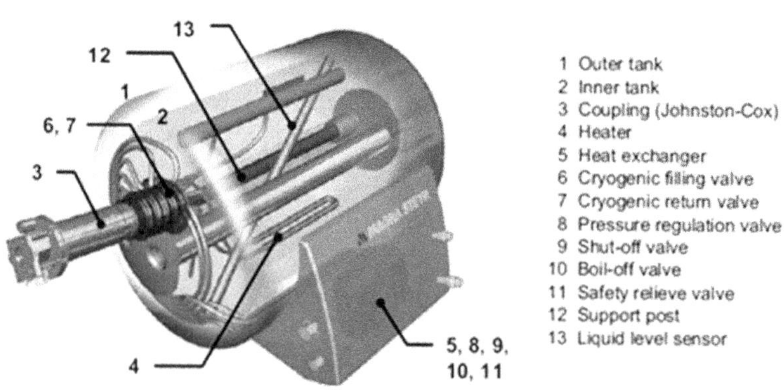

Table 2. Boil-off losses in tank with double-walled vacuum insulation

Per Day Losses (in %)	Tank of Volume (in m^3)
0.40	50
0.20	100
0.06	20000

Source: Züttel (2003)

The term "dormancy" refers to the time period between a vehicle being parked or put into idle mode and the process of venting. Typically, a valve is opened and hydrogen is released once the pressure level reaches about 10 bar. Depending on the application, the vented hydrogen may diffuse into the air, be catalytically burned, or captured. If a vehicle's fuel cell or engine is operating, hydrogen consumption is quicker than the boil-off effects, so venting is not required. But, throughout periods of vehicle inactivity, boil-off causes pressure of tank to rise, eventually requiring venting to control the pressure.

Designing LH_2 transfer lines involves numerous heat transfer considerations. The transfer of LH_2 typically results in two-phase flow due to heat exchange to the fluid, specifically throughout the filling process startup. To prevent the critical reduction in carrying capacity due to the presence of vapor in the transfer line, insulation is necessary to minimize evaporation. However, despite adequate insulation, low-pressure liquid hydrogen tanks still experience an energy loss of up to 8% during fueling. Additionally, cooling losses occur during refilling, and the entire transfer line must be cooled to about 20 K. Even with substantial efforts, these losses cannot be entirely eliminated and remain noteworthy. The insulation methods for the transfer line are analogous to those employed for the storage vessel.

To avoid boil-off effects, the insulation system for tank walls aims to delay them, although it cannot completely eliminate them. Hence, it is essential to vent gaseous hydrogen to regulate the pressure inside the tank (Colozza, 2002). The venting system should be airtight to prevent the entry of air, which can freeze and obstruct the flow of hydrogen. An extra protection layer (vacuum jacket) is also present for cryogenic vessels to ensure safety in case of accidents (Petitpas & Aceves, 2012). Thus, unless ignited, a severe explosion will not occur in case of leakage or tank rupture.

Merits-Demerits

Cryogenic LH_2 tanks are considered safer than high-pressure hydrogen tanks and can be filled relatively quickly. However, maintaining hydrogen at the extremely low temperature required for liquefaction (20 K) is a major challenge. The process of liquefaction is energy-intensive and continuous boil-off during storage is also a significant drawback (Amos, 1998). The cooling technology required for LH_2 is time-consuming and energy-intensive, with almost 30% to 40% of the lower heating value of hydrogen being consumed during the process (Barthelemy et al., 2016). Low pressure LH_2 storage systems are relatively cost-effective (Aceves et al., 2013), but the high cost and energy consumption of liquefaction process have been widely criticized. The LH_2 storage system is typically used in situations where the cost of hydrogen is not a major problem and where the hydrogen is consumed quickly, such as in air and space applications. The intricacy of the LH_2 storage system, along with the challenge of reducing boil-off gas, leads to higher overall LH_2 system costs compared to CGH_2 systems, especially at a large scale.

METAL HYDRIDE (SOLID STATE HYDROGEN STORAGE)

Ensuring both reliability and safety in hydrogen storage and delivery is crucial for the widespread use of hydrogen. High pressure is required for gaseous hydrogen storage, and cryogenic temperatures are necessary for storing it in a liquid form. Alternatively, hydrogen can be stored in solid form by embedding it reversibly in a solid matrix. Metal hydrides are a promising option for solid-state hydrogen storage because they offer several advantages, such as normal storage temperature and pressure, absorption and desorption of hydrogen at a constant pressure (Harries et al., 2012), high volumetric storage capacity (Sakintuna et al., 2007), large number of charge–discharge cycles and most importantly safety. The hydrogen is combined with solid metals or metallic alloys to provide solid-state storage. While some research has yielded higher hydrogen volumetric densities than those achieved with compressed gas or liquid hydrogen, the densities are still lower than future targets (Yoo et al., 2004).

Solid State Hydrogen Storage Reactor

The metal hydride reactor is used to hold the metal hydride materials, and operates as a packed bed vessel that allows for heat and mass exchange with the metal hydride bed. The reactor is responsible for facilitating the hydride formation and decomposition reactions, and its design plays a critical role in achieving optimal performance. The process of reaction kinetics, porous flow, and heat and mass transfer in the reactor is complex and transient, and is affected by various parameters. The reactor configuration is a crucial parameter in reactor design and several different types have been proposed, including tubular, disc, and tank or chamber reactors, as shown in Figure 6.

Figure 6. Schematic of different reactors: (a) tubular reactor, (b) annular disc reactor, (c) tank reactor

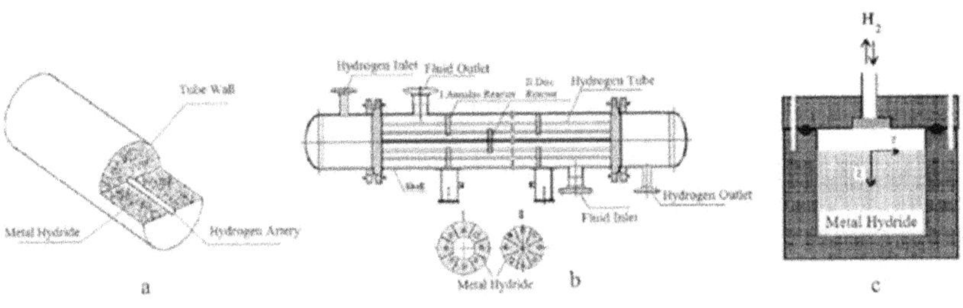

(a) Tubular reactor, (b) Annular disc reactor, (c) Tank reactor

Merits-Demerits

Conventional metal hydrides are commercially existing (Chen et al., 2003) also they are much safer than compressed hydrogen, because of which they can be used in fuel-cell-powered mining locomotives (Story, 2000). Volumetric densities of metal hydride are comparable to that of liquid hydrogen, however low gravimetric density is their key drawback. Reduced storage capacity is another significant disadvantage associated with metal hydrides. This is because hydriding and dehydriding processes occur at the molecular level within the interstitial sites of the metal matrix. There is a chance that impurities present in the tank could get trapped in these interstitial sites, leading to a decrease in the overall storage capacity.

Challenges for Metal Hydride Systems

The advancement of metal hydride systems faces several key challenges, including identifying appropriate materials for development of hydride, controlling hydrogen uptake and release mechanisms, and reducing production as well as operational costs. Research on metal hydrides has concentrated on kinetics and crystal structures, but the technology suffers from two inherent issues, which are poor heat transfer characteristics and inadequate kinetics. Therefore, special consideration must be given to the thermal design of the storage devices. The other major challenge is the slow hydriding and dehydriding rates, which are limited, in part, by the thermal properties of the system. As a result, the rate of reaction in a metal hydride reactor is frequently restricted by the rate at which heat can be added or removed from the hydride rather than the intrinsic kinetics of the hydriding and dehydriding reactions (Harries et al., 2012).

The thermodynamics of hydrides play a critical role in their effectiveness. Metal hydride formation is an exothermic reaction that generates a significant amount of heat and increases the hydride's volume. Inadequate heat removal can lead to a rise in temperature that can cause the process to stall. Additionally, high internal temperatures can cause metal hydrides to sinter, which negatively affects their hydrogen storage capacity (Sandrock et al., 2002). Chemical thermodynamics indicate that a specific temperature is required for the dehydriding process to take place. Without adequate heat input, hydrogen release will cease due to the decrease in the bed temperature.

While heavy metals can release hydrogen at lower temperatures, light metals require significant heating to release hydrogen. Metal alloys that can bind a large amount of hydrogen are typically very heavy, and the stability of metal hydrides is due to their low thermodynamic kinetics (Schuth et al., 2004). However, the stability of metal hydrides also means that more heat is needed to release the stored hydrogen, so efficient internal heat transfer is necessary for optimal system performance and reliability (Gomes et al., 2005). To be useful, the hydride material must be capable of absorbing and releasing hydrogen repeatedly without fatigue (Felderhoff & Bogdanovi, 2009; Rowsell & Yaghi, 2005). Therefore, researchers are focusing on two key areas of study: uptake-release kinetics and the retention of cycling capacity. To achieve high volumetric densities in metal hydride storage systems is another goal of hydride research. The most significant volumetric densities have been achieved with atomic ratios of two hydrogen atoms to one metal atom.

Heat Transfer Improvement (Augmentation/ Intensification Techniques)

Numerous studies have been conducted to investigate the parameters affecting the heat transfer process in hydrides. These studies have explored the properties of hydrides (Huston, 1980), as well as heat and mass transfer, (Ram Gopal & Murthy, 1993; Urunkar & Patil, 2021a; 2021b; 2023), and system simulation (Gopal & Murthy, 1995). Some researchers have developed cooling and heating systems and tested experimentally to improve performance (Inomata et al., 1998). Researchers have attempted various methods to enhance the heat conductivity of the hydrogen absorbing medium, such as incorporating copper wire net structure, inserting aluminum foam, expanded natural graphite (ENG), multilayer sheet structure in waved form used for packing, and metal hydride compact with micro encapsulation (Mellouli et al., 2009). However, it should be noted that while these heat transfer intensification techniques can enhance the rates of hydrogen absorption and desorption reactions, they do not increase the storage capacity. Additionally, such techniques may not be suitable for portable and mobile hydrogen storage applications, as the added weight of high thermal conductivity materials may be counterproductive and lack of contribution to hydrogen sorption.

To enhance the heat conductivity of hydrogen storage devices, several design parameters like geometrical distribution of the alloy, filters, and heat exchangers can be considered. While the geometry of the reactor doesn't affect the process of hydride formation, it has an impact on the reaction rate (Souahlia et al., 2011). By improving the reactor configuration as well as heat exchanger design, the hydrogen storage density and capacity can be increased, and the refueling time can be decreased (Garrison et al., 2012). However, operating pressure and temperature are essential factors to consider when designing heat exchangers (Raju & Kumar, 2012).

To enhance the heat transfer in metal hydride storage systems, various techniques can be classified into three categories. Firstly, methods to increase the thermal conductivity of the hydride bed such as integrating ENG and metal foam into the alloy bed. Secondly, techniques to add heat transfer area by incorporating cooling tubes, fins, etc. Lastly, improvements in operating parameters can also enhance heat transfer.

COMPLEX HYDRIDE

Another option for solid state hydrogen storage with comparatively lighter weight is complex metal hydride. These materials have high theoretical hydrogen storage capacity (Jain et al., 2010), great storage densities as well as low hydrogen release temperatures and pressures (Srinivasan et al., 2008) and are relatively inexpensive. When compared to simple metal hydrides, complex metal hydrides differ in that they form an ionic or covalent compound throughout the hydrogen absorption. However, conventional metal hydrides like intermetallic hydrides have a low gravimetric storage capacity for hydrogen (Leng et al., 2006) due to their composition of heavy elements from the periodic table, with only lithium (Li), sodium (Na), beryllium (Be), and magnesium capable of forming solid state compounds along with hydrogen. Intense research interest has developed in low weight complex hydrides such as alanates [$A1H_4$], borohydrides [BH_4], amides [NH_2], imides[NH], etc.

Complex hydrides are materials that have a salt-like structure, in which the hydrogen atoms are typically situated at the corners of a tetrahedron and create a covalent bond with the central atoms, resulting in a crystal structure composed of complex anions. These compounds are typically represented by the formula $A_xMe_yH_z$, where A is usually an element from the first or second group of the periodic table, and Me is often an element from the boron or aluminum groups. Sodium (Na), Lithium (Li), Magnesium (Mg), Aluminum (Al), and Boron (B), which belong to Groups I, II, and III of the periodic table, have been used to create a variety of metal hydrogen complexes. Complex hydrides in Group III, such as borohydrides and alanates, are capable of bonding four hydrogen atoms with an alkali metal, partially covalent compound, or ionic compound (Zuttel et al., 2003). The number of hydrogen atoms per metal atom is two in several cases. This kind of complexes shows the highest volumetric density and the highest gravimetric density at room temperature. Hydrogen storage capacities of some complex hydrides are shown in Table 3.

Table 3. Complex hydride's storage capacities for hydrogen

Hydride	H_2 (wt. %)
$LiBH_4$	18.0
$Al(BH_4)_3$	17.0
$Be(AlH_4)_2$	11.3
$LiAlH_4$	10.5
$NaBH_4$	10.4
$Mg(AlH_4)_2$	9.3
$Ti(AlH_4)_4$	9.3
$Ca(AlH_4)_2$	7.7
$NaAlH_4$	7.5
$KAlH_4$	5.7

Source: Murthy (2012)

Limitations

The basic understandings of atomic structure as well as the thermodynamics of hydrogen uptake-release are very important in order to form complex hydrides in practical point of view. In contrast to the metallic hydrides, hydrogen is released through multistep chemical reaction from the complex hydrides, which require different conditions at different step. Therefore, this cannot be considered as entirely reversible process unlike in metal hydrides. So, there is a big difference between the theoretical and the practically attainable hydrogen storage capacities. Despite having high energy densities, complex hydrides pose challenges in handling safely and are prone to decompose into stable elements that are difficult to refuel with hydrogen on board vehicles (Ley et al., 2014). Furthermore, they exhibit poor kinetics and high desorption temperatures, making rehydrogenation conditions very stringent. Due to these reasons, utilizing complex hydrides for hydrogen storage is a difficult task because of their low hydrogen releasing process dynamics.

A significant challenge with many complex hydrides is that they do not meet the hydrogen storage targets established by the US Department of Energy, as they have high thermodynamic stability as well as slow hydrogen cycling kinetics (Huang et al., 2006). Unlike simple metal hydrides, the thermal simulation of complex hydrides is more intricate. Nevertheless, these challenges can be partly addressed through thermodynamic destabilization techniques, such as adding a new element or introducing a dopant or catalyst. These modifications can alter or tune the materials to achieve favorable thermodynamic and kinetic properties for hydrogen storage applications (Albanese et al., 2013). To improve the thermodynamics of metal hydrides, various approaches are being implemented, including doping, controlling the particle size of metal hydrides, and combining different reactants. These strategies can be used together, but the number of variables increases, which can make the systems more challenging to study. While alanates, borates, and imides show promising gravimetric hydrogen storage capacity, they encounter issues with high desorption temperatures.

In order to develop effective systems for potential complex hydrides, it is essential to characterize their thermal properties through ongoing research. However, these hydrides are highly sensitive to moisture and can generate hazardous hydrogen gas when exposed to air-borne water vapour (Anton et al., 2007). To ensure safety, experiments must be conducted under inert conditions. It is also important to note that critical thermal properties vary greatly between the hydrogen charging and discharging states, and depend on factors like particle diameter and density. As a result, the thermal properties of tanks that are partially and fully charged will vary significantly, and the heat transfer systems must be adaptable to accommodate these different conditions.

SUMMARY AND CONCLUDING REMARKS

Hydrogen is an attractive alternative source of energy as it is clean, non-toxic and most importantly, renewable. Despite its potential benefits, the hydrogen economy faces significant challenges that require considerable effort to overcome. Hydrogen production, purification, storage, distribution, and end use are all important factors in the hydrogen economy. The biggest challenge is developing hydrogen storage systems that are safe, portable, compact, and cost-effective. This chapter examines various technologies used for hydrogen storage, including storage as gas, liquid, or solid compounds. Various hydrogen storage methods include compression at high pressure, cooling to cryogenic temperatures to liquefy the gas, physisorption, and absorption into solid-state compounds such as metal hydrides and complex hydrides.

Solid-state hydrogen storage seems to be the most viable solution since it is safer and more convenient than other methods.

The previous research has shown that every potential method for storing hydrogen presents significant challenges regarding heat transfer. In certain cases, these challenges could determine the feasibility of a particular storage method. This chapter examines the significant heat transfer techniques and issues associated with various hydrogen storage methods. Effective cooling is critical for compressed and liquid hydrogen storage systems, as maintaining extremely low temperature for extended periods is necessary. Efficient heat transfer is also vital for metal-hydride systems, as it enables rapid hydrogen sorption. Specific issues are inherent to each technique, making it difficult to apply individual improvements to other designs. Each technique has its advantages and disadvantages, so it is not recommended to focus on a single material as a global solution for the hydrogen storage problem.

Heavy metal hydrides have limitations in terms of gravimetric density and relatively high temperature requirements, while light metal hydrides face challenges due to the significant energy required to release hydrogen. Critical design factors for solid-state hydrogen storage devices include its hydrogen storage capacity (gravimetric) and rate of charging-discharging. Most candidate materials release hydrogen at temperatures above 90°C, and materials that discharge hydrogen at lower temperatures are unable to store significant quantities of hydrogen.

Considerable attention must be given to numerous geometric and operational factors in the charging and discharging processes of metal hydride storage tanks, as their properties fluctuate continuously. Various heat transfer techniques are commonly employed, such as fins, metal foam, cooling tubes, water jackets, and ENG. Fins and cooling tubes were the most commonly studied, accounting for over fifty percent of the research. However, the thermal conductivity of the metal hydride appears to be a major limiting factor for the storage tank's performance. Metal foam has been found to be more efficient than fins for heat transfer enhancement, while ENG composites are both cost-effective and perform better than metal foam. Improving only one aspect of the heat transfer process, such as heat transfer coefficient or thermal conductivity or heat transfer area, may not be sufficient to resolve the heat transfer issue. Therefore, it is recommended to integrate several techniques, such as combining mesh and cooling tubes or ENG and coiled tube, to improve the system's performance.

Due to the limited storage capacity of metals and alloys, there has been a suggestion to use complex hydrides like alanates and borates, as well as carbons, as alternative storage media for hydrogen. These complex hydrides have high hydrogen storage capacities, low discharge pressure, and mild temperatures, making them a potential option for solid-state hydrogen storage. Nonetheless, many of these hydrides have a high level of thermodynamic stability and slow kinetics throughout hydrogen sorption. To address these issues, techniques such as nano confinement, thermodynamic destabilization, and catalyst addition can be utilized to improve their performance to some extent.

According to the United States Department of Energy (US-DOE) objectives, there is currently no flawless option for a hydrogen storage material. This chapter can provide valuable insights to professionals, scholars, and scientists working on hydrogen storage material investigations.

REFERENCES

Aceves, S. M., Petitpas, G., Espinosa-Loza, F., Matthews, M. J., & Ledesma-Orozco, E. (2013). Safe, long range, inexpensive and rapidly refuelable hydrogen vehicles with cryogenic pressure vessels. *International Journal of Hydrogen Energy, 38*(5), 2480–2489. doi:10.1016/j.ijhydene.2012.11.123

Al-zareer, M., Dincer, I., & Rosen, M. A. (2018). Analysis and assessment of a hydrogen production plant consisting of coal gasification, thermochemical water decomposition and hydrogen compression systems. *Energy Conversion and Management, 157*, 600–618. doi:10.1016/j.enconman.2017.11.047

Albanese, E., Kalantzopoulos, G. N., Vitillo, J. G., Pinatel, E., Civalleri, B., Deledda, S., Bordiga, S., Hauback, B. C., & Baricco, M. (2013). Theoretical and experimental study on Mg(BH4)2 – Zn(BH4)2 mixed borohydrides. *Journal of Alloys and Compounds, 580*, S282–S286. doi:10.1016/j.jallcom.2013.02.165

Amos, W. A. (1998). *Costs of storing and transporting hydrogen costs of storing and transporting hydrogen*. National Renewable Energy Laboratory.

Anton, D. L., Mosher, D. A., Opalka, S. M., Tang, X., Laube, B. L., Brown, R. J., Thomas, H., Arsenault, S., Wu, R., Strickler, J., Anton, D. L., Zidan, R., Berseth, P., Read, C., Headquarters, D. O. E., Manager, T., Adams, J., Field, D. O. E., & Officer, P. (2008). Complex Hydride Compounds with Enhanced Hydrogen Storage Capacity Final Report. *Proceedings of the 2004 Annual U.S. DOE Hydrogen Program Review.*

Anton, D. L., Mosher, D. A., Tang, X., Brown, R. J., Arsenault, S., Saitta, S., Laube, B. L., Dold, R. H., Anton, D. L., Adams, J., Field, D. O. E., & Officer, P. (2007). High Density Hydrogen Storage System Demonstration Using NaAlH4 Based Complex Compound Hydrides. *Proceedings of the Annual U.S. DOE Hydrogen Program Review.*

Baker, C. R., & Shaner, R. L. (1978). A study of the efficiency of hydrogen liquefaction. *International Journal of Hydrogen Energy, 3*(3), 321–334. doi:10.1016/0360-3199(78)90037-X

Barron, R. f., & Nellis, G. f. (1999). *CryogeniC Heat transfer*. CRC Press. doi:10.1201/b15230

Barthelemy, H., Weber, M., & Barbier, F. (2016). Hydrogen storage : Recent improvements and industrial perspectives. *International Journal of Hydrogen Energy, 42*(11), 7254–7262. doi:10.1016/j.ijhydene.2016.03.178

Basic Research Needs for the Hydrogen Economy. (2003). Report on the Basic Energy Sciences Workshop on Hydrogen Production, Storage, and Use.

Cengel, Y. A., & Boles, M. A. (2002). *Thermodynamics: An Engineering Approach*. McGraw-Hill Higher Education.

Chen, Y., Sequeira, C. A. C., Chen, C., Wang, X., & Wang, Q. (2003). Metal hydride beds and hydrogen supply tanks as minitype PEMFC hydrogen sources. *International Journal of Hydrogen Energy, 28*(3), 329–333. doi:10.1016/S0360-3199(02)00064-2

College of the Desert. (2001). *Hydrogen Fuel Cell Engines and Related Technologies: Report*. Author.

Colozza, A. J. (2002). *Hydrogen Storage for Aircraft Applications: Overview*. Report No. NASA/CR-211867.

Felderhoff, M., & Bogdanovi, B. (2009). High temperature metal hydrides as heat storage materials for solar and related applications. *International Journal of Molecular Sciences, 10*(1), 325–344. doi:10.3390/ijms10010325 PMID:19333448

Garrison, S. L., Hardy, B. J., Gorbounov, M. B., Tamburello, D. A., Corgnale, C., Bart, A., Mosher, D. A., & Anton, D. L. (2012). Optimization of internal heat exchangers for hydrogen storage tanks utilizing metal hydrides. *International Journal of Hydrogen Energy, 37*(3), 2850–2861. doi:10.1016/j.ijhydene.2011.07.044

Gökçek, M., & Kale, C. (2018). Optimal design of a hydrogen refuelling station (HRFS) powered by hybrid power system. *Energy Conversion and Management, 161*, 215–224. doi:10.1016/j.enconman.2018.02.007

Gomes, S., Renaudin, G., Hagemann, H., Yvon, K., Sulic, M. P., & Jensen, C. M. (2005). Effects of milling, doping and cycling of NaAlH4 studied by vibrational spectroscopy and X-ray diffraction. *Journal of Alloys and Compounds, 390*(1-2), 305–313. doi:10.1016/j.jallcom.2004.08.036

Gopal, M. R., & Murthy, S. S. (1995). Performance of a metal hydride cooling system. *International Journal of Refrigeration, 18*(6), 413–420. doi:10.1016/0140-7007(95)98164-G

Harries, B. D. N., Paskevicius, M., Sheppard, D. A., Edward, T., Price, C., & Buckley, C. E. (2012). Concentrating solar thermal heat storage using metal hydrides. *Proceedings of the IEEE, 100*(2), 539–549. doi:10.1109/JPROC.2011.2158509

Hibbeler, R. C. (2018). *Mechanics of materials*. Pearson Education Limited.

Huang, Z. G., Guo, Z. P., Calka, A., Wexler, D., Lukey, C., & Liu, H. K. (2006). Effects of iron oxide (Fe2O3, Fe3O4) on hydrogen storage properties of Mg-based composites. *Journal of Alloys and Compounds, 422*(1-2), 299–304. doi:10.1016/j.jallcom.2005.11.074

Huston, E. L., & Sandrock, G. D. (1980). Engineering properties of metal hydrides. *Journal of the Less Common Metals, 74*(2), 435–443. doi:10.1016/0022-5088(80)90182-4

Inomata, A., Aoki, H., & Miura, T. (1998). Measurement and modelling of hydriding and dehydriding kinetics. *Journal of Alloys and Compounds, 278*(1-2), 103–109. doi:10.1016/S0925-8388(98)00008-5

Jacobsen, R. T., Penoncello, S. G., & Lemmon, E. W. (1997). Thermodynamic Properties of Cryogenic Fluids. Springer. https://doi.org/https://doi.org/10.1007/978-1-4899-1798-0_5

Jain, I. P., Jain, P., & Jain, A. (2010). Novel hydrogen storage materials : A review of lightweight complex hydrides. *Journal of Alloys and Compounds, 503*(2), 303–339. doi:10.1016/j.jallcom.2010.04.250

Krainz, G., Hödl, P., & Hofmeister, F. (2004). Automotive Production of Liquid Hydrogen Storage Systems. *The International German Hydrogen Energy Congress*.

Leng, H., Ichikawa, T., & Fujii, H. (2006). Hydrogen Storage Properties of Li - Mg - N - H Systems with Different Ratios of. *Journal of Physical Chemistry, 110*(26), 12964–12968. doi:10.1021/jp061120h PMID:16805600

Ley, M. B., Jepsen, L. H., Lee, Y., Cho, Y. W., Von Colbe, M. B., Dornheim, M., Rokni, M., Jensen, J. O., Sloth, M., Filinchuk, Y., Jørgensen, J. E., Besenbacher, F., & Jensen, T. R. (2014). Complex hydrides for hydrogen storage – new perspectives. *Materials Today, 17*(3), 122–128. doi:10.1016/j.mattod.2014.02.013

Mahfuz, M. H., Kamyar, A., Afshar, O., Sarraf, M., Anisur, M. R., Kibria, M. A., Saidur, R., & Metselaar, I. H. S. C. (2014). Exergetic analysis of a solar thermal power system with PCM storage. *Energy Conversion and Management*, *78*, 486–492. doi:10.1016/j.enconman.2013.11.016

Mellouli, S., Askri, F., Dhaou, H., Jemni, A., & Ben Nasrallah, S. (2009). Numerical study of heat exchanger effects on charge / discharge times of metal – hydrogen storage vessel. *International Journal of Hydrogen Energy*, *34*(7), 3005–3017. doi:10.1016/j.ijhydene.2008.12.099

Murthy, S. S. (2012). Heat and Mass Transfer in Solid State Hydrogen Storage : A Review. *Journal of Heat Transfer*, *134*(3), 1–11. doi:10.1115/1.4005156

Patil, S. D., & Gopal, M. R. (2013). Entropy generation and its relation to heating and cooling requirements of a metal hydride hydrogen storage reactor. *International Journal of Hydrogen Energy*, *38*(31), 13687–13693. doi:10.1016/j.ijhydene.2013.08.018

Pesonen, O., & Alakunnas, T. (2017). *Energy Storage: a missing piece of the puzzle for the self-sufficient living*. UAS publication series B: research reports and compilations.

Petitpas, G., & Aceves, S. M. (2012). Modeling of sudden hydrogen expansion from cryogenic pressure vessel failure. *International Journal of Hydrogen Energy*, *38*(19), 8190–8198. doi:10.1016/j.ijhydene.2012.03.166

Prabhukhot, P. R., Wagh, M. M., & Gangal, A. C. (2016). A Review on Solid State Hydrogen Storage Material. *Advances in Energy and Power*, *4*(2), 11–22. doi:10.13189/aep.2016.040202

Raju, M., & Kumar, S. (2012). Optimization of heat exchanger designs in metal hydride based hydrogen storage systems. *International Journal of Hydrogen Energy*, *37*(3), 2767–2778. doi:10.1016/j.ijhydene.2011.06.120

Ram Gopal, M., & Murthy, S. S. (1993). Parametric studies on heat and mass transfer in metal hydride beds. *Chemical Engineering and Processing*, *32*(4), 217–223. doi:10.1016/0255-2701(93)80003-Y

Rowsell, J. L. C., & Yaghi, O. M. (2005). Strategies for Hydrogen Storage in Metal-Organic Frameworks. *Angewandte Chemie International Edition*, *44*(30), 4670–4679. doi:10.1002/anie.200462786 PMID:16028207

Sakintuna, B., Lamari-darkrim, F., & Hirscher, M. (2007). Metal hydride materials for solid hydrogen storage : A review. *International Journal of Hydrogen Energy*, *32*(9), 1121–1140. doi:10.1016/j.ijhydene.2006.11.022

Sandi, G. (2004). Hydrogen storage and its limitations. *The Electrochemical Society Interface*, *13*(3), 40–44. doi:10.1149/2.F06043IF

Sandrock, G., Gross, K., Thomas, G., Jensen, C., Meeker, D., & Takara, S. (2002). Engineering considerations in the use of catalyzed sodium alanates for hydrogen storage. *Journal of Alloys and Compounds*, *330*, 696–701. doi:10.1016/S0925-8388(01)01505-5

Schuth, F., Bogdanovic, B., & Felderhoff, M. (2004). Light metal hydrides and complex hydrides for hydrogen storage. *Chemical Communications (Cambridge)*, (20), 2249–2258. doi:10.1039/B406522K PMID:15489969

Souahlia, A., Dhaou, H., Askri, F., Mellouli, S., Jemni, A., & Ben Nasrallah, S. (2011). Experimental study and characterization of metal hydride containers. *International Journal of Hydrogen Energy*, *36*(8), 4952–4957. doi:10.1016/j.ijhydene.2011.01.074

Srinivasan, S., Diego, E., Goswami, Y., & Stefanakos, E. (2008). Effects of catalysts doping on the thermal decomposition behavior of Zn(BH4)2. *International Journal of Hydrogen Energy*, *33*(9), 2268–2272. doi:10.1016/j.ijhydene.2008.02.062

Story, G. C. (2000). Hydride bed/fuelcell proect. *Proceedings of the 2000 Hydrogen Program Review*.

Timmerhaus, K. D., & Flynn, T. M. (1989). Cryogenic Process Engineering (K. Mendelssohn, C. Rizzuto, & Klaus, Eds.). Plenum Press.

Tzimas, E., Filiou, C., Peteves, S. D., & Veyret, J. (2003). *Hydrogen storage : state of the art and future perspective*. Institute for Energy, Directorate General Joint Research Centre.

Urunkar, R. U., & Patil, S. D. (2021a). Enhancement of heat and mass transfer characteristics of metal hydride reactor for hydrogen storage using various nanofluids. *International Journal of Hydrogen Energy*, *46*(37), 19486–19497. doi:10.1016/j.ijhydene.2021.03.090

Urunkar, R. U., & Patil, S. D. (2021b). Numerical study on performance evaluation of multi-tubular sodium alanate hydride reactor by enhancing heat and mass transfer characteristics using nanofluids. *Proceedings of the 26th National and 4th International ISHMT-ASTFE Heat and Mass Transfer Conference December 17-20, 2021, IIT Madras, Chennai-600036, Tamil Nadu, India*, 2613–2618. 10.1615/IHMTC-2021.3950

Urunkar, R. U., & Patil, S. D. (2023). Performance analysis of sodium alanate hydride reactor with different nanofluids. *International Journal of Hydrogen Energy*, xxxx. https://doi.org/10.1016/j.ijhydene.2023.02.105

Wang, H., Lin, H. J., Cai, W. T., Ouyang, L. Z., & Zhu, M. (2016). Tuning kinetics and thermodynamics of hydrogen storage in light metal element based systems – A review of recent progress. *Journal of Alloys and Compounds*, *658*, 280–300. doi:10.1016/j.jallcom.2015.10.090

Wolf, J. (2002). Liquid-hydrogen technology for vehicles. *MRS Bulletin*, *27*(9), 684–687. doi:10.1557/mrs2002.222

Yoo, E., Gao, L., Komatsu, T., Yagai, N., Arai, K., Yamazaki, T., Matsuishi, K., Matsumoto, T., & Nakamura, J. (2004). Atomic hydrogen storage in carbon nanotubes promoted by metal catalysts. *The Journal of Physical Chemistry B*, *108*(49), 18903–18907. doi:10.1021/jp047056q

Zhang, B., & Wu, Y. (2017). Recent advances in improving performances of the lightweight complex hydrides Li-Mg-N-H system. *Progress in Natural Science*, *27*(1), 21–33. doi:10.1016/j.pnsc.2017.01.005

Zheng, J., Liu, X., Xu, P., Liu, P., Zhao, Y., & Yang, J. (2011). Development of high pressure gaseous hydrogen storage technologies. *International Journal of Hydrogen Energy*, *37*(1), 1048–1057. doi:10.1016/j.ijhydene.2011.02.125

Züttel, A. (2003). Materials for hydrogen storage. *Materials Today*, *6*(9), 24–33. doi:10.1016/S1369-7021(03)00922-2

Zuttel, A., Wenger, P., Rentsch, S., Mauron, P., & Emmenegger, C. (2003). LiBH4 a new hydrogen storage material. *Journal of Power Sources*, *118*(1-2), 1–7. doi:10.1016/S0378-7753(03)00054-5

Chapter 9
Hydrogen Safety Issues:
The Inherent Challenges of Hydrogen and Possible Precautions

Ayşenur Öztürk Aydın
Atatürk University, Turkey

Ayşe Bayrakçeken Yurtcan
Atatürk University, Turkey

ABSTRACT

Hydrogen energy is a promising source to meet the global energy demand. The adoption of hydrogen energy by society is possible with the installation and operation of reliable hydrogen infrastructures. Hydrogen has some risk factors inherently. The wide flammability range in air, high flame speed, and low minimum ignition energy cause fire and explosion accidents. In addition, hydrogen embrittlement in metals leads to ruptures in equipment and the formation of leakages. Hydrogen refueling stations, commercial hydrogen facilities, nuclear power plants, and laboratories are places that are open to accidents caused by hydrogen equipment failure or wrong human interventions. Determination of standards for safe operation and making quantitative risk analysis assessments help eliminate accidents or mitigate their damaging consequences. This chapter aims to raise awareness about the safety of hydrogen energy by addressing the risks of hydrogen, past accidents, the lessons learned, and the applicable precautions and recommendations against hazards.

DOI: 10.4018/978-1-6684-6721-3.ch009

INTRODUCTION

The orientation to solar, hydro, wind and geothermal renewable energy sources increases with the start of depletion of fossil fuels and rising energy needs. However, they have limited use because of intermittent energy support and there are no liquid or gas fuel forms for transportation. Hydrogen has become attractive to the energy sector in recent years as a resource that provides clean energy production and is an abundant element. Hydrogen (141.9 kJ/g) has almost three times the heating value of natural gas (54 kJ/g), and since it is high flammability in the air, the energy required for the start of combustion in vehicles is lower than other fuels. Hydrogen is nitrogen and sulfur oxides-free, zero-emission fuel for the transportation sector because of the emergence of water as a byproduct of the hydrogen combustion reaction (Okolie et al., 2021). Some features of hydrogen were summarized in Table 1.

Table 1. The properties of hydrogen

Hydrogen Properties	
Lower heating value (LHV, MJ/kg)	120
Higher heating value (HHV, MJ/kg)	142
Density at 273 K (kg/m^3)	0.09
Boiling point at atmospheric pressure (K)	20.3
Liquid density (kg/m^3)	70.8
Flammability concentration limits in air (vol. %)	4-75
Diffusion coefficient in air (cm^2/s)	0.61

Reproduced from Rivard et al. (2019) with permission from the MDPI

Worldwide, the produced 60 million tons of hydrogen is for industrial applications. Ammonia production (53%) is the first area where hydrogen consumption is the most, followed by the oil industry and methanol synthesis (40%). The remaining small share belongs to the production of polymers and resins (7%). The hydrogen economy is growing, and it promises hope for the future as an energy source in electricity production and transportation, apart from these industries (Brandon & Kurban, 2017). Hydrogen energy systems designed for vehicles have now gained their marketplace. These vehicles will be expected to reach sales rates of 3% by 2030 and 36% by 2050 globally. With the battery system working in parallel with the polymer electrolyte membrane (PEM) fuel cell, Toyota launched the fuel cell vehicle "Mirai" which provides a maximum power density of 144 kW and supplies the power demands of the vehicle during acceleration (Yue et al., 2021). Fuel cells are energy conversion tools that directly convert the chemical energy of hydrogen used as fuel into electricity and are one of the leading products of hydrogen technology. Fuel cell vehicles working with hydrogen energy have been used not only in light-duty but also in heavy-duty vehicles, military defense vehicles, railways, forklifts, maritime transportation, submarines, and aviation. Figure 1 is a schematic picture of the applications of PEM fuel cells. The widespread use of fuel cells in vehicles will be possible by reducing costs, improving catalyst performance and durability, establishing a hydrogen support infrastructure, and safely storing and transferring hydrogen. Hydrogen energy is also preferred to meet the electricity and heat of homes and commercial buildings. Energy efficiency can be increased to the 70-90% range with the utilization of waste heat. Combined Heat and Power Units (CHPs) produce electricity and heat from a single source; the consumers use the electricity for energy needs while the released heat provides the space heating and supplies hot water (Öztürk, Akay, Erkan, & Bayrakçeken Yurtcan, 2020). Hydrogen energy is favorable for mobile applications, but they pose more complex requirements for hydrogen storage than others. For these applications, the gravimetric and volumetric storage capacities are both significant (Abe, Popoola, Ajenifuja, & Popoola, 2019).

Figure 1. The applications of PEM fuel cells
Reproduced from Kandidayeni et al. (2020) with permission from the MDPI

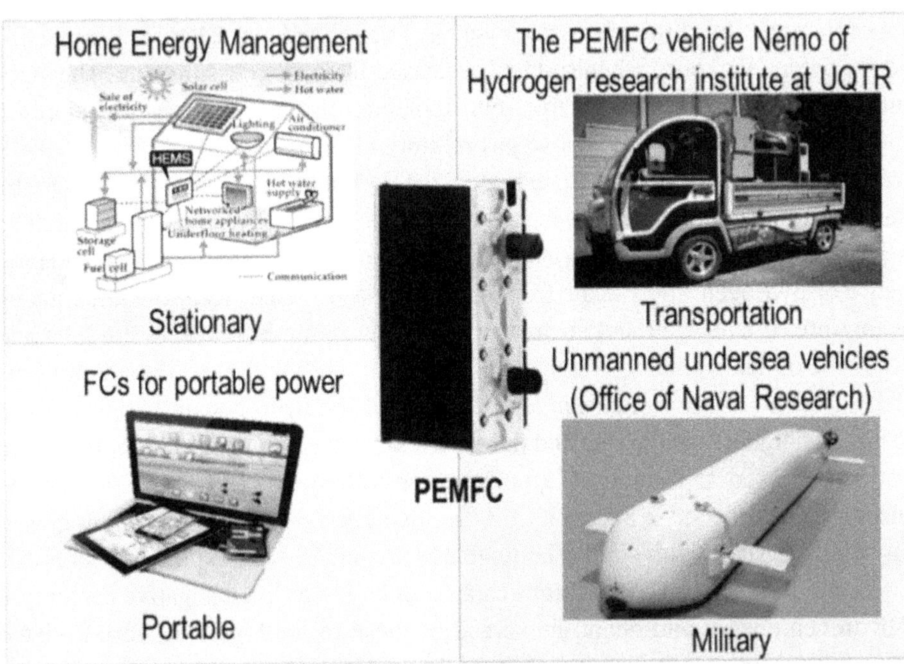

Even though hydrogen has applications in many sectors and is the rising star in energy applications with favorable properties, some hazards and safety issues have to be solved. Storing hydrogen in gaseous or liquid form is challenging because the former requires high pressure (P≥200 bar), while the latter requires very low temperatures (T≥-253 °C). Storing a sufficient amount of fuel must be necessary to cover a reasonable distance in hydrogen vehicles. The hydrogen is pressurized in the range of 350-700 bar in the storage tanks. The pressure does not exceed 700 bar because a further increase reduces the compressibility of the hydrogen. Hydrogen gas stored at 70 bar and 300 K will release maximum mechanical energy of 0.55 kWh/kg if it expands adiabatically to atmospheric pressure. Increasing the vessel pressure to 1000 bar (1400% increase) leads to an insignificant change (10%) in the maximum mechanical energy release (Petitpas & Aceves, 2013). Liquid hydrogen requires more careful work because evaporating liquid hydrogen creates a denser cloud of hydrogen gas, unlike gaseous hydrogen, and is more difficult to disperse into the atmosphere. Since the expansion ratio of liquid hydrogen to gaseous hydrogen is about 848, if a tank filled with liquid hydrogen evaporates completely, the tank's internal pressure will increase significantly (Aziz, 2021). Hydrogen, which carries high risks such as fire and explosion, also easily leaks from tanks and pipelines. Hydrogen makes metals brittle and creates ruptures in pieces of equipment, causing leaks. Since the leaked hydrogen diffuses rapidly, it can cause respiratory distress in a closed environment. Since hydrogen is highly flammable, it has the risk of explosion when mixed with air above a threshold concentration (lower flammability limit (LFL): 3.9 vol. %, @20 °C) (Jeon, Choi, & Kim, 2019) and the potential to burn when suddenly released from the pressure tank. The auto-ignition temperature refers to the minimum temperature required for the auto-ignition of a given quantity of an explosive gas mixture (Kong, Eckhoff, & Alfert, 1995). Auto-ignition at the air-contact surface of the combustible gas mixture occurs by maintaining the temperature rise in the environment for a sufficiently long time. The rise in temperature occurs when a strong shock wave, resulting from the accidental release of hydrogen into the air from a high-pressure tank, heats the air, causing the mixture of hydrogen and air to spontaneously ignite both inside the tube and at the tube outlet (Yamada, Watanabe, Hayashi, & Tsuboi, 2009). Table 2 presents the LFL values and auto-ignition temperatures of the hydrogen and other fuels in the air. The higher burning rate and the colorless flame of hydrogen complicate fire control in a closed environment (Flynn, Ricci, & Bellaby, 2012). The safe storage and transport of hydrogen are necessary for the adaptation of hydrogen technologies to daily life. In systems with highly flammable hydrogen, fire and explosion events that may occur under insecure conditions causes society to create a negative perception of hydrogen energy and delay the spread of these technologies. Material-related issues and hydrogen-handling issues cause the accidents in hydrogen-based systems.

That hydrogen makes metals brittle, the liner bubbles, the risk of fire in composite overwrapped pressure vessels (COPVs), and the degradation of carbon fibers bring along material-related safety problems. Conditions such as the compression process of the hydrogen, the sudden change in temperature during the emptying of the tank, the fluctuation in pressure, the leakage of hydrogen (late detection of the leak as colorless and odorless hydrogen), and contamination are the issues that need to be considered in the processes related to hydrogen (Moradi & Groth, 2019).

Table 2. LFL values and auto-ignition temperatures of some flammable fuels in the air

Flammable Gas	LFL (vol. %, @~20 °C)	Auto-Ignition Temperature (°C)
Hydrogen	3.90	585
Methane	4.95	640
Propane	2.05	500
Diesel	0.60	320
Gasoline	1.30	460
Ammonia	15.50	650
Carbon monoxide	12.30	609

Source: Ishaq et al. (2022), Jeon et al. (2019), Kobayashi et al. (2019), Hankin and Shah (2017), Kondo et al. (2011), Astbury and Hawksworth (2007), Kong et al. (1995)

Establishing safe conditions for the widespread use and operation of hydrogen energy is possible by thoroughly analyzing the risks and consequences arising from the nature of hydrogen. Quantitative Risk Assessment (QRA) is quite necessary to reveal safety codes and standards, and performing these analyzes requires the availability of reliable data for specific conditions. The names of the four platforms that allow the acquisition of safety data for hydrogen systems are as follows (West, Al-Douri, Hartmann, Buttner, & Groth, 2022):

- H_2Tools Lessons Learned
- Hydrogen Incidents and Accidents Database (HIAD)
- National Renewable Energy Lab's (NREL) Composite Data Products (CDPs)
- Center for Hydrogen Safety (CHS)

Risk assessment analyzes, which change depending on application and jurisdiction, are performed to maintain a reliable operation in hydrogen energy-based systems. Risk assessment analyzes can be categorized into three general approaches (Hansen, 2020):

- **Safety Distances:** This approach considers the operation of hydrogen refueling stations (HRS) or some standardized installations at safe distances tabulated by certain jurisdictions.
- **Credible Approach:** The safety distance in this approach is determined by evaluating the maximum credible event or dimensioning accident scenarios. Local conditions affect the risk analysis assessment. The weaknesses of this approach are the different interpretations of the word "credible" and ignoring the low-frequency high-consequence scenarios.
- **Probabilistic Study:** This approach is a comprehensive quantitative analysis method in which risk experts or authorities evaluate all possible accident scenarios together with the frequency/consequence relationship. In this method, where distances to fatality frequencies, barrier failure frequencies, and costs are estimated, uncertainties are eliminated or mitigated including low-frequency high-consequence events.

In this book chapter, the authors aim to examine the risks of hydrogen, the causes of hydrogen-based accidents, and the precautions to prevent these accidents. Risk analysis evaluations for the safe operation of hydrogen processes are another heading in the text.

CASE STUDY

A Historical View of Hydrogen Accidents

Although hydrogen is a clean energy source, it has the potential to harm the environment if not handled carefully. Failure to take adequate precautions and not assessing the existing risks in detail causes occurring catastrophic accidents. The hydrogen-based accidents are principally due to the high flammability and rapid diffusivity of hydrogen and the release of a large amount of energy in burning or exploding cases (Mohammadfam & Zarei, 2015). HRSs, transportation and storage processes, and plants operating with hydrogen and nuclear reactors are private areas where different risk assessments are present for the projection of accidents. Taking precautions will, of course, alleviate or eliminate the occurrence of many situations. However, estimating accident risks can be challenging as thermal and hydraulic conditions, working compartment conditions, many physical events, and environmental factors change the way hydrogen behaves. The accident type and the interventions to repair the damages also complicate understanding of the nature of the accident (Liu, Sun, Ding, & Bian, 2021). There have also been accidents with hydrogen in history, which had severe consequences.

In May 1937, the German-origin passenger zeppelin Hindenburg, which contained 7 million cubic feet of hydrogen gas, ignited within seconds as a result of leakage of hydrogen gas for a probable reason and crashed to the ground at Naval Air Station in Lakehurst. Hydrogen gas leaking from the electrostatic charge caught fire, and the world's biggest zeppelin (803.8 ft) caught fire in just 35 seconds. This accident resulted in the death of 36 crew and passengers in total (DiLisi, 2017). In water-cooled nuclear reactors, reasons such as malfunctions in the cooling system, equipment breakthroughs, and mismanagement of operators cause the reactor core to overheat and even melt, resulting in the release of a gas mixture containing abundant hydrogen. The pressure that occurs due the formation of sufficient conditions for the hydrogen to mix with the air or to burn exceeds the pressure limit of the protective equipment, resulting in an explosion. As a result of the deterioration of the reactor integrity, high-temperature mixed gases and radioactive substances are emitted into the environment (F. Liu et al., 2021). The Fukushima Daiichi Nuclear Power Plant accident occurred in March 2011. The severe earthquake that occurred in Japan and the following gigantic tsunami damaged the electrical system of the nuclear reactor, and the power of the system that cools the nuclear core decreased due to this damage. This failure caused the accumulation of hydrogen in the reactor roof and subsequently the explosion of reactors 1 and 3. The area of radioactive pollution released into the environment has spread to the northwest region of Fukushima Prefecture, about 60 km, with the effect of snowfall (Hirose, 2012).

H_2 tools is a beneficial public platform funded by The U.S. Department of Energy (DOE) that presents possible causes of hydrogen-related accidents in the past, contributing factors to the accident, how the accident resulted in damage, and lessons learned from the accident. Table 3 presents the selected cases among 221 accidents reported on the platform.

Table 3. The hydrogen accidents

Year	Setting	Accident	Probable Cause	Contributing Factor	Equipment	Injury	Case/Lesson
2019	Government Facility		-Material Incompatibility -Inadequate Inspection	-Fatigue	-Compressor		Pressure Sensor Diaphragm Rupture on H_2 Compressor/Verify H_2 compatibility of components and their specifications
2018	Commercial Facility		-Flammable mixture	-Abnormal Operations -Failure to Follow Procedure	-Hydrogen Production		Unauthorized Field Modified Equipment, Drain Lines Cause Control Cabinet Explosion/Separate hydrogen drain trap lines with other drain lines
2012	Fueling Station		-Equipment Failure -Material Incompatibility	-Deficiency in Procedures -Communication	-Burst Disk -Hydrogen tubes -Vent line		Pressure Relief Device Fails at Fueling Station/Provide adequate quality control procedures during the design
2010	Refinery		-Equipment Failure -Hazard Not Identified in Advance -Incorrect Engineering Hazard Analysis Calculation	-Design Flaw -Equipment Failure -Inadequate System Monitoring/ Oversight	-Piping -Heat Exchanger		Rupture of Refinery Heat Exchanger/Process hazards analysis (PHA) and damage mechanism hazard reviews (DMHRs) need to carefully consider all assumptions
2008	Commercial Facility		-Equipment Failure	-Inadequate Inspection	-Piping		Hydrogen Fire at an Aircraft Parts Manufacturing Facility/Replace the existing copper and carbon steel hydrogen pipeline with ¾-inch schedule 40 stainless steel.
2005	Laboratory		-Human Error	-Design Flaw -Incorrect Protocol -Training Issue	-Valve -Electrical Equipment		Hydrogen Lab Fire/Lock the valves when working in gas manifolds. Maintain a clear path to the exit in case of an emergency.
1995	Municipal Refuse Incineration Facility		-Deficiency in Procedures -Human Error	Flammable Mixture in Confined Area Training Issue	-Chisel -Refuse Incinerator		Hydrogen Gas Explosion In Municipal Refuse Incineration Facility/Aluminum should have been separated from the refuse.
1984	Chemical Plant		-Equipment Failure -Hazard Not Identified in Advance -Inadequate Equipment	-Flammable Mixture in Confined Area -Lack of Hydrogen Detection Equipment	-Compressor -Gasket -Flange		Hydrogen Explosion In A Styrene Plant/Process hazard analysis was introduced.

continues on following page

Table 3. Continued

1974	Chemical Plant		-Equipment Failure -Human Error -Improper Fire Fighting Technique	-Abnormal Operations -Decision Making -Individual Action	-Burst Disk -Vessel -Vent line		Liquid Hydrogen Tank Boiling Liquid Expanding Vapor Explosion (Bleve) Due To Water-Plugged Vent Stack/An additional secondary backup vent stack was added to liquid hydrogen tanks.
1969	Hydrogen Storage/Use Facility		-Improper Purging Procedure -Failure to Follow Standard Operating Procedures	-Decision Making -Human Error -Inadequate System Monitoring/ Oversight -Situational Awareness	-Piping -Valve		Hydrogen Tube Trailer Explosion/Personnel must be motivated to adhere to established policies and procedures.

leakage	ignition	fire	explosion	injury	death	lost time injury	property damage	facility closure

Source: H2Tools.org (2022)

CHALLENGES OF HYDROGEN

Hydrogen-Related Problems

Accident possibilities of production, storage, transportation, and operation of hydrogen harden safe utilization of hydrogen energy. Of the 120 hydrogen accidents recorded between 1999-2019 mostly occurred at laboratories (38.3%), hydrogen refueling stations (10.6%), and hydrogen-related commercial facilities (9.0%). Equipment failures and human-induced errors have a larger share in laboratory accidents. Equipment breakdown is generally due to pipe ruptures and connection/valve failures. Damage to the valves mounted on the cylinder outlets or the connections that secure the hydrogen cylinder is the most common failure (Yang et al., 2021). Brief explanations of some hydrogen-related problems are as the following in the below sub-headings:

- **Hydrogen Embrittlement:** Hydrogen embrittlement, induced by the electrochemical reaction and high-pressure gaseous hydrogen environment, results from the loss of mechanical properties following the crack growth and onset fracture in high-strength materials. Austenitic stainless steel-based equipment is more suitable against hydrogen embrittlement in hydrogen processes. Mechanisms such as hydrogen-enhanced decohesion mechanism

(HEDE), adsorption-induced dislocation emission (AIDE), and hydrogen-enhanced localized plasticity (HELP) are known phenomena of hydrogen embrittlement. The hydrogen atoms interact with the material during cathodic charge, electrolytic coating, and welding processes initiating the formation of cracks that lead to fatigue with the microcracks created along the slip bands and cause brittleness. The parameters that affect the mechanical properties of the steel in the presence of hydrogen are divided into four groups (Dwivedi & Vishwakarma, 2018), these are:

- ◦ Chemical composition
- ◦ Hydrogen charging conditions
- ◦ Microstructure of steel
- ◦ Testing conditions

- **Hydrogen Diffusion:** The diffusion rate of hydrogen in air is four times that of methane and twelve times that of gasoline, and its diffusion coefficient in the air is 0.61 cm^2/s at normal temperature and pressure. Although the flame velocity of hydrogen in the air is high, the concentration of hydrogen decreases to a safe level rapidly due to its high diffusion rate and the effect of buoyancy (Najjar, 2013). Hydrogen causes embrittlement in metals, usually by triggering crack formation at the grain boundaries and propagating into the grain. Austenite stainless steel is a popular, high-resistant material against hydrogen embrittlement. It has a single phase of austenite, which causes hydrogen to exhibit low diffusion and high solubility. Kim et al. (Kim, Kang, & Kim, 2021) stated that the carbon included in the structure at the rate of 0.1-0.3% by mass improves the stability of austenite and limits the propagation of cracks along the grain.
- **Hydrogen Leakage:** Hydrogen leakage, which is a common reason for the occurrence of hydrogen accidents, starts with the formation of a hydrogen jet firstly as a result of high-pressure hydrogen leaking out. The hydrogen jet occurs in various forms changing with the pressure ratio of the gas source and the surrounding atmosphere (Yang et al., 2021):
  - ◦ Subsonic jet (gas flow fully expands)
  - ◦ Critical-state jet (gas flow reaches local sonic velocity)
  - ◦ Under-expanded jet (gas flow expands and accelerates more when released into the atmosphere)

The considerable low temperature of liquid hydrogen and the evaporation process combined with the different leakage conditions can result in unexpected hazardous areas. In the modeling study of Sun et al. (Sun, 2022) for HRS, a small amount of liquid hydrogen condensed around the hole after it leaked, and a large amount evaporated quickly and formed massive hydrogen clouds. A cloud of flammable hydrogen vapor can spread rapidly, traveling a distance of about 3.5 m downwind, and upsoar to much higher areas under the influence of buoyancy. No leakage in hydrogen storage tanks is crucial for safe operation in nuclear power plants. Leakage factors such as rupture size, pressure, and gas storage tank location, as well as environmental factors such as wind speed and direction, atmospheric stability, and ground roughness, affect the extent of damage from any leak condition. Positioning the nuclear reactor at a safe distance from the location of hydrogen storage tanks is a necessary precaution in the layout of nuclear power plants. Gao et al. (Gao, Wang, Peng, Zhang, & Chen, 2022) stated that the minimum separation distance is 338 m for the safe operation of the nuclear reactor under harsh conditions with high wind speed, and 30% of this distance is due to the transmission of shock waves created by the explosion.

- **Hydrogen Self-Ignition:** Because hydrogen requires nominal ignition energy (0.017 mJ), pressurized hydrogen released from a vessel or pipeline can ignite spontaneously and cause fire or explosion accidents without an ignition source. The mechanism of the self-ignition phenomenon is the shock wave produced as a result of the sudden leakage of high-pressure hydrogen into the air first. The oxidizer is heated up between the shock wave and the hydrogen expansion jet. Then, the heat flow from the oxidizer causes an increase in the temperature of the combustible mixture formed by diffusion, resulting in self-ignition (Duan, Zeng, Jin, Wang, & Sun, 2022). The short length of the pipelines and the low release pressure of hydrogen reduce the possibility of self-ignition (Pan et al., 2019). Listed below are the names of some accepted mechanisms that trigger the self-ignition of hydrogen (Astbury & Hawksworth, 2007):
    - Reverse Joule-Thomson effect
    - Electrostatic ignition
    - Diffusion ignition
    - Sudden adiabatic compression
    - Hot surface ignition
- **Hydrogen Explosion:** Leakage of hydrogen, which is inherently flammable and explosive in a closed environment, causes serious accidents. The rapid release of hydrogen into the environment creates a large explosive cloud, and its ignition leads to a turbulent flare at a very high speed, creating excessive pressure. In such hydrogen-air mixtures, detonation can occur with easy way. When hydrogen release is slow, there is enough time for the hydrogen to diffuse rapidly, and a relatively harmless explosive hydrogen/air mixture forms. The pressure bears of storage tanks should be designed to cover the constant volume explosion pressure of the stored hydrogen. The tanks, which

have larger length/diameter ratio, may prone to high-speed deflagration up to flame acceleration and explosion. The rapid release of hydrogen and the dangers that may occur in pressurized limited-volume storage tanks must be avoided as much as possible (Ng & Lee, 2008).

In 2011, the Fukushima Daiichi nuclear reactor accident began with a hydrogen leak, progressing with a failure to supply coolant, failure of the relief valve, and human error with the devastating result that led to a massive nuclear explosion. The partial oxidation of the cladding with liquid water and steam produced 890 kg of hydrogen which was included directly in the reactor contents. The release of 8-10% hydrogen and noble gases into the environment was enough to cause a nuclear disaster. Since the risk of hydrogen-induced explosions is high, purveying qualified personnel for the operation of nuclear reactors, designing the control room considering user errors, having an emergency plan, and establishing a reliable monitoring and management system are vital (Gharari, Kazeminejad, Kojouri, & Hedayat, 2018).

Hydrogen Storage

Storage is essential for the widespread use of hydrogen in fuel cell-powered vehicles and other areas. Although the requirements for different situations vary, the expected features of an ideal storage method are high gravimetric/volumetric energy density, operation in room conditions, fast filling and emptying of fuel, safe operation, and low cost. Especially for mobile applications, the gravimetric/volumetric energy density should be high and the size and weight as low as possible. The hydrogen storage methods differ according to the hydrogen forms or used chemicals: compressed gas, liquefied hydrogen, cold/cryo compression, metal-organic framework (MOFs), metal hydrides, metal borohydrides, carbon nanostructures, Kubas type hydrogen, liquid organic hydrogen carriers (LOHCs), chemical hydrogen. Each storage method has its energy density, kinetics, and efficiency values. Complex thermal management systems, boil-off, poor efficiency and stability, expensive catalysts, slow response rates, high operating pressures, low energy densities, and emerging violent and uncontrolled spontaneous reactions are the main problems of current hydrogen storage methods (Rivard et al., 2019).

The hydrogen liquefaction requires going down to pretty low temperatures, and extra insulation is necessary to prevent the hydrogen boiling at -253 °C from being lost by evaporation. The liquefaction and cooling processes of hydrogen are also energy-requiring processes. In methods where hydrogen is physically stored, the design of high-pressure containers, the use of extra energy in the compression process, and the possibility of gas/liquid hydrogen leakage from the vessel lead to risky situations. Ignition or explosion of high-pressure hydrogen tanks can result in tragic accidents. Since the physical storage of hydrogen in liquid or gaseous form involves some serious risks, there is grand trend towards material-based technologies in which chemical or physical adsorption processes happen for the hydrogen storage. The risks of leakage or explosion mitigate with the adsorbed hydrogen. However, although metal hydrides are a safe form of storage so that hydrogen does not leak out, large amounts of metal powder are required to replace systems that use compressed hydrogen gas with metal hydrides (Dewangan, Mohan, Kumar, Sharma, & Ahn, 2022).

In recent years, underground hydrogen storage systems have been proposed as an economical way to store large amounts of gaseous hydrogen. Salt caverns, aquifers, and depleted natural gas/oil reservoirs are the largest hydrogen storage areas and are considered more reliable than above-ground systems as they prevent direct contact of hydrogen with atmospheric oxygen. The mechanical properties and chemical stability of the salt and the geological features of the place make salt caverns suitable for hydrogen storage. The high salt concentration in the environment also prevents the consumption of hydrogen by microscopic living things. The underground water layer, whose pores are filled with fresh or saline water, can be considered a hydrogen storage area if it has a good reservoir structure and an impermeable layer that will not leak gaseous stored hydrogen. As a result of biochemical reactions or the reaction of hydrogen with minerals in rocks, it is powerful to store hydrogen in its pure form in aquifers. The tight caprock that maintains its integrity in depleted natural gas/oil reservoirs, the well-established geological features, and the established necessary infrastructure above and below the surface make these sites a favorite for hydrogen storage. In addition, the residual gas in these reservoirs acts as a cushion gas against the risks of hydrogen (Zivar, Kumar, & Foroozesh, 2021).

Hydrogen Transport

Establishing a suitable transport system for the use of hydrogen in different industries is an important issue. Since hydrogen exists in various states according to temperature and pressure, knowing the properties of hydrogen is essential to protect against explosion and leakage risks. Compressed gas cylinders, cryogenic liquid tankers, pipelines, and natural gas blending are known techniques for transporting hydrogen. Moderate amounts of hydrogen compressed at pressures ranging from 200-500 bar are transferred by trucks in gas cylinders or tubes. For the large amounts of hydrogen transfer, vehicles called compressed gas hydrogen tube trailers, which contain a protective frame for safety, are used. Transporting hydrogen without liquefaction with tube trailers is more economically viable at $2.86 per kg H_2 delivered to a small-scale power plant (Faye, Szpunar, & Eduok, 2022).

High-pressure composite cylinders made using high molecular weight polymers or aluminum liners to prevent hydrogen permeability are also used for hydrogen storage. The pressure load resistance of the tubes is increased by over-wrapping the filament-wound carbon epoxy composite structure on the liner (Hu, Chen, Sundararaman, Chandrashekhara, & Chernicoff, 2008). Composite hydrogen transport cylinders carry higher risks than conventional methods, as they bring unknown/inexperienced phenomena. The pressure-volume product of the cylinders is a parameter that should be considered to increase reliability (Becker & Mair, 2017). The flammability of polymers leads to the failure of high-pressure composite tubes in the event of a fire. The hardness and strength of the composite, which changes depending on the temperature, cause serious accidents when high pressure occurs (34.5-70 MPa). For safety working with hydrogen, how well the high-pressure composite hydrogen storage cylinders maintain their structural integrity under fire exposure has to be evaluated (Hu et al., 2008). The Joule Thomson effect and the heat released by gas compression resulting from the rapid filling of hydrogen cause a serious temperature increase in the cylinder; the mechanical strength of the carbon fiber/epoxy composite is seriously destroyed due to the continuous high pressure and temperature cycling (L. Wang, Zheng, Wei, Wang, & Wei, 2015).

Although the transfer of hydrogen by trucks is current practice, it is more economically feasible to lay pipelines when considering long-term use. Designing pipelines with suitable materials and ensuring normal operating conditions are vital for sustainable and safe hydrogen energy use. In risk analysis assessments, pipeline disruption scenarios and possible consequences are evaluated regarding the effect of factors such as air temperature, heat flow, and thermal damages on pipeline disruptions (Medeiros, Alencar, & de Almeida, 2016). The implementation of a new technology may be possible with some adaptations applied to the existing system. For example, it is economically advantageous to use existing natural gas lines for the initial phases where the need for hydrogen is relatively low (Guler, Gecici, & Erdogan, 2021). Injecting some hydrogen into the natural gas line may cause changes in chemical and physical properties, change in material strength, and brings up some safety issues. The risks posed by hydrogen vary according to the volume percentage injected into the natural gas grid. Hydrogen makes metals brittle and diffuses rapidly, increasing the likelihood of leaks in pipelines. The escaping hydrogen tends to show deflagration to detonation transition when the atmospheric conditions are sufficient for the explosion. 28.65 g of trinitrotoluene (TNT) releases the same amount of energy as 1 g of hydrogen, so strict precautions are vital to avoid hydrogen-induced explosions (Messaoudani, Rigas, Hamid, & Hassan, 2016).

Hydrogen Refueling Stations

With the developing hydrogen economy, HRSs need to be installed to provide hydrogen support to fuel cell-powered vehicles. Figure 2 shows the principal components in an HRS configuration schematically. Compressor, distributor, high-pressure storage vessel, water chiller, heat exchanger, etc. equipment such as HRSs are components of HRSs, and the whole system must be well designed and operated in terms of performance and cost. HRSs can be classified into different categories such as hydrogen production location, hydrogen storage status, construction form, and construction content (Tian, Lv, Zhou, Zhang, & He, 2022). There are classifications named with different colors according to the production way of hydrogen in HRSs. These are as follows: Production of hydrogen with renewable energy sources (RESs)-based energy, storage, and transfer to HRS (blue arrow), on-site production of hydrogen at HRS with RESs and transfer to customers (green arrow), production of hydrogen by residual capacity and transfer to customers (red arrow), on-site production of hydrogen at HRS by steam methane reforming and transfer to customers (gray arrow) (Kuvvetli, 2020).

Figure 2. Basic components of HRS
Source: Reproduced from Rei (2020) with permission from the Scientific Research

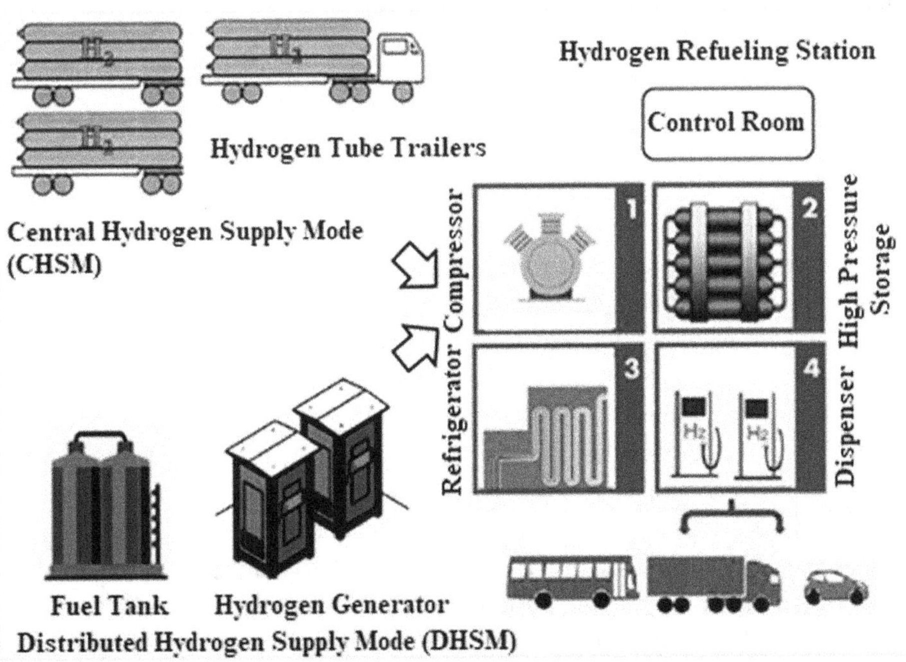

Against the risks inherent in hydrogen, it is necessary to be conscious of possible accidents at HRSs and take strict precautions. In a recent example, the HRS accident in Sandvika, Norway, on 10 June 2019, resulted from an incorrectly inserted plug causing a gas leakage following an explosion. Hydrogen, which has compelling properties such as wide flammability limits (4-75%), low minimum ignition energy, and high laminar burning speed, causes combustion and explosion events in such installations. Metallic components such as storage containers, pipes, and valves lose their strength due to hydrogen embrittlement, and hydrogen leakage occurs through cracks (Ade, Wilhite, & Goyette, 2020). The risks that arise in situations where there is an undesired release of hydrogen are affected by the volume and initial pressure of the hydrogen storage tank, the diameter and location of the hole where the hydrogen is released, the ambient wind effect, and the ventilation situation (Qian, Li, Gao, & Jin, 2020). Liang et al. (Liang, Pan, Zhang, Xie, & Liu, 2019) conducted a simulation study about the leakages and explosions in HRS. They found that the 90 MPa tube causes the highest destruction among the hydrogen storage tanks with 20, 45, and 90 MPa pressures; they also found that in the event of an explosion, if the leak and wind direction are the same, the furthest damaging distance is 35.7 m and the furthest lethal distance is 18.8 m.

SOLUTIONS AND RECOMMENDATIONS

Standards and Codes

DOE predicts that by 2040, cars and light-duty trucks will need 150 megatons of hydrogen per year to power them with fuel cells. It is necessary to establish safe infrastructures for the hydrogen economy at the social level and to operate them within the framework of decisive policies and regulations to supply this demand. Standards regulators, such as government institutions or relatively independent agencies, plan and regulate certain aspects like designing and operating services and setting working conditions to ensure reliable operations in various industries. The standards and regulations set for the safe use of hydrogen energy by untrained people also include scientific approaches, community expectations and experiences gained from past accidents (Apak, Atay, & Tuncer, 2012). Hydrogen regulations, codes, and standards frame the topics: physics of high-pressure hydrogen emissions, quantitative risk assessment, hydrogen compatibility of materials, and hydrogen fuel quality. Created by Sandia National Laboratories (SNL), Hydrogen Risk Assessment Models (HyRAM) is a QRA platform that performs risk assessments for hydrogen infrastructure. This platform provides design decisions, justifies changes in the existing code rules, and uses scientific developments in code revisions (Marchi et al., 2017).

The codes, such as American Society of Mechanical Engineers (ASME), National Fire Protection Association (NFPA2), and Society of Automotive Engineers (SAE), set the rules for hydrogen storage operations. The ASME VIII Boiler and Pressure Vessel Code is related to the structural design, and the NFPA2 Code specifies the testing and operating processes in hydrogen storage systems. The production of composite cylinders has to comply with various standards like ISO 11119-1, ISO 11119-2, and ISO 11119-3. In addition, the ISO 11623:2015 standard is used, which contains requirements for the periodic inspection and testing of composite portable gas cylinders for compressed, liquefied, or dissolved gases under pressure (Hassan, Ramadan, Saleh, & Hissel, 2021). STORHY (Hydrogen storage systems for automotive application and network of excellence), HYSAFE (Safety of hydrogen as an energy carrier), NESSHY (Novel efficient solid storage for hydrogen), and HYTRAIN (Hydrogen storage research training network) are the leading projects implemented by the European Commission to support the on-site storage of hydrogen.

Countries such as America, Germany, France, and Russia have determined the acceptance criteria of detected failures (during construction and inspection) with different technical codes according to national regulations to ensure safety in nuclear reactor containers. The American ASME Boiler and Pressure Vessel (B&PV) code is a well-established technical code that includes a set of standards, specifications, design formulas, and criteria used in the design, manufacture, assembly, inspection, and certification of pressure vessels. The German Nuclear Safety Standards Commission (KTA) is a comprehensive quality assurance system that includes standards for the production, construction, and operation of nuclear power plants and safety criteria. Design and Construction Rules of The Mechanical Components of Pressurized Water Reactors of The Nuclear Islands (RCC-M) French-origin code concerns the safety and usability of mechanical components such as tanks, supports, containers, vessels, reactor cores, heat exchangers, pumps, valves, pipes in the reactor of nuclear power plants. Later versions of the Rules and Norms in Nuclear Industry (PNAE G) code set by Russia added experience gained in versions of the ASME B&PV code (Rodriguez-Prieto, Camacho, Mendoza, Kickhofel, & Lomonaco, 2021).

Material, design, installation, and pressure testing, welding and heat treatment, operation, maintenance, and inspection of hydrogen pipelines are put into operation and inspected under the standards determined by ASME b31.12–2019 'Hydrogen Piping and Pipelines' and Compressed Gas Association (CGA) g5.6-r2013 'Hydrogen Pipeline Systems'. ASME standards apply to pure or mixed hydrogen piping systems meeting the following issues: hydrogen content (>10% vol.), pressures (<21 MPa), temperatures (-62°C-232°C), and moisture content (<20ppm). The CGA G-5.6-2013 standard is valid for systems whose temperature falls in the range of (-40°C - 175 °C)

(H. Wang, Tong, Z., Zhou, G., Zhang, C., Zhou, H., Wang, Y., Zheng, W., 2022). To prevent performance degradation due to impurities in the hydrogen content in fuel cells, hydrogen fuel standards (ISO 14687-2 and SAE J2719) have determined the levels of 4 ppb for H_2S, 200 ppm for CO, and 100 ppb for NH_3, which are critical pollutants (Marchi et al., 2017).

Risk Assessment Analysis

QRA is an assessment method that uses science and engineering models and data to facilitate the design and control of safe infrastructure for hydrogen facilities as well as provide a framework for the development and revision of hydrogen regulations, codes, and standards. The QRA process provides the user with the information, methodology, data, and models used to conduct the analysis, these evaluations are more meaningful when they contain a large amount of information covering more than one discipline. Defining, integrating, and operating appropriate physical models in QRAs when determining hydrogen codes and standards requires the knowledge of experts working in this field. HyRAM is the best-known software toolkit with a standard methodology in QRA for assessing the security of hydrogen refueling and storage infrastructure (Groth & Hecht, 2017). QRA helps decision makers determine an acceptable risk level by evaluating actual hazards and risks that may occur, based on consequences analysis and frequency analysis, considering the possible relevant system conditions and failures (West et al., 2022). Figure 3 presents the basic steps of risk assessment for hydrogen fuel stations using hazard identification.

There are also qualitative risk assessments for hazard identification, such as hazard and operability (HAZOP) studies, Failure Mode and Impact Analysis (FMEA), or risk matrices for off-site or on-site HRSs. Qualitative analysis methods are not adequate one by self for determining safety distances between activities involving hazardous materials, risk reduction measures, and decision-making for land use planning; therefore, QRA based on results and frequency scenarios provides more meaningful information. Frequency analyses for HRSs have problems due to insufficient statistical data reported for faults. For this reason, the frequency data bank is obtained by making use of other events in the past (Suzuki et al., 2021).

An example of the application of QRAs is the determination of safe distances (to allow people to escape in case of danger) between processes associated with the mechanisms of the hydrogen vapor cloud determined by QRA in HRSs (Y. L. Liu et al., 2021). Froeling et al. (Froeling, Droge, Nane, & Van Wijk, 2021) determined the flame characteristics of a hydrogen jet, which causes a dangerous fire, employing models they created using state-of-the-art software and also taking into account geometric factors. Skob et al. (Skob, Ugryumov, & Granovskiy, 2021) developed the calculation method for the probability of damage of varying degrees of severity to the personnel exposed to a possible hydrogen explosion shock wave in a coal mine. As in these studies, it is possible to create safer conditions by combining various QRAs on many parameters for different applications.

Figure 3. Risk assessment flowchart of HRS
Source: Reproduced from Sakamoto et al. (2018) with permission from the MDPI

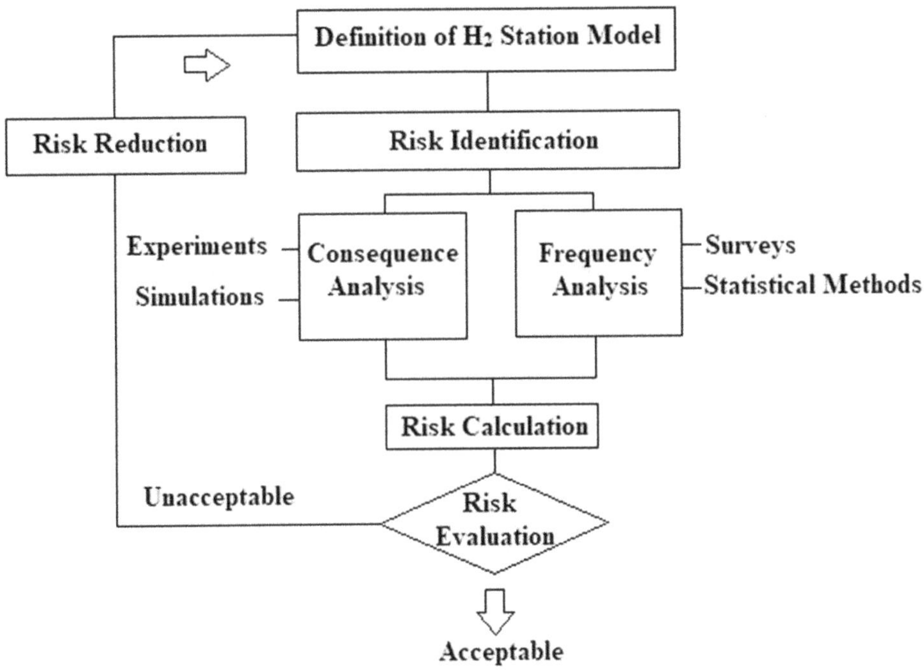

Recommendations

Life safety, property protection, and environmental protection are the main topics that have to be considered in terms of safety in closed areas working with hydrogen. Within the Hyindoor Europe Project led by the EU Fuel Cells and Hydrogen Joint Venture, the safety actions to mitigate or eliminate the dangerous consequences of hydrogen release in confined environments are listed below (Fuster et al., 2017):

- Reduce the hydrogen supply pipeline diameter and operating pressure to a workable minimum level and use flow restrictors if it is necessary
- Design the hydrogen infrastructure to destroy jet formation to prevent reaching flammable hydrogen concentration

- Identify possible ignition sources and reduce their number
- Use of fire resistant storage tanks to allow people to evacuate
- Minimize hydrogen inventory to inhibit creating flammable mixture even inside the closed enclosure after hydrogen released
- Determine hazard distances before QRA
- Ensure the use of side vents instead of roof vents to activate vertical ventilation
- Use of several vents with a height difference between them instead of a single vent in the same area and set the vents throughout the building for being independent of the wind direction
- Keep ventilation pipes and ducts out as much as possible.

Material-based measures against embrittlement caused by hydrogen in metals are listed below (Foorginezhad et al., 2021; Kovac, Paranos, & Marcius, 2021):

- Use the chromium-rich steels, chromium-molybdenum alloys, and composite reinforced plastic
- Add titanium and aluminum alloys to the base material
- Choose an amorphous structure material
- Apply zinc and nickel plating techniques
- Apply graphene and niobium coatings
- Add an oxide, nitrogen, or carbon diffusion layer to create a barrier for hydrogen penetration

According to Hadef et al. (2020), some of the recommendations determined according to the risk analysis assessments for hydrogen production facilities are listed below:

- Using sensors to detect hydrogen leakages
- Perform periodic inspections of hydrogen sensors
- Install another independent system to ensure that the emergency shutdown is partial or complete
- Locate batteries outside the building to avoid contact with hydrogen
- Isolate the operator from the EMS control system and equipment so that does not create a domino effect in the event of an accident
- Make all system changes or new equipment integration in the presence of qualified experts to avoid wrong design basis and update the risk analysis

HRSs must become widespread since hydrogen is envisioned as fuel in vehicles soon. The safe operation of HRSs facilitates the public adoption of hydrogen energy. The accidents on HRSs are due to leakages from dispensers or ruptures in tube trailers. The following precautions are necessary for a safe operation (Kikukawa, Mitsuhashi, & Miyake, 2009):

- Design earthquake resistant storage tanks
- Monitor pressure of thermally insulated vacuum layer in storage tanks
- Ensure installation of fire protection wall along station boundaries
- Lay the pipes with trenches
- Equip the storage tank, pipes, dispenser hoses and dispenser to handle cryogenic liquids to handle the cryogenic temperatures required by liquid hydrogen
- Have collision protection on the dispenser
- Install heater and orifice in secondary caps of valve and manual line valves
- Connect the relief valve to the vent line

In every process involving hydrogen, operators must be aware of the risks, know the actions to be taken against accidents and comply with safety rules. Everyone involved in the hydrogen industry needs to learn lessons from the accidents so far.

FUTURE RESEARCH DIRECTIONS

The scientific studies will continue to enhance the features of the equipment used for hydrogen production, storage, and transportation in a way that will provide more safety. The hydrogen sensitivity and stability of the sensors will improve to understand whether hydrogen is leaking or not. More robust mouth valves will be integrated into the hydrogen storage tank to make the hydrogen supply of vehicles in more reliable conditions. Temperature control units will be developed to prevent sudden temperature rise during the rapid filling of hydrogen into vehicles. The HRSs will have layout designs at the correct geometries. Various safety facilities such as flame alarm detectors, emergency radiation devices, automatic spray cooling systems, and nitrogen purge systems will be components of the HRS layouts. Mechanical components such as hydrogen filling pipelines, valves, gaskets, and storage tanks will be strengthened against hydrogen embrittlement using different alloys and composite materials (Zhang et al., 2022). The numerical simulations of leakage and diffusion in hydrogen storage tanks will be extensible in terms of a grid, turbulence model, and reaction kinetic conditions. It would be helpful to elucidate the hydrogen damage mechanisms that occur with the synergistic effect between hydrogen and common impurities such as sulfides, carbon monoxide, and carbon dioxide (Li et al., 2022). The more effective safety regulations will be valid regarding the distances and physical locations between nuclear power plants, hydrogen production facilities, and transportation systems (Clarke, 2022). According to the information obtained from hydrogen-based accidents, the experts will evaluate the possible risk factors and decide on the most suitable designs and operating conditions with different QRA methods. As hydrogen takes more place in our lives, society's awareness of security will increase, and a hydrogen energy culture will be born.

CONCLUSION

Hydrogen is accepted as the solution to the energy crisis in many areas in the future. Although hydrogen is a clean energy carrier with a high energy density, there are challenges in adapting and spreading it to daily life due to inherent dangers. Factories producing hydrogen, hydrogen filling stations, nuclear power plants, laboratories, which are hydrogen areas, should be designed and operated considering these risks. Hydrogen has high flammability when mixed with air above a limit concentration, and its flame rate is also high. Because the hydrogen flame is colorless, it is hard to discern the area from which it spreads. Since it is a very small-sized element, it easily leaks from storage tanks and pipelines and also causes brittleness by penetrating between atoms in metals and renders equipment unusable. For these reasons, people's safety should be kept at a high level in terms of infrastructure and working conditions in hydrogen plants. Hydrogen infrastructures must be manufactured based on the safety standards determined by the leading organizations of the countries, and the personnel must have information about potential threats, security measures, and emergency action plans. Entrepreneurs and operators must evaluate the potential risks detected by the quantitative risk analysis assessment software created by experts in the field and lessons learned from fatal accidents in the past. Equipment failure or human-induced factors mainly cause accidents. The equipments are improved according to the results obtained from the risk analysis assessments, and people are made more conscious for preventing the accidents. New durable composite materials are tested to increase the safety of hydrogen storage tanks. By establishing safe distances between different processes in hydrogen plants, catastrophic situations are prevented in the event of a possible accident. In addition, the operation of the processes is followed with sensitive sensors, alarms, and improved inspection and monitoring systems. If hydrogen energy is available in safe conditions, it will be advantageous regarding energy needs. The adoption of hydrogen energy by society will increase in parallel with the provision of safe hydrogen energy support. More scientific studies by many disciplines and experienced people are needed to remove the obstacles in front of hydrogen energy.

ACKNOWLEDGMENT

This research received no specific grant from any funding agency in the public, commercial, or not-for-profit sectors.

REFERENCES

Abe, J. O., Popoola, A. P. I., Ajenifuja, E., & Popoola, O. M. (2019). Hydrogen energy, economy and storage: Review and recommendation. *International Journal of Hydrogen Energy*, *44*(29), 15072–15086. doi:10.1016/j.ijhydene.2019.04.068

Ade, N., Wilhite, B., & Goyette, H. (2020). An integrated approach for safer and economical design of Hydrogen refueling stations. *International Journal of Hydrogen Energy*, *45*(56), 32713–32729. doi:10.1016/j.ijhydene.2020.08.232

Apak, S., Atay, E., & Tuncer, G. (2012). Renewable hydrogen energy regulations, codes and standards: Challenges faced by an EU candidate country. *International Journal of Hydrogen Energy*, *37*(7), 5481–5497. doi:10.1016/j.ijhydene.2012.01.005

Astbury, G. R., & Hawksworth, S. J. (2007). Spontaneous ignition of hydrogen leaks: A review of postulated mechanisms. *International Journal of Hydrogen Energy*, *32*(13), 2178–2185. doi:10.1016/j.ijhydene.2007.04.005

Aziz, M. (2021). Liquid Hydrogen: A Review on Liquefaction, Storage, Transportation, and Safety. *Energies*, *14*(18). doi:ARTN 591710.3390/en14185917

Becker, B., & Mair, G. W. (2017). Risks and safety level of composite cylinders. *International Journal of Hydrogen Energy*, *42*(19), 13810–13817. doi:10.1016/j.ijhydene.2017.01.145

Brandon, N. P., & Kurban, Z. (2017). Clean energy and the hydrogen economy. *Philosophical Transactions of the Royal Society a-Mathematical Physical and Engineering Sciences*, *375*(2098). doi:ARTN 2016040010.1098/rsta.2016.0400

Clarke, M. C. (2022). Can the hydrogen economy concept be the solution to the future energy crisis? *Australian Journal of Multi-Disciplinary Engineering*, *18*(1), 70–84. doi:10.1080/14488388.2022.2046325

Dewangan, S. K., Mohan, M., Kumar, V., Sharma, A., & Ahn, B. (2022). A comprehensive review of the prospects for future hydrogen storage in materials-application and outstanding issues. *International Journal of Energy Research*, *46*(12), 16150–16177. Advance online publication. doi:10.1002/er.8322

DiLisi, G. A. (2017). The Hindenburg Disaster: Combining Physics and History in the Laboratory. *The Physics Teacher, 55*(5), 268–273. doi:10.1119/1.4981031

Duan, Q. L., Zeng, Q., Jin, K. Q., Wang, Q. S., & Sun, J. H. (2022). Mechanism of self-ignition and flame propagation during high-pressure hydrogen release through a rectangular tube. *Process Safety and Environmental Protection, 164*, 283–290. doi:10.1016/j.psep.2022.06.017

Dwivedi, S. K., & Vishwakarma, M. (2018). Hydrogen embrittlement in different materials: A review. *International Journal of Hydrogen Energy, 43*(46), 21603–21616. doi:10.1016/j.ijhydene.2018.09.201

Faye, O., Szpunar, J., & Eduok, U. (2022). A critical review on the current technologies for the generation, storage, and transportation of hydrogen. *International Journal of Hydrogen Energy, 47*(29), 13771–13802. doi:10.1016/j.ijhydene.2022.02.112

Flynn, R., Ricci, M., & Bellaby, P. (2012). Ambiguity, complexity and uncertainty surrounding the hazards of hydrogen and public views of emergent risks. *Journal of Risk Research, 15*(4), 373–387. doi:10.1080/13669877.2011.634517

Foorginezhad, S., Mohseni-Dargah, M., Falahati, Z., Abbassi, R., Razmjou, A., & Asadnia, M. (2021). Sensing advancement towards safety assessment of hydrogen fuel cell vehicles. *Journal of Power Sources, 489*. doi:ARTN 22945010.1016/j.jpowsour.2021.229450

Froeling, H. A. J., Droge, M. T., Nane, G. F., & Van Wijk, A. J. M. (2021). Quantitative risk analysis of a hazardous jet fire event for hydrogen transport in natural gas transmission pipelines. *International Journal of Hydrogen Energy, 46*(17), 10411–10422. doi:10.1016/j.ijhydene.2020.11.248

Fuster, B., Houssin-Agbomson, D., Jallais, S., Vyazmina, E., Guy, D. N., Bernard-Michel, G., ... Kinderen, J. D. (2017). Guidelines and recommendations for indoor use of fuel cells and hydrogen systems. *International Journal of Hydrogen Energy, 42*(11), 7600–7607. doi:10.1016/j.ijhydene.2016.05.266

Gao, Q. X., Wang, L. J., Peng, W., Zhang, P., & Chen, S. Z. (2022). Safety analysis of leakage in a nuclear hydrogen production system. *International Journal of Hydrogen Energy, 47*(7), 4916–4931. doi:10.1016/j.ijhydene.2021.11.099

Gharari, R., Kazeminejad, H., Kojouri, N. M., & Hedayat, A. (2018). A review on hydrogen generation, explosion, and mitigation during severe accidents in light water nuclear reactors. *International Journal of Hydrogen Energy, 43*(4), 1939–1965. doi:10.1016/j.ijhydene.2017.11.174

Groth, K. M., & Hecht, E. S. (2017). HyRAM: A methodology and toolkit for quantitative risk assessment of hydrogen systems. *International Journal of Hydrogen Energy*, *42*(11), 7485–7493. doi:10.1016/j.ijhydene.2016.07.002

Guler, M. G., Gecici, E., & Erdogan, A. (2021). Design of a future hydrogen supply chain: A multi period model for Turkey. *International Journal of Hydrogen Energy*, *46*(30), 16279–16298. doi:10.1016/j.ijhydene.2020.09.018

H2Tools.org. (2022). *Lessons Learned*. Retrieved September 1, 2022, from https://h2tools.org/lessons?search_api_fulltext=

Hadef, H., Negrou, B., Ayuso, T. G., Djebabra, M., & Ramadan, M. (2020). Preliminary hazard identification for risk assessment on a complex system for hydrogen production. *International Journal of Hydrogen Energy*, *45*(20), 11855–11865. doi:10.1016/j.ijhydene.2019.10.162

Hankin, A., & Shah, N. (2017). Process exploration and assessment for the production of methanol and dimethyl ether from carbon dioxide and water. *Sustainable Energy & Fuels*, *1*(7), 1541–1556. doi:10.1039/C7SE00206H

Hansen, O. R. (2020). Hydrogen infrastructure-Efficient risk assessment and design optimization approach to ensure safe and practical solutions. *Process Safety and Environmental Protection*, *143*, 164–176. doi:10.1016/j.psep.2020.06.028

Hassan, I. A., Ramadan, H. S., Saleh, M. A., & Hissel, D. (2021). Hydrogen storage technologies for stationary and mobile applications: Review, analysis and perspectives. *Renewable & Sustainable Energy Reviews, 149*. doi:ARTN 11131110.1016/j.rser.2021.111311

Hirose, K. (2012). 2011 Fukushima Dai-ichi nuclear power plant accident: Summary of regional radioactive deposition monitoring results. *Journal of Environmental Radioactivity*, *111*, 13–17. doi:10.1016/j.jenvrad.2011.09.003 PMID:22119330

Hu, J., Chen, J., Sundararaman, S., Chandrashekhara, K., & Chernicoff, W. (2008). Analysis of composite hydrogen storage cylinders subjected to localized flame impingements. *International Journal of Hydrogen Energy*, *33*(11), 2738–2746. doi:10.1016/j.ijhydene.2008.03.012

Ishaq, H., Dincer, I., & Crawford, C. (2022). A review on hydrogen production and utilization: Challenges and opportunities. *International Journal of Hydrogen Energy*, *47*(62), 26238–26264. doi:10.1016/j.ijhydene.2021.11.149

Jeon, J., Choi, W., & Kim, S. J. (2019). A flammability limit model for hydrogen-air-diluent mixtures based on heat transfer characteristics in flame propagation. *Nuclear Engineering and Technology*, *51*(7), 1749–1757. doi:10.1016/j.net.2019.05.005

Kandidayeni, M., Macias, A., Boulon, L., & Trovao, J. P. F. (2020). Online Modeling of a Fuel Cell System for an Energy Management Strategy Design. *Energies, 13*(14). doi:ARTN 371310.3390/en13143713

Kikukawa, S., Mitsuhashi, H., & Miyake, A. (2009). Risk assessment for liquid hydrogen fueling stations. *International Journal of Hydrogen Energy*, *34*(2), 1135–1141. doi:10.1016/j.ijhydene.2008.10.093

Kim, K. S., Kang, J. H., & Kim, S. J. (2021). Carbon effect on hydrogen diffusivity and embrittlement in austenitic stainless steels. *Corrosion Science, 180*. doi:ARTN 10922610.1016/j.corsci.2020.109226

Kobayashi, H., Hayakawa, A., Somarathne, K. D. K. A., & Okafor, E. C. (2019). Science and technology of ammonia combustion. *Proceedings of the Combustion Institute*, *37*(1), 109–133. doi:10.1016/j.proci.2018.09.029

Kondo, S., Takizawa, K., Takahashi, A., & Tokuhashi, K. (2011). On the temperature dependence of flammability limits of gases. *Journal of Hazardous Materials*, *187*(1-3), 585–590. doi:10.1016/j.jhazmat.2011.01.037 PMID:21288637

Kong, D., Eckhoff, R. K., & Alfert, F. (1995). Auto-Ignition of Ch4 Air, C3h8 Air, Ch4 C3h8 Air and Ch4 Co2 Air Using an Ignition Bomb. *Journal of Hazardous Materials*, *40*(1), 69–84. doi:10.1016/0304-3894(94)00082-R

Kovac, A., Paranos, M., & Marcius, D. (2021). Hydrogen in energy transition: A review. *International Journal of Hydrogen Energy*, *46*(16), 10016–10035. doi:10.1016/j.ijhydene.2020.11.256

Kuvvetli, Y. (2020). Multi-objective and multi-period hydrogen refueling station location problem. *International Journal of Hydrogen Energy*, *45*(55), 30845–30858. doi:10.1016/j.ijhydene.2020.08.126

Li, H., Cao, X. W., Liu, Y., Shao, Y. B., Nan, Z. L., Teng, L., Peng, W., & Bian, J. (2022). Safety of hydrogen storage and transportation: An overview on mechanisms, techniques, and challenges. *Energy Reports*, *8*, 6258–6269. doi:10.1016/j.egyr.2022.04.067

Liang, Y., Pan, X. M., Zhang, C. M., Xie, B., & Liu, S. J. (2019). The simulation and analysis of leakage and explosion at a renewable hydrogen refuelling station. *International Journal of Hydrogen Energy*, *44*(40), 22608–22619. doi:10.1016/j.ijhydene.2019.05.140

Liu, F., Sun, Z. N., Ding, M., & Bian, H. Z. (2021). Research progress of hydrogen behaviors in nuclear power plant containment under severe accident conditions. *International Journal of Hydrogen Energy*, *46*(73), 36477–36502. doi:10.1016/j.ijhydene.2021.08.151

Liu, Y. L., Liu, Z., Wei, J. J., Lan, Y. Q., Yang, S. Y., & Jin, T. (2021). Evaluation and prediction of the safe distance in liquid hydrogen spill accident. *Process Safety and Environmental Protection*, *146*, 1–8. doi:10.1016/j.psep.2020.08.037

Marchi, C. S., Hecht, E. S., Ekoto, I. W., Groth, K. M., LaFleur, C., Somerday, B. P., ... James, C. W. (2017). Overview of the DOE hydrogen safety, codes and standards program, part 3: Advances in research and development to enhance the scientific basis for hydrogen regulations, codes and standards. *International Journal of Hydrogen Energy*, *42*(11), 7263–7274. doi:10.1016/j.ijhydene.2016.07.014

Medeiros, C. P., Alencar, M. H., & de Almeida, A. T. (2016). Hydrogen pipelines: Enhancing information visualization and statistical tests for global sensitivity analysis when evaluating multidimensional risks to support decision-making. *International Journal of Hydrogen Energy*, *41*(47), 22192–22205. doi:10.1016/j.ijhydene.2016.09.113

Messaoudani, Z. L., Rigas, F., Hamid, M. D. B., & Hassan, C. R. C. (2016). Hazards, safety and knowledge gaps on hydrogen transmission via natural gas grid: A critical review. *International Journal of Hydrogen Energy*, *41*(39), 17511–17525. doi:10.1016/j.ijhydene.2016.07.171

Mohammadfam, I., & Zarei, E. (2015). Safety risk modeling and major accidents analysis of hydrogen and natural gas releases: A comprehensive risk analysis framework. *International Journal of Hydrogen Energy*, *40*(39), 13653–13663. doi:10.1016/j.ijhydene.2015.07.117

Moradi, R., & Groth, K. M. (2019). Hydrogen storage and delivery: Review of the state of the art technologies and risk and reliability analysis. *International Journal of Hydrogen Energy*, *44*(23), 12254–12269. doi:10.1016/j.ijhydene.2019.03.041

Najjar, Y. S. H. (2013). Hydrogen safety: The road toward green technology. *International Journal of Hydrogen Energy*, *38*(25), 10716–10728. doi:10.1016/j.ijhydene.2013.05.126

Ng, H. D., & Lee, J. H. S. (2008). Comments on explosion problems for hydrogen safety. *Journal of Loss Prevention in the Process Industries, 21*(2), 136–146. doi:10.1016/j.jlp.2007.06.001

Okolie, J. A., Patra, B. R., Mukherjee, A., Nanda, S., Dalai, A. K., & Kozinski, J. A. (2021). Futuristic applications of hydrogen in energy, biorefining, aerospace, pharmaceuticals and metallurgy. *International Journal of Hydrogen Energy, 46*(13), 8885–8905. doi:10.1016/j.ijhydene.2021.01.014

Öztürk, A., Akay, R. G., Erkan, S., & Bayrakçeken Yurtcan, A. (2020). Introduction to fuel cells. In Direct Liquid Fuel Cells Fundamentals, Advances and Future (pp. 1-47). Academic Press.

Pan, X. H., Yan, W. Y., Jiang, Y. M., Wang, Z. L., Hua, M., Wang, Q. Y., & Jiang, J. C. (2019). Experimental Investigation of the Self-Ignition and Jet Flame of Hydrogen Jets Released under Different Conditions. *ACS Omega, 4*(7), 12004–12011. doi:10.1021/acsomega.9b01214 PMID:31460312

Petitpas, G., & Aceves, S. M. (2013). Modeling of sudden hydrogen expansion from cryogenic pressure vessel failure. *International Journal of Hydrogen Energy, 38*(19), 8190–8198. doi:10.1016/j.ijhydene.2012.03.166

Qian, J. Y., Li, X. J., Gao, Z. X., & Jin, Z. J. (2020). A numerical study of unintended hydrogen release in a hydrogen refueling station. *International Journal of Hydrogen Energy, 45*(38), 20142–20152. doi:10.1016/j.ijhydene.2020.05.063

Rei, M.-H., Kao, Y.-L., Yeh, G.-T., & Chen, S.-C. (2020). The Influence of the Hydrogen Supply Modes on a Hydrogen Refueling Station. *Modern Economy, 11*(4), 867–880. doi:10.4236/me.2020.114065

Rivard, E., Trudeau, M., & Zaghib, K. (2019). Hydrogen Storage for Mobility: A Review. *Materials, 12*(12). doi:ARTN 197310.3390/ma12121973

Rodriguez-Prieto, A., Camacho, A. M., Mendoza, C., Kickhofel, J., & Lomonaco, G. (2021). Evolution of Standardized Specifications on Materials, Manufacturing and In-Service Inspection of Nuclear Reactor Vessels. *Sustainability, 13*(19). doi:ARTN 1051010.3390/su131910510

Sakamoto, J., Misono, H., Nakayama, J., Kasai, N., Shibutani, T., & Miyake, A. (2018). Evaluation of Safety Measures of a Hydrogen Fueling Station Using Physical Modeling. *Sustainability, 10*(11). doi:ARTN 384610.3390/su10113846

Skob, Y. A., Ugryumov, M. L., & Granovskiy, E. A. (2021). Numerical assessment of hydrogen explosion consequences in a mine tunnel. *International Journal of Hydrogen Energy*, *46*(23), 12361–12371. doi:10.1016/j.ijhydene.2020.09.067

Sun, R., Pu, L., Yu, H., Dai, M., & Li, Y. (2022). Modeling the diffusion of flammable hydrogen cloud under different liquid hydrogen leakage conditions in a hydrogen refueling station. *International Journal of Hydrogen Energy*, *47*(61), 25849–25863. doi:10.1016/j.ijhydene.2022.05.303

Suzuki, T., Shiota, K., Izato, Y., Komori, M., Sato, K., Takai, Y., Ninomiya, T., & Miyake, A. (2021). Quantitative risk assessment using a Japanese hydrogen refueling station model. *International Journal of Hydrogen Energy*, *46*(11), 8329–8343. doi:10.1016/j.ijhydene.2020.12.035

Tian, Z., Lv, H., Zhou, W., Zhang, C. M., & He, P. F. (2022). Review on equipment configuration and operation process optimization of hydrogen refueling station. *International Journal of Hydrogen Energy*, *47*(5), 3033–3053. doi:10.1016/j.ijhydene.2021.10.238

Wang, H., Tong, Z., Zhou, G., Zhang, C., Zhou, H., Wang, Y., & Zheng, W. (2022). Research and demonstration on hydrogen compatibility of pipelines: A review of current status and challenges. *International Journal of Hydrogen Energy*, *47*(66), 28585–28604. doi:10.1016/j.ijhydene.2022.06.158

Wang, L., Zheng, C. X., Wei, S., Wang, B. C., & Wei, Z. X. (2015). Thermo-mechanical investigation of composite high-pressure hydrogen storage cylinder during fast filling. *International Journal of Hydrogen Energy*, *40*(21), 6853–6859. doi:10.1016/j.ijhydene.2015.04.018

West, M., Al-Douri, A., Hartmann, K., Buttner, W., & Groth, K. M. (2022). Critical review and analysis of hydrogen safety data collection tools. *International Journal of Hydrogen Energy*, *47*(40), 17845–17858. doi:10.1016/j.ijhydene.2022.03.244

Yamada, E., Watanabe, S., Hayashi, A. K., & Tsuboi, N. (2009). Numerical analysis on auto-ignition of a high pressure hydrogen jet spouting from a tube. *Proceedings of the Combustion Institute*, *32*(2), 2363–2369. doi:10.1016/j.proci.2008.07.001

Yang, F. Y., Wang, T. Z., Deng, X. T., Dang, J., Huang, Z. Y., Hu, S., Li, Y., & Ouyang, M. G. (2021). Review on hydrogen safety issues: Incident statistics, hydrogen diffusion, and detonation process. *International Journal of Hydrogen Energy*, *46*(61), 31467–31488. doi:10.1016/j.ijhydene.2021.07.005

Yue, M. L., Lambert, H., Pahon, E., Roche, R., Jemei, S., & Hissel, D. (2021). Hydrogen energy systems: A critical review of technologies, applications, trends and challenges. *Renewable & Sustainable Energy Reviews, 146*. doi:ARTN 11118010.1016/j.rser.2021.111180

Zhang, C. Z., Cao, X. J., Bujlo, P., Chen, B., Zhang, X., Sheng, X. F., & Liang, C. (2022). Review on the safety analysis and protection strategies of fast filling hydrogen storage system for fuel cell vehicle application. *Journal of Energy Storage, 45*. doi:ARTN 10345110.1016/j.est.2021.103451

Zivar, D., Kumar, S., & Foroozesh, J. (2021). Underground hydrogen storage: A comprehensive review. *International Journal of Hydrogen Energy*, *46*(45), 23436–23462. doi:10.1016/j.ijhydene.2020.08.138

KEY TERMS AND DEFINITIONS

Diffusivity: It is the diffusion rate of a substance, fluid, or heat from the high-density to the low-density region, which varies according to the media.

Explosion: It is a devastating event that occurs as a result of rapid expansion in the volume due to the propensity to the powerful release of accumulated energy sourced from high temperatures or high-pressure gas into the environment.

Flammability: Flammability is a parameter that indicates the ease of ignition of material and the intensity of heat release after being ignited.

Hydrogen: Hydrogen is a flammable substance with the smallest atomic number among chemical elements, releasing water when burned with oxygen.

Hydrogen Embrittlement: It means breaking the strength of the metal as a result of reduced energy required for the formation of cracks via the very small-sized hydrogen atoms that can easily penetrate the metals.

Hydrogen Refueling Station: This infrastructure provides fuel for vehicles that use hydrogen and, in some layouts, also includes hydrogen production units within the same body.

Hydrogen Storage: This process is stockpiling hydrogen for later use by techniques such as compressed gas, cryogenic liquid, or entrapment in certain chemical compounds.

Leakage: It is the amount of liquid or gas that escapes from a pipeline or container through the formation of a rupture, deterioration of mechanical parts, or easy evaporation.

Minimum Ignition Energy: This term is used for flammable gases and refers to the minimum discharge energy required for gas to ignite.

Risk Assessment Analysis: Risk assessment is to identify risk factors and potential hazards in the current operation, take appropriate measures against hazards to eliminate or mitigate the damage that may occur, and prepare action plans.

Safety: The whole of the conditions creates confidence to prevent injury, loss of life, or property damage caused by possible accident risks.

Transportation: It is the transfer of produced hydrogen to different demand points by pipelines, cryogenic liquid tank trucks, or gaseous tube trailers.

Chapter 10
Comparative Analysis Regarding the Performance Criteria of Hydrogen Fuel Cells Used in Green Electricity Generation for Mobility

Raluca Andreea Felseghi
Technical University of Cluj-Napoca, Romania

ABSTRACT

Today, hydrogen is recognized as a non-polluting energy carrier because it does not contribute to global warming if it is produced from renewable energy resources. This chapter presents one of the possibilities of using hydrogen fuel cells in the energy support of electric vehicle charging stations. A comparative study was presented for five situations in which fuel cells have different nominal powers, a study whose objective is to investigate the equipment capability, assuming their use as the first and only source of electricity in a future economy based on hydrogen, in which case the hydrogen fuel used by the fuel cell is available to the final consumer through a centralized distribution infrastructure. The main decision criteria taken into account include electricity generated by fuel cell, degree of autonomy from the classic electricity grid, total hydrogen consumption, energy purchased from the grid, excess energy, total CO_2 *emissions, initial investment, electricity unit prices, total cost calculated for 25 years of system life.*

DOI: 10.4018/978-1-6684-6721-3.ch010

BACKGROUND

This chapter comes to present one of the possibilities of using the hydrogen to electricity conversion technology in the energy support of electric vehicle charging stations (Dash et al., 2022; Mastoi et al., 2022; Ravi & Aziz, 2022), approaching a comparative study on five situations in which the fuel cells have different rated powers, a study that investigates the capability and energy efficiency of the equipment, assuming their use as the first and only source of electricity generation in a future hydrogen-based economy (Badea et al., 2021), in which case the fuel used by the fuel cell is available to the final consumer through a distribution infrastructure centralized (Ali et al., 2022), hydrogen production is also outsourced and independent of the energy generation system (Akarsu & Genç, 2022, Turoń et al., 2020).

The technical-economic and ecological study carried out aims to determine and highlight a set of general criteria that allow the characterization of these types of systems (Felseghi et al., 2021; Longden et al., 2022; Venkatasatish & Dhanamjayulu, 2022), but also to approach this energy solution of using the fuel cell in the field of green mobility from the point of view of sustainability (Farias et al., 2022; Kovač et al., 2021; Park et al., 2022), based on a multi-criteria comparative analysis carried out on the five fuel cells of different powers.

For the development and substantiation of the study, simulations were carried out in operation of energy systems with fuel cells having nominal powers of 1 kW, 2 kW, 3 kW, 4 kW, 5 kW, and the results of these simulations were presented in the form of a comparative analysis.

An electric vehicle charging station type consumer, was used as input data, with a total electricity demand of 6759 kWh/year. As an initial condition, it was established that for the configuration of the studied energy systems, the consumer is also connected to the electricity distribution network, and that excess energy storage in batteries is not foreseen, in order to create the possibility to quantify and capitalize this excess as a form of green energy that can be exported to the grid, but also to have the electricity available to ensure the start-ups for the component equipment of the systems (Aşchilean et al., 2021; Badea et al., 2015, Enescu et al., 2023).

Given the particular virtual conditions of this case, the operating principle assumes that the energy system uses the fuel cell supplied with hydrogen from the centralized network to generate electricity and supports the consumer energetically by ensuring the total energy demand through the electrochemical conversion of hydrogen (Ahmad et al., 2022; Karaca & Dincer, 2023; Jones et al., 2020; Rojas et al., 2022). This type of system was further named Energy System with Fuel Cell powered by Hydrogen from the Distribution Network (SE - FC_H).

Based on the input data of the energy requirement and the particularities of the energy systems, the computational simulations regarding the operation of these types of systems were carried out using iHOGA version 2.5 software (Dufo-López & Bernal-Agustín, 2020), and as global optimization criteria, aspects related to the minimization of emissions were initially set of carbon dioxide, uncovered energy and excess energy (Badea et al., 2019; Felseghi & Papp, 2014; Filote et al., 2020; Li & Taghizadeh-Hesary, 2022). The obtained value results are presented in detail in the following.

CONFIGURATION AND PERFORMANCE CRITERIA OF HYDROGEN FUEL CELLS USED IN GREEN ELECTRICITY GENERATION

Fuel Cell With Rated Power of 1 kW (SE - FC_H 1kW)

Table 1. Performance indicators: Annual averages - SE - FC_H 1 kW system

Performance Indicator	Value	Unit
Total energy demand	6.759,00	kWh/ year
Energy generated by the fuel cell	6776,70	kWh/ year
Energy demand not covered by the system	773,96	kWh/ year
Energy purchased from the grid	773,96	kWh/ year
Excess energy	0	kWh/ year
Energy exported to the grid	0	kWh/ year
Total hydrogen consumption	593,30	kg
Total carbon dioxide emissions	502,00	kg CO_2 / year

In a global calculation regarding the provision of electricity for the total energy demand of 6759 kWh/year, the energy system with a 1 kW fuel cell generates a percentage of 88.55%, the remaining 11.45% of electricity being purchased from the centralized national grid of electricity distribution.

The performance indicators of the 1 kW fuel cell power generation system, after one year of operation, have the annual average values shown in Table 1.

Energy production and hydrogen consumption, graphically presented in Figure 1 highlights the consumption of hydrogen within the fuel cell, which depends on its nominal power, and the real power generated in the system is influenced by the consumption and efficiency diagrams.

Taking into account the lower calorific value of hydrogen of 33.3 kWh/kg, in the case of the 1 kW fuel cell, the values of hydrogen consumption and electricity production were obtained, presented synthetically in Figure 1. Thus, during the period a year the 1 kW fuel cell consumes 593.3 kg of hydrogen to produce 6776 kWh of electricity.

Figure 1. Hydrogen consumption in the electricity generation: SE - FC_H 1kW system

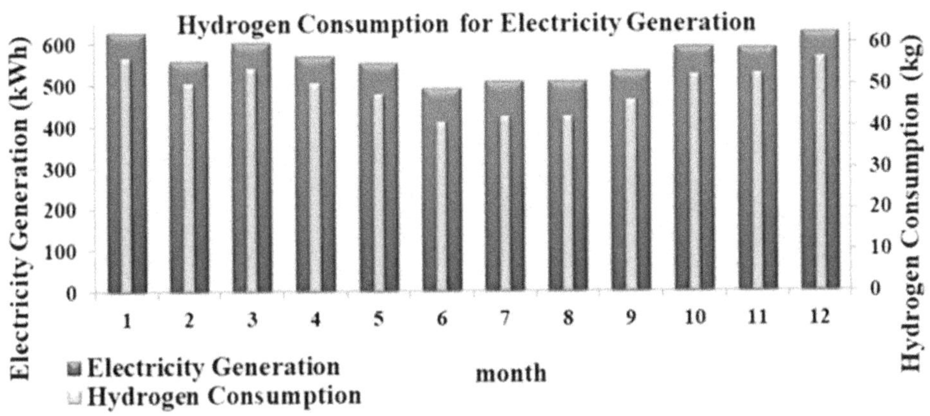

The financial indicators of the 1 kW fuel cell energy generation system are presented in value in Table 2 and graphically, in percentage in Figure 2 summarizing information on the following financial aspects.

Table 2. System costs: SE - FC_H 1 kW

Financial Indicator	Cost [Euro]
Initial investment	8664
Total system cost / 25 years of operation	192232
Cost - fuel cell	67422
Cost - inverter	2632
Fuel cost (hydrogen)	116363
Cost - Energy purchased from the grid	4276
Cost - Energy exported to the grid	0
Other costs	1539
Energy unit price	1,14 euro/kWh

The initial investment costs for the implementation of such an energy system are relatively low, being 8664 euros, but with a total lifetime of 25 years, the system costs 192232 euros. It is observed that the largest share in the total cost is represented by the purchase of hydrogen fuel, with a percentage of 60.53%, the fuel cell representing 35.07% of the total cost, and the other costs regarding the converters and inverters, as well as the purchase from the network of electricity not covered by the system totals 4.39%.

Figure 2. Percentage costs of component equipment and fuel: SE - FC_H 1 kW

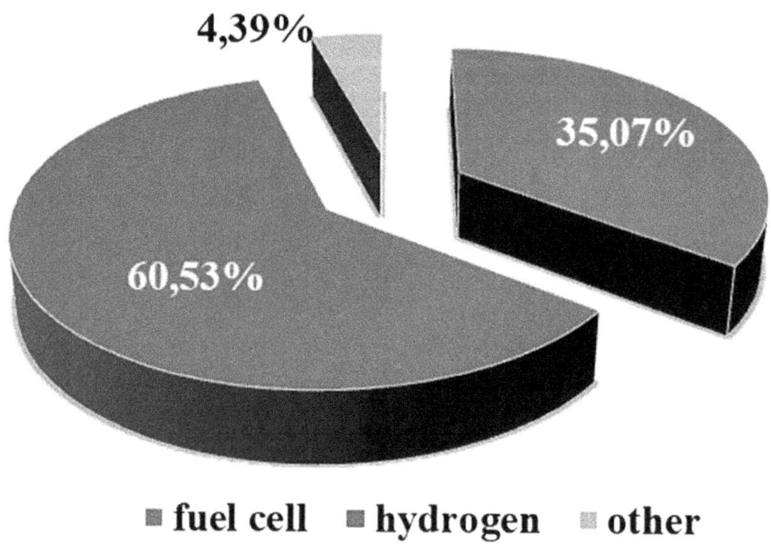

All these costs were calculated for a lifetime of the energy system of 25 years. It is observed that the unit price of energy, defined as the ratio between the total annual cost of the energy generation system and the total annual amount of electricity produced by this system, registers a value equal to 1.14 euros/kWh.

Fuel Cell With Rated Power of 2 kW (SE - FC_H 2kW)

Given the fact that the energy demand of the analyzed consumer is not fully covered by the 1 kW fuel cell system, the computational simulation of an energy system equipped with a fuel cell with a higher rated power, namely 2 kW, was carried out.

The performance indicators of the 2 kW fuel cell power generation system, after one year of operation, have the total annual average values shown in Table 3.

It is observed that after one year of operation, the total power generated by the fuel cell is 7677.50 kWh/year, of which, after the conversion performed by the inverter, 88.7% of it is actually used for the energy support of the consumer.

Table 3. Performance indicators: Annual averages - SE - FC_H 2 kW system

Performance Indicator	Value	Unit
Total energy demand	6.759,00	kWh/ year
Energy generated by the fuel cell	7.677,50	kWh/ year
Energy demand not covered by the system	2,10	kWh/ year
Energy purchased from the grid	2,10	kWh/ year
Excess energy	0	kWh/ year
Energy exported to the grid	0	kWh/ year
Total hydrogen consumption	578,45	kg
Total carbon dioxide emissions	386,00	kg CO_2 / year

From the tabular value data, it can be seen that if a 2 kW fuel cell energy generation system is chosen, the energy requirement requested by the consumer cannot be fully ensured, for a relatively small amount of energy, respectively 0.1%. Although the annual average of the energy generated by the 2 kW fuel cell registers a value that exceeds the energy consumption, it still results in this uncovered energy. Energy production and hydrogen consumption are shown graphically in Figure 3.

Figure 3. Hydrogen consumption in the electricity generation: SE - FC_H 2kW system

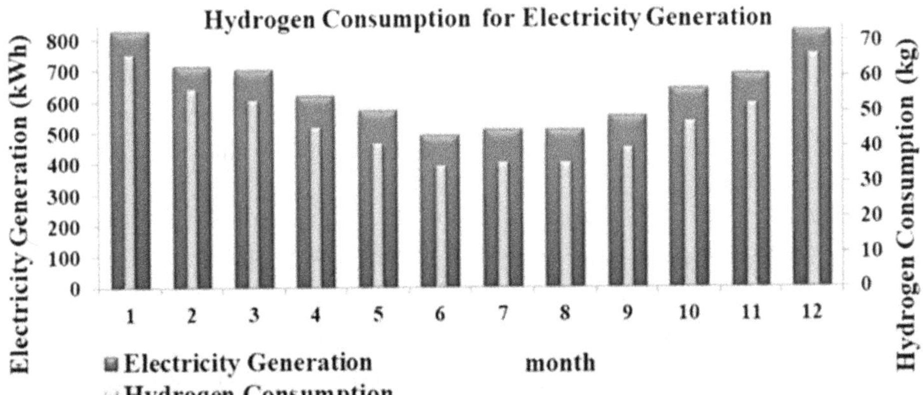

Hydrogen consumption within the fuel cell depends on its nominal power and the real power generated in the system, this real power being influenced by the consumption and efficiency charts. The average annual values calculated for these parameters are: 578.45 kg of hydrogen to produce 7677.50 kWh of electricity.

The financial indicators of the 2 kW fuel cell energy generation system are presented in value in Table 4 and graphically, in percentage in Figure 4, summarizing information on financial aspects.

Table 4. System costs: SE - FC_H 2 kW

Financial Indicator	Cost [Euro]
Initial investment	13764
Total system cost / 25 years of operation	209401
Cost - fuel cell	91036
Cost - inverter	2632
Fuel cost (hydrogen)	113452
Cost - Energy purchased from the grid	11
Cost - Energy exported to the grid	0
Other costs	4913
Energy unit price	1,24 euro/kWh

In this case, the unit price of energy registers an increase compared to the previous case studied, the value being equal to 1.24 euros/kWh.

It is observed that the largest share in the total cost is the purchase of hydrogen fuel, with a percentage of 54.18%, at the same time, an increase in the costs of the hydrogen conversion equipment compared to the first system studied, implicitly, are also increasing the other global expenses related to the operation of this energy generation system related to a lifetime of 25 years, the period proposed for analysis.

Instead, the expenses related to the purchase of electricity from the centralized distribution system decrease, along with the decrease in the value of the energy not covered by the system.

Fuel Cell With Rated Power of 3 kW (SE - FC_H 3kW)

Taking into account the small percentage of non-coverage of the energy requirement in the case of the 2 kW cell, the operation simulation of a fuel cell with a higher nominal power, namely of 3 kW, was carried out. In this case, a 100% degree of coverage of the energy demand is found.

Figure 4. Percentage costs of component equipment and fuel: SE - FC_H 2 kW

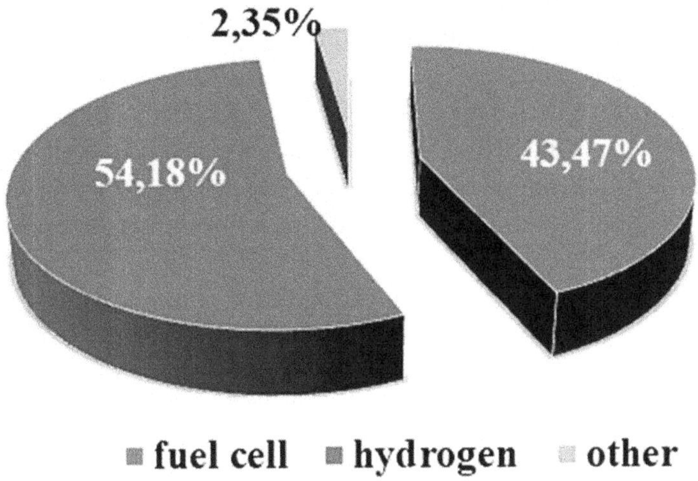

There was no lack of energy coverage during any period of operation, but neither was there an excess of energy. The performance indicators of the 3 kW fuel cell power generation system, after one year of operation, have the total annual average values shown in Table 5.

Table 5. Performance indicators: Annual averages - SE - FC_H 3 kW system

Performance Indicator	Value	Unit
Total energy demand	6.759,00	kWh/ year
Energy generated by the fuel cell	7680,10	kWh/ year
Energy demand not covered by the system	0	kWh/ year
Energy purchased from the grid	0	kWh/ year
Excess energy	0	kWh/ year
Energy exported to the grid	0	kWh/ year
Total hydrogen consumption	549,01	kg
Total carbon dioxide emissions	578,00	kg CO_2 / year

From the tabular value data, it can be seen that if a 3 kW fuel cell energy generation system is chosen, the energy requirement requested by the consumer is fully covered, the hydrogen to electricity conversion technology can support energetically the electric vehicle charging station that was subjected to the study.

The diagram of annual average values emphasizes the aspect regarding the degree of coverage of 100% of the total required energy that can be provided by the studied energy system. The annual average of the energy generated by the 3 kW fuel cell registers a value that exceeds the energy consumption by 12% previously explained aspect and due to the difference obtained from the efficiency of the inverter.

Energy production and hydrogen consumption are shown in Figure 5.

Hydrogen consumption within the fuel cell depends on its nominal power and the real power generated in the system, this real power being influenced by the consumption and efficiency charts. Numerically, the values calculated according to the previous details, a consumption of 549.01 kg of hydrogen was obtained to produce 7680.10 kWh of electricity.

Figure 5. Hydrogen consumption in the electricity generation: SE - FC_H 3kW system

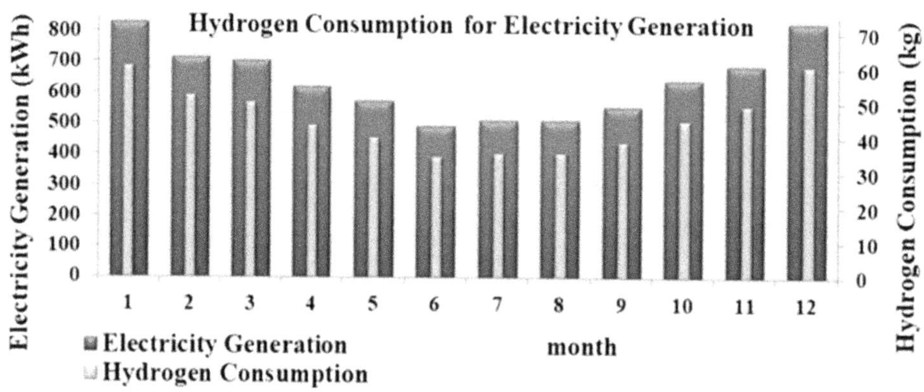

The financial indicators of the energy generation system with a 3 kW fuel cell are presented in value in Table 6 and graphically, in percentage in Figure 6, summarizing information regarding financial aspects.

In this case, the unit price of energy also registers an increase compared to the 2 kW fuel cell system, the value being equal to 1.29 euros/kWh.

It is observed that the largest share in the total cost is represented by the purchase of hydrogen fuel, with a percentage of 49.34%, closely followed by the costs regarding the hydrogen conversion equipment with 48.21%, implicitly the other global expenses related to of the operation of this energy generation system related to a lifetime of 25 years, a period proposed for analysis, but the expenses regarding the purchase of electricity from the centralized distribution system are no longer recorded, along with the incomplete satisfaction of the energy demand.

Table 6. System costs: SE - FC_H 3 kW

Financial Indicator	Cost [Euro]
Initial investment	16824
Total system cost / 25 years of operation	218227
Cost - fuel cell	105204
Cost - inverter	2632
Fuel cost (hydrogen)	107682
Cost - Energy purchased from the grid	0
Cost - Energy exported to the grid	0
Other costs	5341
Energy unit price	1,29 euro/kWh

Figure 6. Percentage costs of component equipment and fuel: SE - FC_H 3 kW

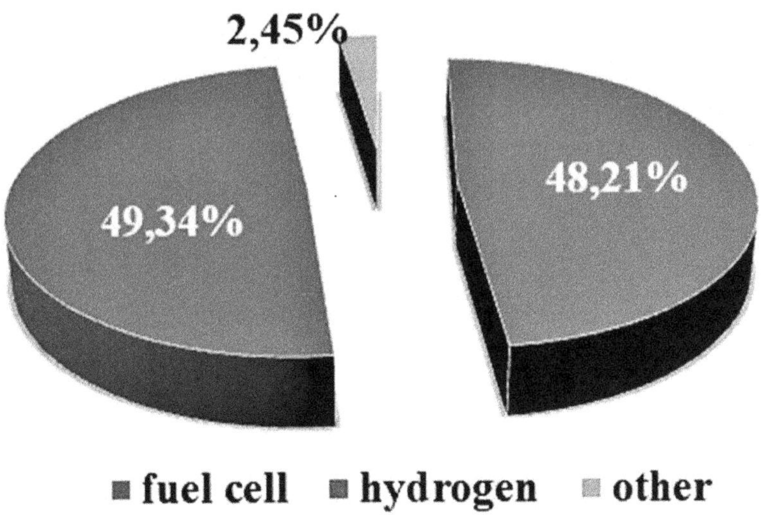

Fuel Cell With Rated Power of 4 kW (SE - FC_H 4kW)

If until now the goal has been to fully cover the energy demand through the energy generation system with fuel cells, in the present sub-chapter it is proposed to study the hypothesis of using a fuel cell with a higher rated power (4 kW) in the system, having as objective the parametric study regarding the influence of this equipment on the global energy efficiency of the energy system of this type. Energy performance indicators of the 4 kW fuel cell power generation system after one year of operation.

Table 7. Performance indicators: Annual averages - SE - FC_H 4 kW system

Performance Indicator	Value	Unit
Total energy demand	6.759,00	kWh/ year
Energy generated by the fuel cell	7736,80	kWh/ year
Energy demand not covered by the system	0	kWh/ year
Energy purchased from the grid	0	kWh/ year
Excess energy	56,70	kWh/ year
Energy exported to the grid	56,70	kWh/ year
Total hydrogen consumption	556,30	kg
Total carbon dioxide emissions	770,00	kg CO_2 / year

From the tabular data (Table 7) it can be seen that if a 4 kW fuel cell energy generation system is chosen, the energy required by the consumer is fully provided, but there is also an excess of energy throughout the year. From the diagram of annual average values, it can be seen that the total annual energy requirement is covered in 100% percentage, while also obtaining an excess of energy in a small percentage of 0.85%. The annual average of the energy generated by the 4 kW fuel cell registers a value that exceeds the energy consumption and excess energy combined, a problem due to the efficiency of the inverter. Energy production and hydrogen consumption are shown graphically in Figure 7.

Figure 7. Hydrogen consumption in the electricity generation: SE - FC_H 4kW system

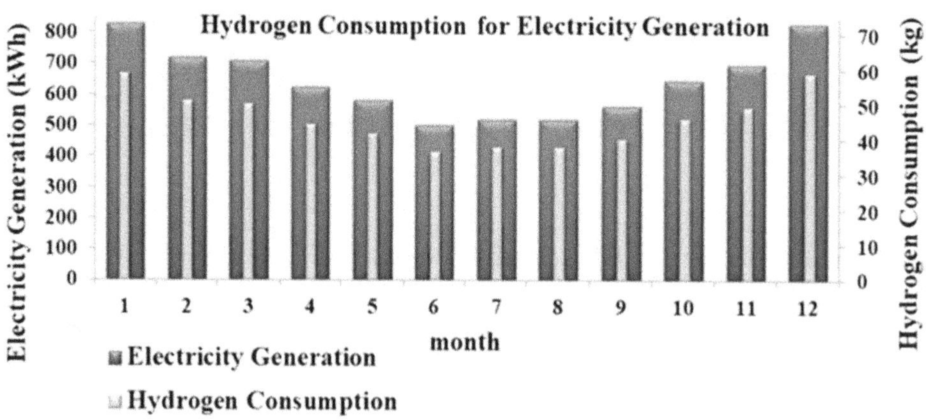

The values with reference to the consumption of hydrogen involved in energy generation were calculated according to the prescriptions in the previous subsections, and as results the following annual values were obtained: a consumption in the amount of 556.30 kg of hydrogen involved in the production of 7736.80 kWh of electricity.

The financial indicators of the energy generation system with a 4 kW fuel cell are presented in value in Table 8 and graphically, in percentage in Figure 8, summarizing information on financial aspects.

Table 8. System costs: SE - FC_H 4 kW

Financial Indicator	Cost [Euro]
Initial investment	19374
Total system cost / 25 years of operation	231676
Cost - fuel cell	117011
Cost - inverter	2632
Fuel cost (hydrogen)	109108
Cost - Energy purchased from the grid	0
Cost - Energy exported to the grid	149
Other costs	5706
Energy unit price	1,37 euro/kWh

In this case, the unit price of energy also registers an increase, the value being equal to 1.37 euros/kWh.

It is observed that the largest share in the total cost is this time represented by the costs regarding the hydrogen conversion equipment with 50.47%, and the purchase of hydrogen fuel takes second place, with a percentage of 47.06%, as well as the other elements financials are higher compared to the previously studied power generation system (*SE - FC_H 3kW*). As a new element in this analysis, the cost regarding the energy exported to the network appears, a cost generated by the appearance of a supplement of electricity delivered by the system, and the expenses regarding the purchase of electricity from the centralized distribution system are no longer recorded, along with the uneven satisfaction of the energy demand.

Figure 8. Percentage costs of component equipment and fuel: SE - FC_H 4 kW

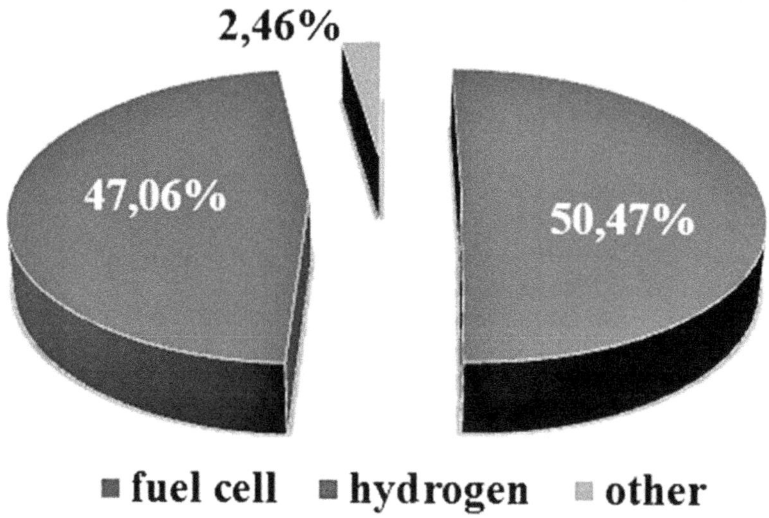

Fuel Cell With Rated Power of 5 kW (SE - FC_H 5kW)

As a final hypothesis of this study, computational simulations were carried out on an energy generation system that integrates a 5 kW fuel cell. The energy performance indicators of the 5 kW fuel cell energy generation system, after one year of operation, have the monthly average values shown in Table 9.

Compared to the previous system, there is an increase in the amount of electricity generated by the fuel cell, which implicitly leads to a greater excess of energy, and at the same time to an increased fuel consumption compared to the situation analyzed in the case of *SE - FC_H 4kW*. The performance indicators of the 5 kW fuel cell power generation system, after one year of operation, have the total annual average values shown in Table 9.

From the diagram of annual average values, it can be seen that the total annual energy requirement is covered in 100% percentage, while also obtaining an excess of energy in a small percentage of 2.95%, increasing compared to the previous system studied. The annual average energy generated by the 5 kW fuel cell exceeds the combined energy consumption and excess energy, a problem discussed earlier, due to the efficiency of the inverter.

Energy production and hydrogen consumption are presented graphically in Figure 9.

Table 9. Performance indicators: Annual averages - SE - FC_H 5 kW system

Performance Indicator	Value	Unit
Total energy demand	6.759,00	kWh/ year
Energy generated by the fuel cell	7878,20	kWh/ year
Energy demand not covered by the system	0	kWh/ year
Energy purchased from the grid	0	kWh/ year
Excess energy	198,10	kWh/ year
Energy exported to the grid	198,10	kWh/ year
Total hydrogen consumption	583,05	kg
Total carbon dioxide emissions	963,00	kg CO_2 / year

Figure 9. Hydrogen consumption in the electricity generation: SE - FC_H 5kW system

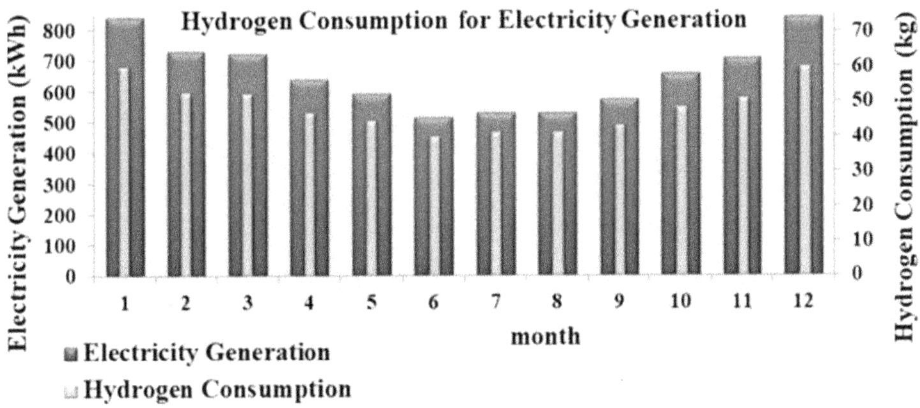

The values with reference to the consumption of hydrogen involved in energy generation were also calculated, and as results the following annual values were obtained: a consumption in the amount of 583.05 kg of hydrogen involved in the production of 7878, 20 kWh of electricity.

The financial indicators of the 5 kW fuel cell power generation system are presented in value in Table 10 and graphically, as a percentage in Figure 10, summarizing information on financial aspects defined in detail in the previous subsections.

In this case, the unit price of energy is 1.57 euros/kWh, being higher than in the case of the previously studied energy system.

Table 10. System costs: SE - FC_H 5 kW

Financial Indicator	Cost [Euro]
Initial investment	21924
Total system cost / 25 years of operation	265905
Cost - fuel cell	146000
Cost - inverter	2632
Fuel cost (hydrogen)	114356
Cost - Energy purchased from the grid	0
Cost - Energy exported to the grid	523
Other costs	6072
Energy unit price	1,57 euro/kWh

Figure 10. Percentage costs of component equipment and fuel: SE - FC_H 5 kW

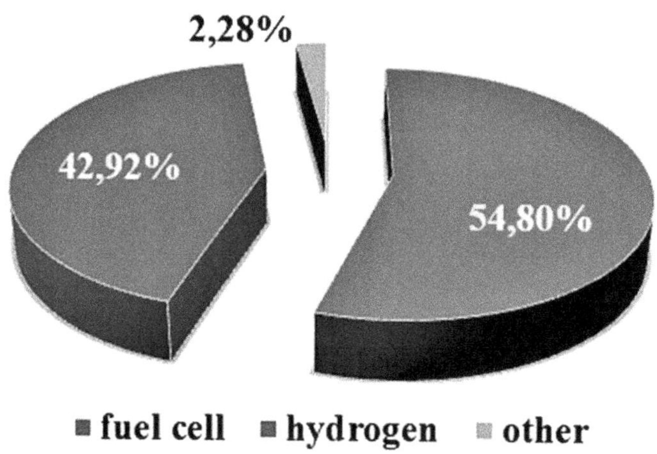

It is found that the largest share in the total cost is also this time represented by the costs regarding the hydrogen conversion equipment with 54.80%, and the purchase of hydrogen fuel being in second place, with a percentage of 42.92%, as well as the other financial elements are on the rise compared to the previous 4kW fuel cell power generation system.

COMPARATIVE ANALYSIS OF THE RESULTS

In order to carry out an analysis of the performance parameters that characterize the five simulated situations, in the following, the results were graphically illustrated in comparative diagrams in order to substantiate the decision of choosing the optimal equipment for the energy system that uses the fuel cell as the first and only source of electricity generation in a future hydrogen-based economy.

Comparative data were presented regarding the energy generated by the externally powered hydrogen fuel cell for the five types of fuel cells of different nominal powers, the excess energy resulting from operation and the provision of the energy requirement was sought. The annual average values of the energy parameters were presented graphically in Figure 11, and the main financial indicators were presented comparatively in Figure 13.

Figure 11. Comparative diagram regarding the average annual values of energy parameters

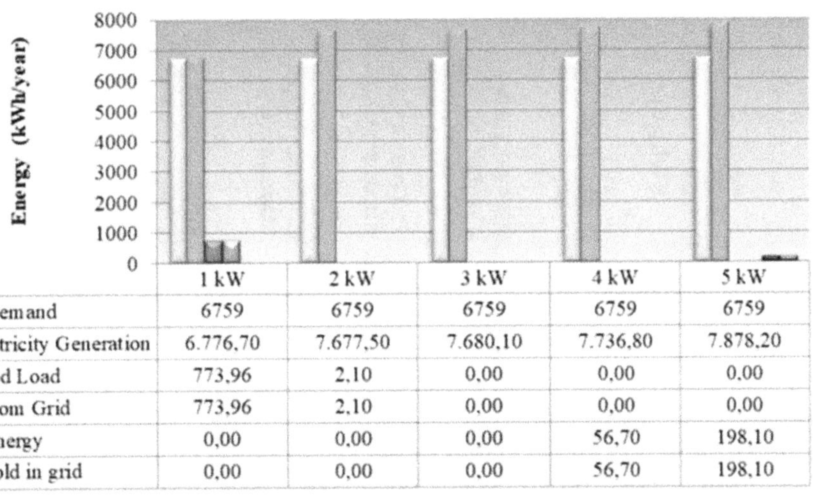

	1 kW	2 kW	3 kW	4 kW	5 kW
Energy Demand	6759	6759	6759	6759	6759
FC - Electricity Generation	6.776,70	7.677,50	7.680,10	7.736,80	7.878,20
Uncovered Load	773,96	2,10	0,00	0,00	0,00
Energy from Grid	773,96	2,10	0,00	0,00	0,00
Excess Energy	0,00	0,00	0,00	56,70	198,10
Energy sold in grid	0,00	0,00	0,00	56,70	198,10

With regard to the energy generated by the fuel cell, a linear increase is observed in direct proportion to the increase in the nominal power of the hydrogen conversion equipment. In the same sense, if in the case of the 1 kW fuel cell it is found that the energy requirement is not covered in a percentage of 11.45%, for the situation of the 2 kW cell there is a small uncovered load of 1%, and in the hypothesis of the 3 kW fuel, all energy consumption for the analyzed consumer is satisfied, also for fuel cell systems with higher nominal powers. At the same time, for the systems with the 4 kW and 5 kW fuel cell, the systems deliver an excess amount of energy, which is very small for the 4 kW, respectively 3% in the case of the 5 kW.

The hydrogen consumption of the fuel cell depends on its nominal power and the real power generated in the system, and the degree of energy efficiency of the fuel cell is given by the ratio between the energy generated by the equipment and the product of its hydrogen consumption and the lower calorific power of hydrogen (Felseghi et al. 2019; Razi & Dincer, 2022; Trapp et al., 2022). Comparatively analyzing the parameters related to the consumption of hydrogen in the generation of electricity, it is observed that a minimum consumption was achieved in the case of the 3 kW fuel cell. This consumption related to the amount of energy generated, while taking into account the fact that the total energy demand of the consumer is covered by the system in this case, can be considered the optimal consumption of hydrogen for the energy support of the studied consumer.

Figure 12. Comparative diagram regarding the average annual values of hydrogen consumption in electricity production

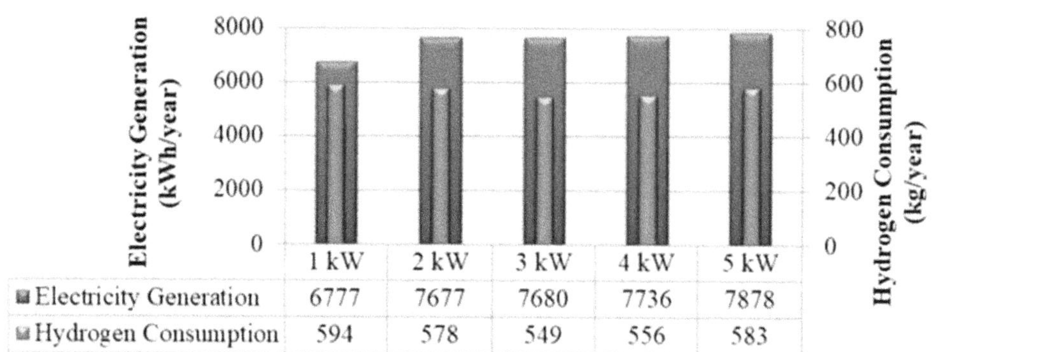

If the *SE - FC$_H$ 1kW* case is analyzed, it is found that this system consumes 8.2% more hydrogen and produces 11.75% less electricity than in the *SE - FC$_H$ 3kW* hypothesis, and the *SE - FC$_H$ 2kW* consumes with 5.3% more hydrogen and produces electricity in a smaller quantity by only 3 kWh/year than *SE - FC$_H$ 3kW*. Regarding the fuel cells with energy generation capacity higher than that considered optimal, a 1.3% higher consumption is found in the case of *SE - FC$_H$ 4kW*, respectively 6.2% higher in the case of *SE - FC$_H$ 5kW*, consumption which, however, is lower than in the case of the 1 kW fuel cell, emphasizing the fact that in these last two situations excess energy is obtained in addition to the degree of 100% coverage of the energy requirement.

From the perspective of the two parameters, hydrogen consumption and energy generated by the fuel cell (Figure 12), for the energy support of the studied electric vehicle charging station, it can be concluded that the 3 kW fuel cell is optimal. The equipment with lower nominal powers do not correspond to the energy needs of the consumer, also registering an increased hydrogen consumption due to the low degree of energy efficiency of the equipment in relation to the energy demand. The equipment with higher nominal powers ensure the energy needs of the consumer, at the same time realizing an unjustified excess of energy in the present case.

The financial indicators that characterize the costs of the simulated systems in operation are presented comparatively in Figure 13. The linear increase of the total costs, but also of the purchase costs of the hydrogen conversion equipment is observed, directly proportional to the increase in the nominal power of the fuel cell.

Figure 13. Comparative diagram of the main financial indicators

	Total	Hydrogen Fuel	Fuel Cell
1 kW	192232	116363	67422
2 kW	209401	113452	91036
3 kW	218227	107682	105204
4 kW	231676	109108	117011
5 kW	265905	114356	146000

Given the previously determined hydrogen consumption to be optimal in the case of the 3 kW fuel cell and the aspects discussed in the previous analysis regarding the hydrogen consumption of each type of fuel cell, it is observed that these are reflected in the fuel purchase costs. Regarding the purchase of fuel - hydrogen, the lowest cost is recorded in the case of the 3 kW fuel cell.

The initial investment is schematically presented in the comparative diagram in Figure 14. Given that this indicator is directly influenced by the nominal capacity of the electrochemical hydrogen conversion equipment, i.e. the purchase price of the fuel cell or in other words the price of the technological transfer from the equipment manufacturer to the final consumer, it increases linearly with the increase in nominal power of it, this increasing trend can be observed in the comparative diagram of investment costs. The 1 kW fuel cell has the lowest purchase cost of 8664 euros, the 2 kW one being 58.9% higher, the 3 kW one 94.2% higher than the first and 22.3% than the second one. Compared to the 3 kW fuel cell, the investment costs for those with higher nominal powers are 15.5% higher in the case of the 4 kW cell, respectively by 30.3% in the case of the 5 kW.

Figure 14. Comparative diagram of investment costs

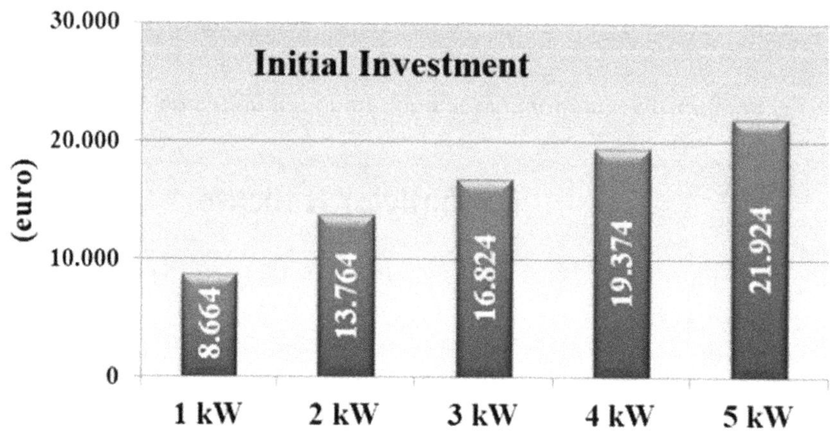

The unit price of electricity generated through energy production systems based on fuel cells of different powers is schematically presented in the comparative diagram in Figure 15.

As discussed in the previous analyses, with the increase in the nominal power of the fuel cell, the investment costs and the total operating costs of the system increase compared to 25 years of life, which is implicitly reflected in the unit price of the energy delivered by these systems. The lowest unit price is obtained in the base SE - FC_H *1kW* of 1.14 euros/kWh. For SE - FC_H *2kW* there is an 8.8% increase in the unit tariff, then for the 3 kW fuel cell it is higher by 13.2% compared to the first one. Analyzing compared to SE - FC_H *3kW* fuel cells with higher nominal powers, a unit price is obtained 6.2% higher in the case of the 4 kW cell and 21.7% higher for the 5 kW one.

Figure 15. Comparative diagram regarding unit prices of electricity

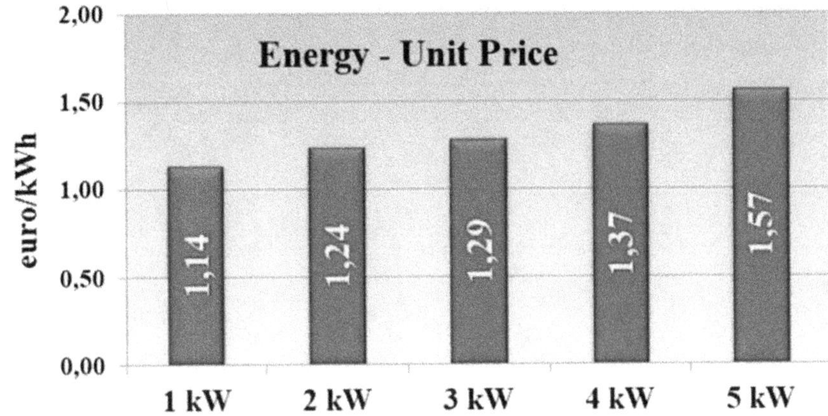

As environmental protection parameters, the aspects related to the carbon dioxide emissions obtained as a result of the operation of these electricity generation systems were analyzed in the present study: Valoric emissions are graphically illustrated in comparison in Figure 16.

The lowest value of carbon dioxide emissions is recorded in the case of the 2 kW fuel cell, respectively 386 kg CO_2/year. In the case of the energy system with a 1 kW fuel cell, emissions are 30% higher, and for the one with a 3 kW cell, the emissions are 49.75% higher. In the case of the other two systems, SE - FC_H $4kW$ registers increased emissions by 100%, and SE - FC_H $5kW$ 150% compared to the reference system SE - FC_H $2kW$. If this performance indicator is compared to the average value calculated for the carbon footprint (5.23 tons of CO_2/year) of charging stations for electric vehicles supported energetically by classic systems, a significant decrease is found in the case of SE - FC_H, an extremely relevant in the context of the sustainability of energy generation systems.

From the comparative presentation and analysis of the results obtained from the simulations, the conclusions detailed above were obtained, motivated by pro and con arguments regarding each of the five analyzed systems. Aspects related to the rated power and the capability of the fuel cell technology integrated into energy systems with the role of the first and only source of electricity generation in a future hydrogen-based economy, directly influence the energy efficiency of these systems in ensuring the energy demand of to the consumer under analysis.

Figure 16. Comparative diagram regarding CO_2 emissions

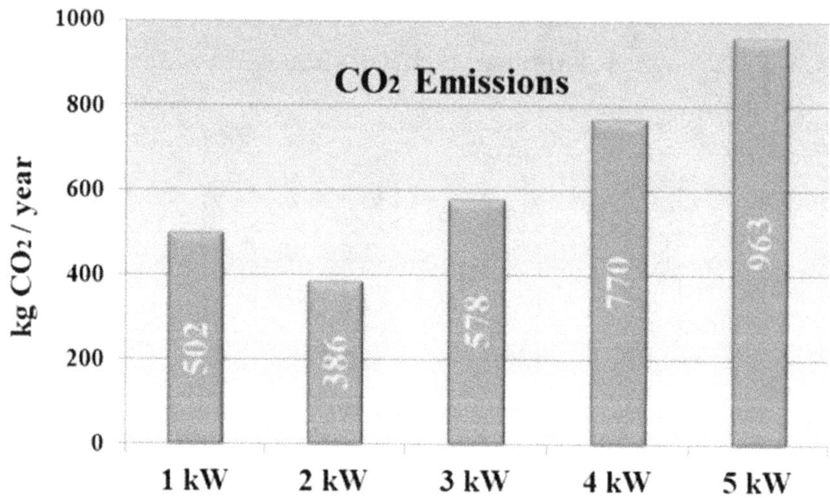

CONCLUSION

In this chapter, the results of simulations of energy systems with fuel cells of different powers were presented, which were carried out separately for five proposed types, with the aim of demonstrating the capability of this equipment to energetically support an electric vehicle charging station under the conditions uses hydrogen as the first and only source of electricity generation.

The objectives pursued in this hypothesis are: the full coverage of the energy demand and the study of the achieved performances, simultaneously with the establishment of the optimal configuration of the energy generation system under conditions of minimizing the unmet need, excess energy and CO_2 emissions.

The results of the computational simulations were presented in comparative diagrams, which highlighted the 3 kW fuel cell as the optimal equipment due to the arguments that are favorable to this system. Among the five situations analyzed, the SE - $FC_H\,3kW$ system has the highest energy efficiency, a statement supported by the fact that in this case a 100% coverage of the energy requirement was achieved without resulting in excess energy, and the hydrogen consumption compared to the production energy has the lowest value compared to the other four systems under analysis.

For the scientific substantiation of this decision, a multi-criteria analysis was carried out through which the conditions according to which the studied variants are ranked were expanded. In this analysis, the global optimization criteria refer to: unmet energy demand, energy generated by the fuel cell, excess energy, hydrogen consumption, investment cost, total system cost, energy unit price, CO_2 emissions.

FUTURE RESEARCH DIRECTIONS

The studies, analyzes and results of research carried out and presented in this chapter, as well as the problems and technical limitations encountered, allow the identification and establishment of future research directions in the field of the topic addressed:

- Expanding the comparative study and carrying out a comparative multicriteria analysis based on dedicated tools, specifically the Application of Analytic Hierarchy Process;
- The comprehensive study of high-power energy systems based on hydrogen energy, with practical applicability in the mobility domain;
- Analysis regarding the wide acceptance of hydrogen energy as a fuel for mobility domain.

REFERENCES

Ahmad, F., Iqbal, A., Ashraf, I., Marzband, M., & khan, I. (2022). Optimal location of electric vehicle charging station and its impact on distribution network: A review. *Energy Reports*, *8*, 2314–2333. doi:10.1016/j.egyr.2022.01.180

Akarsu, B., & Genç, M. S. (2022). Optimization of electricity and hydrogen production with hybrid renewable energy systems. *Fuel*, *324*, 124465. doi:10.1016/j.fuel.2022.124465

Ali, A., Shakoor, R., Raheem, A., Muqeet, H. A. U., Awais, Q., Khan, A. A., & Jamil, M. (2022). Latest energy storage trends in multi-energy standalone electric vehicle charging stations: A comprehensive study. *Energies*, *15*(13), 4727. doi:10.3390/en15134727

Aşchilean, I., Cobîrzan, N., Bolboaca, A., Boieru, R., & Felseghi, R. A. (2021). Pairing solar power to sustainable energy storage solutions within a residential building: A case study. *International Journal of Energy Research*, *45*(10), 15495–15511. doi:10.1002/er.6982

Badea, G., Felseghi, R., Aschilean, I., Bolboaca, A., Muresan, D., Soimosan, T., ... Raboaca, S. (2015, August). Energen system for power supply of passive house: Case study. In *2015 Second International Conference on Mathematics and Computers in Sciences and in Industry (MCSI)* (pp. 24-31). IEEE. 10.1109/MCSI.2015.31

Badea, G., Felseghi, R. A., & Aşchilean, I. (2021). Hydrogen-Energy Vector Within a Sustainable Energy System for Stationary Applications. In *Hydrogen Fuel Cell Technology for Stationary Applications* (pp. 1–21). IGI Global. doi:10.4018/978-1-7998-4945-2.ch001

Badea, G., Felseghi, R. A., Varlam, M., Filote, C., Culcer, M., Iliescu, M., & Răboacă, M. S. (2018). Design and simulation of romanian solar energy charging station for electric vehicles. *Energies*, *12*(1), 74. doi:10.3390/en12010074

Dash, S. K., Chakraborty, S., Roccotelli, M., & Sahu, U. K. (2022). Hydrogen Fuel for Future Mobility: Challenges and Future Aspects. *Sustainability (Basel)*, *14*(14), 8285. doi:10.3390u14148285

Dufo-López, R., & Bernal-Agustín, J. L. (2020). *iHOGA PRO+ Software Version 2.5*. Electrical Engineering Department, University of Zaragoza. Available online: https://ihoga.unizar.es/

Enescu, F. M., Birleanu, F. G., Raboaca, M. S., Raceanu, M., Bizon, N., & Thounthong, P. (2023). Electric Vehicle Charging Station Based on Photovoltaic Energy with or without the Support of a Fuel Cell–Electrolyzer Unit. *Energies*, *16*(2), 762. doi:10.3390/en16020762

Farias, C. B. B., Barreiros, R. C. S., da Silva, M. F., Casazza, A. A., Converti, A., & Sarubbo, L. A. (2022). Use of hydrogen as fuel: A trend of the 21st century. *Energies*, *15*(1), 311. doi:10.3390/en15010311

Felseghi, R. A., Aşchilean, I., Cobîrzan, N., Bolboacă, A. M., & Raboaca, M. S. (2021). Optimal Synergy between Photovoltaic Panels and Hydrogen Fuel Cells for Green Power Supply of a Green Building—A Case Study. *Sustainability (Basel)*, *13*(11), 6304. doi:10.3390u13116304

Felseghi, R. A., Carcadea, E., Raboaca, M. S., Trufin, C. N., & Filote, C. (2019). Hydrogen fuel cell technology for the sustainable future of stationary applications. *Energies*, *12*(23), 4593. doi:10.3390/en12234593

Felseghi, R. A., & Papp, A. (2014). Considerations regarding hybrid systems of power generator from renewable energy sources. *Studia Universitatis Babes-Bolyai, Ambientum, 59*.

Filote, C., Felseghi, R. A., Raboaca, M. S., & Aşchilean, I. (2020). Environmental impact assessment of green energy systems for power supply of electric vehicle charging station. *International Journal of Energy Research*, *44*(13), 10471–10494. doi:10.1002/er.5678

Jones, J., Genovese, A., & Tob-Ogu, A. (2020). Hydrogen vehicles in urban logistics: A total cost of ownership analysis and some policy implications. *Renewable & Sustainable Energy Reviews*, *119*, 109595. doi:10.1016/j.rser.2019.109595

Karaca, A. E., & Dincer, I. (2023). Development and evaluation of a solar based integrated hydrogen energy system for mobile applications. *Energy Conversion and Management*, *280*, 116808. doi:10.1016/j.enconman.2023.116808

Kovač, A., Paranos, M., & Marciuš, D. (2021). Hydrogen in energy transition: A review. *International Journal of Hydrogen Energy*, *46*(16), 10016–10035. doi:10.1016/j.ijhydene.2020.11.256

Li, Y., & Taghizadeh-Hesary, F. (2022). The economic feasibility of green hydrogen and fuel cell electric vehicles for road transport in China. *Energy Policy*, *160*, 112703. doi:10.1016/j.enpol.2021.112703

Longden, T., Beck, F. J., Jotzo, F., Andrews, R., & Prasad, M. (2022). 'Clean'hydrogen?–Comparing the emissions and costs of fossil fuel versus renewable electricity based hydrogen. *Applied Energy*, *306*, 118145. doi:10.1016/j.apenergy.2021.118145

Mastoi, M. S., Zhuang, S., Munir, H. M., Haris, M., Hassan, M., Usman, M., Bukhari, S. S. H., & Ro, J. S. (2022). An in-depth analysis of electric vehicle charging station infrastructure, policy implications, and future trends. *Energy Reports*, *8*, 11504–11529. doi:10.1016/j.egyr.2022.09.011

Park, C., Lim, S., Shin, J., & Lee, C. Y. (2022). How much hydrogen should be supplied in the transportation market? Focusing on hydrogen fuel cell vehicle demand in South Korea: Hydrogen demand and fuel cell vehicles in South Korea. *Technological Forecasting and Social Change*, *181*, 121750. doi:10.1016/j.techfore.2022.121750

Ravi, S. S., & Aziz, M. (2022). Clean hydrogen for mobility–Quo vadis? *International Journal of Hydrogen Energy*, *47*(47), 20632–20661. doi:10.1016/j.ijhydene.2022.04.158

Razi, F., & Dincer, I. (2022). Challenges, opportunities and future directions in hydrogen sector development in Canada. *International Journal of Hydrogen Energy*, *47*(15), 9083–9102. doi:10.1016/j.ijhydene.2022.01.014

Rojas, J. M. L., Lopez, R. D., & Navarro, J. A. D. (2022). *Genetic Optimization Techniques for Sizing and Management of Modern Power Systems*. Elsevier.

Trapp, C. T., Kanbach, D. K., & Kraus, S. (2022). Sector coupling and business models towards sustainability: The case of the hydrogen vehicle industry. *Sustainable Technology and Entrepreneurship*, *1*(2), 100014. doi:10.1016/j.stae.2022.100014

Turoń, K. (2020). Hydrogen-powered vehicles in urban transport systems–current state and development. *Transportation Research Procedia*, *45*, 835–841. doi:10.1016/j.trpro.2020.02.086

Venkatasatish, R., & Dhanamjayulu, C. (2022). Reinforcement learning based energy management systems and hydrogen refuelling stations for fuel cell electric vehicles: An overview. *International Journal of Hydrogen Energy*, *47*(64), 27646–27670. doi:10.1016/j.ijhydene.2022.06.088

KEY TERMS AND DEFINITIONS

Clean Energy Generation System: A system designed primarily to provide clean energy services to end users.

Electric Vehicle or Electromobile: Is a zero-emission vehicle powered by an electric motor, with power from an electricity source.

Fuel Cell: Is an electric cell, which, unlike battery cells, can be continuously supplied with fuel, so that the electric power from the output of this electric cell can be maintained indefinitely. Thus, the fuel cell converts hydrogen or hydrogen-based fuels directly into electricity and heat through the electrochemical reaction of hydrogen with oxygen.

Hydrogen as a Carrier of Green Energy: Etymologically, the word hydrogen is a combination of two Greek words, meaning "to make water". Produced from non-fossil sources and raw materials, through the use of different forms of alternative energy (solar, wind, hydroelectric, geothermal, biomass, etc.), hydrogen is considered to be a primary fuel in achieving the supply of so-called "green energy". Thus, systems that operate with hydrogen as fuel can be considered to be among the best solutions for accelerating and ensuring stability at the global level.

Hydrogen Economy: Represents all the energy infrastructures at the national and international level that have hydrogen as an energy resource. Building an infrastructure that allows, easily and efficiently, the transport and delivery of hydrogen energy is a critical step towards a future hydrogen-based economy.

Hydrogen Energy: Today, hydrogen is recognized as a non-polluting energy carrier because it does not contribute to global warming, if it is produced from renewable sources. In addition, hydrogen is the only secondary energy carrier that lends itself to wide application on the market. In the center of attention is the fact that hydrogen can be obtained from a wide range of primary energies. It can be used advantageously for a wide range of applications, from transport and portable to stationary. In addition, hydrogen can also be used in decentralized systems without emitting carbon dioxide.

Multicriteria Comparative Analysis: Describes any structured approach used to determine overall preferences among several alternative options, which options lead to the fulfillment of a number of objectives.

Sustainable Mobility: Decarbonisation of the transport sector. The transport sector must undergo a transformation that will require a 90% reduction in greenhouse gas emissions by 2050, while ensuring affordable solutions for people.

Compilation of References

Abanades. (2019). Metal Oxides Applied to Thermochemical Water-Splitting for Hydrogen Production Using Concentrated Solar Energy. *ChemEngineering*, *3*(3), 63. doi:10.3390/chemengineering3030063

Abbott, D. (2009). Keeping the energy debate clean: How do we supply the world's energy needs? *Proceedings of the IEEE*, *98*(1), 42–66. doi:10.1109/JPROC.2009.2035162

Abd, A. A., Naji, S. Z., Thian, T. C., & Othman, M. R. (2021). Evaluation of hydrogen concentration effect on the natural gas properties and flow performance. *International Journal of Hydrogen Energy*, *46*(1), 974–983. doi:10.1016/j.ijhydene.2020.09.141

Abdel-Basset, M., Gamal, A., Chakrabortty, R. K., & Ryan, M. J. (2021). Evaluation of sustainable hydrogen production options using an advanced hybrid MCDM approach: A case study. *International Journal of Hydrogen Energy*, *46*(5), 4567–4591. doi:10.1016/j.ijhydene.2020.10.232

Abdelkareem, M. A., Elsaid, K., Wilberforce, T., Kamil, M., Sayed, E. T., & Olabi, A. (2020). Environmental aspects of fuel cells: A review. *The Science of the Total Environment*, *141803*. Advance online publication. doi:10.1016/j.scitotenv.2020.1418 PMID:32889267

Abdollahi Haghghi, M., Shamsaiee, M., Ghazanfari Holagh, S., Chitsaz, A., & Rosen, M. A. (2019). Thermodynamic, exergoeconomic, and environmental evaluation of a new multi-generation system driven by a molten carbonate fuel cell for production of cooling, heating, electricity, and freshwater. *Energy Conversion and Management*, *199*, 112040. doi:10.1016/j.enconman.2019.112040

Abe, J. O., Popoola, A. P. I., Ajenifuja, E., & Popoola, O. M. (2019). Hydrogen energy, economy and storage: Review and recommendation. *International Journal of Hydrogen Energy*, *44*(29), 15072–15086. doi:10.1016/j.ijhydene.2019.04.068

Acar, C., Beskese, A., & Temur, G. T. (2022). Comparative Fuel Cell Sustainability Assessment With A Novel Approach. *International Journal of Hydrogen Energy*, *47*(1), 575–594. doi:10.1016/j.ijhydene.2021.10.034

Aceves, S. M., Petitpas, G., Espinosa-Loza, F., Matthews, M. J., & Ledesma-Orozco, E. (2013). Safe, long range, inexpensive and rapidly refuelable hydrogen vehicles with cryogenic pressure vessels. *International Journal of Hydrogen Energy, 38*(5), 2480–2489. doi:10.1016/j.ijhydene.2012.11.123

Ade, N., Wilhite, B., & Goyette, H. (2020). An integrated approach for safer and economical design of Hydrogen refueling stations. *International Journal of Hydrogen Energy, 45*(56), 32713–32729. doi:10.1016/j.ijhydene.2020.08.232

AFHYPAC & FNCCR. (2018). *Déployer les stations hydrogène dans votre territoire*. Available online: http://www.afhypac.org/documents/divers/GUIDE-STATION-HYDROGENE-WEB.pdf

AFHYPAC. (2013). *Les données de base physico-chimiques sur l'hydrogène. Mémento de l'Hydrogène*. Available online:http://www.afhypac.org/documents/toutsavoir/fiche_1.2_donnees_physicochimiques_rev. mars_2013.pdf

Ahamed, M. I., Shakeel, N., Anwar, N., Khan, A., Asiri, A. M., & Dzudzevic-Cancar, H. (2021). Graphene-based nanocomposite for hydrogen storage application. In *Nanomaterials for Hydrogen Storage Applications* (pp. 57–78). Elsevier. doi:10.1016/B978-0-12-819476-8.00006-2

Ahmad, F., Iqbal, A., Ashraf, I., Marzband, M., & khan, I. (2022). Optimal location of electric vehicle charging station and its impact on distribution network: A review. *Energy Reports, 8*, 2314–2333. doi:10.1016/j.egyr.2022.01.180

Aili, D., Henkensmeier, D., Martin, S., Singh, B., Hu, Y., Jensen, J. O., ... Li, Q. (2020). *Polybenzimidazole-Based High-Temperature Polymer Electrolyte Membrane Fuel Cells: New Insights and Recent Progress*. Electrochemical Energy Reviews., doi:10.100741918-020-00080-5

Aithal, P. S., & Acharya, S. (2016). Impact of Green Energy on Global Warming-A Changing Scenario. *International Journal of Scientific Research and Modern Education.*

Akarsu, B., & Genç, M. S. (2022). Optimization of electricity and hydrogen production with hybrid renewable energy systems. *Fuel, 324*, 124465. doi:10.1016/j.fuel.2022.124465

Alaswad, A., Baroutaji, A., Achour, H., Carton, J., Al Makky, A., & Olabi, A. G. (2016). Developments In Fuel Cell Technologies In The Transport Sector. *International Journal Of Hydrogen Energy, 41*(37), 16499–1650. doi:10.1016/j.ijhydene.2016.03.164

Alaswad, A., Omran, A., Sodre, J. R., Wilberforce, T., Pignatelli, G., Dassisti, M., & Olabi, A. G. (2020). Technical And Commercial Challenges Of Proton-Exchange Membrane (Pem) Fuel Cells. *Energies, 14*(1), 144. doi:10.3390/en14010144

Albanese, E., Kalantzopoulos, G. N., Vitillo, J. G., Pinatel, E., Civalleri, B., Deledda, S., Bordiga, S., Hauback, B. C., & Baricco, M. (2013). Theoretical and experimental study on Mg(BH4)2 – Zn(BH4)2 mixed borohydrides. *Journal of Alloys and Compounds, 580*, S282–S286. doi:10.1016/j.jallcom.2013.02.165

Al-Breiki, M., & Bicer, Y. (2021). Comparative life cycle assessment of sustainable energy carriers including production, storage, overseas transport and utilization. *Journal of Cleaner Production, 279*, 123481. doi:10.1016/j.jclepro.2020.123481

Ali, A., Shakoor, R., Raheem, A., Muqeet, H. A. U., Awais, Q., Khan, A. A., & Jamil, M. (2022). Latest energy storage trends in multi-energy standalone electric vehicle charging stations: A comprehensive study. *Energies, 15*(13), 4727. doi:10.3390/en15134727

Al-zareer, M., Dincer, I., & Rosen, M. A. (2018). Analysis and assessment of a hydrogen production plant consisting of coal gasification, thermochemical water decomposition and hydrogen compression systems. *Energy Conversion and Management, 157*, 600–618. doi:10.1016/j.enconman.2017.11.047

Amos, W. A. (1998). *Costs of storing and transporting hydrogen costs of storing and transporting hydrogen*. National Renewable Energy Laboratory.

Andersson, J., & Grönkvist, S. (2019). Large-scale storage of hydrogen. *International Journal of Hydrogen Energy, 44*(23), 11901–11919. doi:10.1016/j.ijhydene.2019.03.063

Anton, D. L., Mosher, D. A., Tang, X., Brown, R. J., Arsenault, S., Saitta, S., Laube, B. L., Dold, R. H., Anton, D. L., Adams, J., Field, D. O. E., & Officer, P. (2007). High Density Hydrogen Storage System Demonstration Using NaAlH4 Based Complex Compound Hydrides. *Proceedings of the Annual U.S. DOE Hydrogen Program Review.*

Anton, D. L., Mosher, D. A., Opalka, S. M., Tang, X., Laube, B. L., Brown, R. J., Thomas, H., Arsenault, S., Wu, R., Strickler, J., Anton, D. L., Zidan, R., Berseth, P., Read, C., Headquarters, D. O. E., Manager, T., Adams, J., Field, D. O. E., & Officer, P. (2008). Complex Hydride Compounds with Enhanced Hydrogen Storage Capacity Final Report. *Proceedings of the 2004 Annual U.S. DOE Hydrogen Program Review.*

Aoyama, T., Iiyama, A., Shinohara, K., Kamegaya, S., Yamamoto, S., & Ban, Y. (2009). Status of FCV development at Nissan and future issues. *SAE International Journal of Engines, 1*(1), 314–323. doi:10.4271/2008-01-0423

Apak, S., Atay, E., & Tuncer, G. (2012). Renewable hydrogen energy regulations, codes and standards: Challenges faced by an EU candidate country. *International Journal of Hydrogen Energy, 37*(7), 5481–5497. doi:10.1016/j.ijhydene.2012.01.005

Apostolou, D., Enevoldsen, P., & Xydis, G. (2019). Supporting green urban mobility – the case of a small-scale autonomous hydrogen refuelling station. *International Journal of Hydrogen Energy, 44*(20), 9675–9689. doi:10.1016/j.ijhydene.2018.11.197

Apostolou, D., & Xydis, G. (2019). A literature review on hydrogen refuelling stations and infrastructure. current status and future prospects. *Renewable & Sustainable Energy Reviews, 113*, 109292. doi:10.1016/j.rser.2019.109292

Appleby, A. J., & Foulkes, F. R. (1993). *Fuel cell handbook. Van Norstand Reinhold, New York, NY, (1989).* Republished by Krieger Publishing Company.

Arico, A. S., Baglio, V., & Antonucci, V. (2009). *Direct methanol fuel cells: history, status and prespectives*. Wiley-VCH.

Armaroli, N., & Balzani, V. (2011). The hydrogen issue. *ChemSusChem*, *4*(1), 21–36. doi:10.1002/cssc.201000182 PMID:21226208

Aşchilean, I., Cobîrzan, N., Bolboaca, A., Boieru, R., & Felseghi, R. A. (2021). Pairing solar power to sustainable energy storage solutions within a residential building: A case study. *International Journal of Energy Research*, *45*(10), 15495–15511. doi:10.1002/er.6982

Asif, U., & Schmidt, K. (2021). Fuel Cell Electric Vehicles (Fcev): Policy Advances To Enhance Commercial Success. *Sustainability (Basel)*, *13*(9), 5149. doi:10.3390u13095149

Aso, S., Kizaki, M., & Mizuno, H. (2009). Development Progress Of The Toyota Fuel Cell Hybrid Vehicle. *Sae International Journal Of Engines*, *1*(1), 296–303. doi:10.4271/2008-01-0420

Astbury, G. R., & Hawksworth, S. J. (2007). Spontaneous ignition of hydrogen leaks: A review of postulated mechanisms. *International Journal of Hydrogen Energy*, *32*(13), 2178–2185. doi:10.1016/j.ijhydene.2007.04.005

Atems, B., Mette, J., Lin, G., & Madraki, G. (2023). Estimating and forecasting the impact of nonrenewable energy prices on US renewable energy consumption. *Energy Policy*, *173*, 113374. doi:10.1016/j.enpol.2022.113374

Ayers, K., Danilovic, N., Ouimet, R., Carmo, M., Pivovar, B., & Bornstein, M. (2019). Perspectives on low-temperature electrolysis and potential for renewable hydrogen at scale. *Annual Review of Chemical and Biomolecular Engineering*, *10*(1), 219–239. doi:10.1146/annurev-chembioeng-060718-030241 PMID:31173524

Ayol, A. (2022). *ScienceDirect Yeast industry wastewater treatment with microbial fuel cells : Effect of electrode materials and reactor configurations*. doi:10.1016/j.ijhydene.2022.05.277

Azam, W., Khan, I., & Ali, S. A. (2023). Alternative energy and natural resources in determining environmental sustainability: A look at the role of government final consumption expenditures in France. *Environmental Science and Pollution Research International*, *30*(1), 1949–1965. doi:10.100711356-022-22334-z PMID:35925458

Aziz, M. (2021). Liquid Hydrogen: A Review on Liquefaction, Storage, Transportation, and Safety. *Energies*, *14*(18). doi:ARTN 591710.3390/en14185917

Badea, G., Felseghi, R. A., & Aschilean, I. (2021). Hydrogen fuel cell technology for stationary applications. *Hydrogen Fuel Cell Technology for Stationary Applications*, 1–332. doi:10.4018/978-1-7998-4945-2

Badea, G., Felseghi, R. A., Aşchilean, I., Răboacă, S. M., & Şoimoşan, T. (2017, December). The role of hydrogen as a future solution to energetic and environmental problems for residential buildings. In AIP Conference Proceedings (Vol. 1917, pp. 030004). AIP Publishing LLC. doi:10.1063/1.5018277

Badea, G., Felseghi, R. A., & Aşchilean, I. (2021). Hydrogen-Energy Vector Within a Sustainable Energy System for Stationary Applications. In *Hydrogen Fuel Cell Technology for Stationary Applications* (pp. 1–21). IGI Global. doi:10.4018/978-1-7998-4945-2.ch001

Badea, G., Felseghi, R. A., Varlam, M., Filote, C., Culcer, M., Iliescu, M., & Răboacă, M. S. (2018). Design and simulation of romanian solar energy charging station for electric vehicles. *Energies*, *12*(1), 74. doi:10.3390/en12010074

Badea, G., Felseghi, R., Aschilean, I., Bolboaca, A., Muresan, D., Soimosan, T., & Raboaca, S. (2015, August). Energen system for power supply of passive house: case study. In *2015 Second International Conference on Mathematics and Computers in Sciences and in Industry (MCSI)* (pp. 24-31). IEEE. 10.1109/MCSI.2015.31

Bae, C., & Kim, J. (2017). Alternative Fuels For Internal Combustion Engines. *Proceedings Of The Combustion Institute*, *36*(3), 3389–3413. doi:10.1016/j.proci.2016.09.009

Bailey, D. (2015). *How The Automobile Changed History.* Abdo.

Bai, S., Tian, K., Meng, J. C., Zhao, Y., Sun, J., Zhang, K., Feng, Y., Luo, R., Li, D., & Chen, A. (2021). Reduced graphene oxide decorated SnO2/BiVO4 photoanode for photoelectrochemical water splitting. *Journal of Alloys and Compounds*, *855*, 156780. doi:10.1016/j.jallcom.2020.156780

Baker, C. R., & Shaner, R. L. (1978). A study of the efficiency of hydrogen liquefaction. *International Journal of Hydrogen Energy*, *3*(3), 321–334. doi:10.1016/0360-3199(78)90037-X

Banerjee, S., Nigam, S., Pillai, C. G. S., & Majumder, C. (2012). Hydrogen storage on Ti decorated SiC nanostructures: A first principles study. *International Journal of Hydrogen Energy*, *37*(4), 3733–3740. doi:10.1016/j.ijhydene.2011.05.078

Barckholtz, T. A., Taylor, K. M., Narayanan, S., Jolly, S., & Ghezel-Ayagh, H. (2022). Molten carbonate fuel cells for simultaneous CO2 capture, Power Generation, and H2 generation. *Applied Energy*, *313*, 118553. doi:10.1016/j.apenergy.2022.118553

Bargal, M. H. S., Abdelkareem, M. A. A., Tao, Q., Li, J., Shi, J., & Wang, Y. (2020a). Liquid cooling techniques in proton exchange membrane fuel cell stacks: A detailed survey. *Alexandria Engineering Journal*, *59*(2), 635–655. doi:10.1016/j.aej.2020.02.005

Baroutaji, A., Carton, J., Stokes, J., & Olabi, A. G. (2014). Design And Development Of Proton Exchange Membrane Fuel Cell Using Open Pore Cellular Foam As Flow Plate Material. *Journal Of Energy Challenges And Mechanics*, *1*(7).

Barrett, S. (2014). Dissecting 'HySA.' *Renewable Energy Focus*, *15*(1), 40–44. doi.org/10.1016/S1755-0084(14)70024-9

Barron, R. f., & Nellis, G. f. (1999). *CryogeniC Heat transfer.* CRC Press. doi:10.1201/b15230

Barthelemy, H., Weber, M., & Barbier, F. (2016). Hydrogen storage : Recent improvements and industrial perspectives. *International Journal of Hydrogen Energy*, *42*(11), 7254–7262. doi:10.1016/j.ijhydene.2016.03.178

Basic Research Needs for the Hydrogen Economy. (2003). Report on the Basic Energy Sciences Workshop on Hydrogen Production, Storage, and Use.

Basic research needs for the hydrogen economy. (2004). www.sc.doe.gov/bes/hydrogen.pdf

Bauman, J., & Kazerani, M. (2008). A Comparative Study Of Fuel-Cell–Battery, Fuel-Cell–Ultracapacitor, And Fuel-Cell–Battery–Ultracapacitor Vehicles. *Ieee Transactions On Vehicular Technology*, *57*(2), 760–769. doi:10.1109/TVT.2007.906379

Becker, B., & Mair, G. W. (2017). Risks and safety level of composite cylinders. *International Journal of Hydrogen Energy*, *42*(19), 13810–13817. doi:10.1016/j.ijhydene.2017.01.145

Bellocchi, S., Colbertaldo, P., Manno, M., & Nastasi, B. (2023). Assessing the effectiveness of hydrogen pathways: A techno-economic optimisation within an integrated energy system. *Energy*, *263*, 126017. doi:10.1016/j.energy.2022.126017

Bessarabov, D., Human, G., Kruger, A. J., Chiuta, S., Modisha, P. M., du Preez, S. P., Oelofse, S. P., Vincent, I., Van Der Merwe, J., Langmi, H. W., Ren, J., & Musyoka, N. M. (2017). South African hydrogen infrastructure (HySA infrastructure) for fuel cells and energy storage: Overview of a projects portfolio. *International Journal of Hydrogen Energy*, *42*(19), 13568–13588. doi:10.1016/j.ijhydene.2016.12.140

Bessette, N. F. II, & Wepfer, W. J. (1996). Prediction of on-design and off-design performance for a solid oxide fuel cell power module. *Energy Conversion and Management*, *37*(3), 281–293. doi:10.1016/0196-8904(95)00188-3

Bhandari, R., Trudewind, C. A., & Zapp, P. (2014). Life cycle assessment of hydrogen production via electrolysis – A Review. *Journal of Cleaner Production, 85*, 151–163. doi:10.1016/j.jclepro.2013.07.048

Bhaskaran, R., Abraham, B. G., & Chetty, R. (2021). Recent advances in electrocatalysts, mechanism, and cell architecture for direct formic acid fuel cells. *WIREs Energy and Environment*, *11*(2). Advance online publication. doi:10.1002/wene.419

Bhatia, S. K., & Myers, A. L. (2006). Optimum conditions for adsorptive storage. *Langmuir*, *22*(4), 1688–1700. doi:10.1021/la0523816 PMID:16460092

Bicer, Y., & Dincer, I. (2017). Comparative life cycle assessment of hydrogen, methanol and electric vehicles from well to wheel. *International Journal of Hydrogen Energy*, *42*(6), 3767–3777. doi:10.1016/j.ijhydene.2016.07.252

Bicer, Y., & Dincer, I. (2018). Life Cycle Environmental Impact Assessments And Comparisons Of Alternative Fuels For Clean Vehicles. *Resources, Conservation And Recycling*, *132*, 141–157. doi:10.1016/j.resconrec.2018.01.036

Blomen, L. J. M. J., & Mugerwa, M. N. (2013). *Fuel cell systems*. Springer Science & Business Media.

Bockris, J. O. M. (2013). The Hydrogen Economy: Its History. *International Journal Of Hydrogen Energy, 38*(6), 2579–2588. doi:10.1016/j.ijhydene.2012.12.026

Bogdanović, B., & Schwickardi, M. (1997). Ti-doped alkali metal aluminium hydrides as potential novel reversible hydrogen storage materials. *Journal of Alloys and Compounds, 253*, 1–9. doi:10.1016/S0925-8388(96)03049-6

Bose, S., Kuila, T., Nguyen, T. X. H., Kim, N. H., Lau, K., & Lee, J. H. (2011). Polymer membranes for high temperature proton exchange membrane fuel cell: Recent advances and challenges. *Progress in Polymer Science, 36*(6), 813–843. doi:10.1016/j.progpolymsci.2011.01.003

Brandon, N. P., & Kurban, Z. (2017). Clean energy and the hydrogen economy. *Philosophical Transactions of the Royal Society a-Mathematical Physical and Engineering Sciences, 375*(2098). doi:ARTN 2016040010.1098/rsta.2016.0400

Budde, B., Alkemade, F., & Hekkert, M. (2015). On The Relation Between Communication And Innovation Activities: A Comparison Of Hybrid Electric And Fuel Cell Vehicles. *Environmental Innovation And Societal Transitions, 14*, 45–59. doi:10.1016/j.eist.2013.11.003

Cahill, A. E., Aiello-Lammens, M. E., Fisher-Reid, M. C., Hua, X., Karanewsky, C. J., Yeong Ryu, H., . . . Wiens, J. J. (2013). How does climate change cause extinction? *Proceedings of the Royal Society B: Biological Sciences, 280*(1750), 20121890. 10.1098/rspb.2012.1890

Carpetis, C. (1982). Estimation of storage costs for large hydrogen storage facilities. *International Journal of Hydrogen Energy, 7*(2), 191–203. doi:10.1016/0360-3199(82)90146-X

Carrette, L., Friedrich, K. A., & Stimming, U. (2000). Fuel Cells: Principles, Types, Fuels, and Applications. *ChemPhysChem, 1*(4), 162–193. doi:10.1002/1439-7641(20001215)1:4<162::AID-CPHC162>3.0.CO;2-Z PMID:23696319

Carrillo, R. J., & Scheffe, J. R. (2017). Advances and trends in redox materials for solar thermochemical fuel production. *Solar Energy, 156*, 3–20. doi:10.1016/j.solener.2017.05.032

Cawthorn, R. G. (2010). The platinum group element deposits of the Bushveld Complex in South Africa. *Platinum Metals Review, 54*(4), 205–215. doi:10.1595/147106710X520222

Cengel, Y. A., & Boles, M. A. (2002). *Thermodynamics: An Engineering Approach*. McGraw-Hill Higher Education.

CernatA.PanaC.NegurescuN.NutuC.FuiorescuD.LazaroiuG. Aspects of an Experimental Study of a Hydrogen Use at Automotive Diesel Engine. *Available at* SSRN 4315127. doi:10.1016/j.heliyon.2023.e13889

Cerniauskas, S., Junco, A. J. C., Grube, T., Robinius, M., & Stolten, D. (2020). Options of natural gas pipeline reassignment for hydrogen: Cost assessment for a Germany case study. *International Journal of Hydrogen Energy, 45*(21), 12095–12107. doi:10.1016/j.ijhydene.2020.02.121

Chan, C. C. (2002). The State Of The Art Of Electric And Hybrid Vehicles. *Proceedings Of The Ieee, 90*(2), 247–275. doi:10.1109/5.989873

Chan, C. C., & Wong, Y. S. (2004). Electric Vehicles Charge Forward. *Ieee Power & Energy Magazine*, *2*(6), 24–33. doi:10.1109/MPAE.2004.1359010

Change, I. C. (2013). The physical science basis. *Contribution of working group I to the fifth assessment report of the intergovernmental panel on climate change, 1535.*

Chapelle, D., Maynadier, A., Bebon, L., & Thiébaud, F. (2021). Hydrogen Storage: Different Technologies, Challenges and Stakes. Focus on TiFe Hydrides. In *Advances in Renewable Hydrogen and Other Sustainable Energy Carriers* (pp. 1–10). Springer. doi:10.1007/978-981-15-6595-3_1

Chau, K. T., & Wong, Y. S. (2002). Overview Of Power Management In Hybrid Electric Vehicles. *Energy Conversion And Management*, *43*(15), 1953–1968. doi:10.1016/S0196-8904(01)00148-0

Chaurasia, A. K., & Mondal, P. (2022). Enhancing biohydrogen production from sugar industry wastewater using Ni, Ni–Co and Ni–Co–P electrodeposits as cathodes in microbial electrolysis cells. *Chemosphere, 286*(November). doi:10.1016/j.chemosphere.2021.131728

Chaurasia, A. K., Mohapatra, S., Shankar, R., & Thakur, L. S. (2022). Technologies for the clean and renewable energy production for the sustainable environment. *Clean Technologies and Sustainable Development in Civil Engineering*, 141–178. doi:10.4018/978-1-7998-9810-8.ch007

Chaurasia, A. K., Goyal, H., & Mondal, P. (2020). Hydrogen gas production with Ni, Ni–Co and Ni–Co–P electrodeposits as potential cathode catalysts by microbial electrolysis cells. *International Journal of Hydrogen Energy*, *45*(36), 18250–18265. doi:10.1016/j.ijhydene.2019.07.175

Chaurasia, A. K., Shankar, R., & Mondal, P. (2021). Effects of nickle, nickle-cobalt and nickle-cobalt-phosphorus nanocatalysts for enhancing biohydrogen production in microbial electrolysis cells using paper industry wastewater. *Journal of Environmental Management*, *298*, 113542. doi:10.1016/j.jenvman.2021.113542 PMID:34426219

Chaurasia, A. K., Siwach, P., Shankar, R., & Mondal, P. (2021). Effect of pre-treatment on mesophilic anaerobic co-digestion of fruit, food and vegetable waste. *Clean Technologies and Environmental Policy*, *1*, 1–14. doi:10.100710098-021-02218-5

Chen, Y., Sequeira, C. A. C., Chen, C., Wang, X., & Wang, Q. (2003). Metal hydride beds and hydrogen supply tanks as minitype PEMFC hydrogen sources. *International Journal of Hydrogen Energy*, *28*(3), 329–333. doi:10.1016/S0360-3199(02)00064-2

Chi, J., & Yu, H. (2018). Water electrolysis based on renewable energy for hydrogen production. *Cuihua Xuebao. Chinese Journal of Catalysis*, *39*(3), 390–394. doi:10.1016/S1872-2067(17)62949-8

Chisalita, D. A., Petrescu, L., Galusnyak, S. C., & Cormos, C. C. (2023). Environmental evaluation of hydrogen production employing innovative chemical looping technologies–A Romanian case study. *International Journal of Hydrogen Energy*, *48*(32), 12112–12128. doi:10.1016/j.ijhydene.2022.06.029

Choi, W., Yoo, E., Seol, E., Kim, M., & Song, H. H. (2020). Greenhouse Gas Emissions Of Conventional And Alternative Vehicles: Predictions Based On Energy Policy Analysis In South Korea. *Applied Energy*, *265*, 114754. doi:10.1016/j.apenergy.2020.114754

Cindrella, L., Kannan, A. M., Lin, J. F., Saminathan, K., Ho, Y., Lin, C. W., & Wertz, J. (2009). Gas diffusion layer for proton exchange membrane fuel cells—A review. *Journal of Power Sources*, *194*(1), 146–160. doi:10.1016/j.jpowsour.2009.04.005

Ciura, A. (2019). *Sustainable development of the public transport sector in EU Countries.*

Clarke, M. C. (2022). Can the hydrogen economy concept be the solution to the future energy crisis? *Australian Journal of Multi-Disciplinary Engineering*, *18*(1), 70–84. doi:10.1080/14488388.2022.2046325

Cloete, S., Ruhnau, O., & Hirth, L. (2021). On capital utilization in the hydrogen economy: The quest to minimize idle capacity in renewables-rich energy systems. *International Journal of Hydrogen Energy*, *46*(1), 169–188. doi:10.1016/j.ijhydene.2020.09.197

Cobîrzan, N., Thalmaier, G., Balog, A. A., Constantinescu, H., Timiş, I., & Streza, M. (2018). Thermophysical properties of fired clay bricks with waste ceramics and paper pulp as pore-forming agent. *Journal of Thermal Analysis and Calorimetry*, *134*(1), 843–851. doi:10.100710973-018-7092-3

Colbertaldo, P., Agustin, S. B., Campanari, S., & Brouwer, J. (2019). Impact of hydrogen energy storage on California electric power system: Towards 100% renewable electricity. *International Journal of Hydrogen Energy*, *44*(19), 9558–9576. doi:10.1016/j.ijhydene.2018.11.062

College of the Desert. (2001). *Hydrogen Fuel Cell Engines and Related Technologies: Report.* Author.

Colozza, A. J. (2002). *Hydrogen Storage for Aircraft Applications: Overview.* Report No. NASA/CR-211867.

Conte, M., Iacobazzi, A., Ronchetti, M., & Vellone, R. (2001). Hydrogen Economy For A Sustainable Development: State-Of-The-Art And Technological Perspectives. *Journal Of Power Sources*, *100*(1-2), 171–187. doi:10.1016/S0378-7753(01)00893-X

Cornwell, O., Radford, S. E., Ashcroft, A. E., & Ault, J. R. (2018). Comparing hydrogen deuterium exchange and fast photochemical oxidation of proteins: A structural characterisation of wild-type and ΔN6 β2-microglobulin. *Journal of the American Society for Mass Spectrometry*, *29*(12), 2413–2426. doi:10.100713361-018-2067-y PMID:30267362

Coyle, E. D., & Simmons, R. A. (2014). *Understanding the global energy crisis*. Purdue University Press. doi:10.26530/OAPEN_469619

Cramer, L. A. (2001). The extractive metallurgy of South Africa's platinum ores. *Journal of the Minerals Metals & Materials Society*, *53*(10), 14–18. doi:10.100711837-001-0048-1

Cui, C., Li, S., Gong, J., Wei, K., Hou, X., Jiang, C., & Ma, J. (2021). Review of molten carbonate-based direct carbon fuel cells. *Materials for Renewable and Sustainable Energy*, *10*(2), 12. Advance online publication. doi:10.100740243-021-00197-7

Cui, Y., Zhang, J., He, H., & Qian, G. (2018). Photonic functional metal–organic frameworks. *Chemical Society Reviews*, *47*(15), 5740–5785. doi:10.1039/C7CS00879A PMID:30046798

Cunanan, C., Tran, M. K., Lee, Y., Kwok, S., Leung, V., & Fowler, M. (2021). A Review Of Heavy-Duty Vehicle Powertrain Technologies: Diesel Engine Vehicles, Battery Electric Vehicles, And Hydrogen Fuel Cell Electric Vehicles. *Clean Technologies*, *3*(2), 474–489. doi:10.3390/cleantechnol3020028

Dankers, R., & Feyen, L. (2008). Climate change impact on flood hazard in Europe: An assessment based on high-resolution climate simulations. *Journal of Geophysical Research*, *113*(D19), D19105. doi:10.1029/2007JD009719

Das, D., & Veziroğlu, T. N. (2001). Hydrogen Production By Biological Processes: A Survey Of Literature. *International Journal Of Hydrogen Energy*, *26*(1), 13–28. doi:10.1016/S0360-3199(00)00058-6

Daş, E., & Yurtcan, A. B. (2016). Effect Of Carbon Ratio In The Polypyrrole/Carbon Composite Catalyst Support On Pem Fuel Cell Performance. *International Journal Of Hydrogen Energy*, *41*(30), 13171–13179. doi:10.1016/j.ijhydene.2016.05.167

Das, H. S., Tan, C. W., & Yatim, A. H. M. (2017). Fuel Cell Hybrid Electric Vehicles: A Review On Power Conditioning Units And Topologies. *Renewable & Sustainable Energy Reviews*, *76*, 268–291. doi:10.1016/j.rser.2017.03.056

Dash, S. K., Chakraborty, S., Roccotelli, M., & Sahu, U. K. (2022). Hydrogen Fuel For Future Mobility: Challenges And Future Aspects. *Sustainability (Basel)*, *14*(14), 8285. doi:10.3390u14148285

Dawood, F., Anda, M., & Shafiullah, G. M. (2020). Hydrogen production for energy: An overview. *International Journal of Hydrogen Energy*, *45*(7), 3847–3869. doi:10.1016/j.ijhydene.2019.12.059

de Carvalho Rocha, W. F., & Sheen, D. A. (2019). Determination of physicochemical properties of petroleum derivatives and biodiesel using GC/MS and chemometric methods with uncertainty estimation. *Fuel*, *243*, 413–422. doi:10.1016/j.fuel.2018.12.126

De Las Heras, N., Roberts, E. P. L., Langton, R., & Hodgson, D. R. (2009). A Review Of Metal Separator Plate Materials Suitable For Automotive Pem Fuel Cells. *Energy & Environmental Science*, *2*(2), 206–214. doi:10.1039/B813231N

De Vries, J. (1994). The industrial revolution and the industrious revolution. *The Journal of Economic History*, *54*(2), 249–270. doi:10.1017/S0022050700014467

Demirbas, A. Y. H. A. N., & Arin, G. O. N. E. N. C. (2004). Hydrogen from biomass via pyrolysis: Relationships between yield of hydrogen and temperature. *Energy Sources*, *26*(11), 1061–1069. doi:10.1080/00908310490494568

Department-of-Energy-US. (2022). *How Do Fuel Cell Electric Vehicles Work Using Hydrogen?* Energy Efficiency and Renewable Energy.

Desideri, U., & Yan, J. (2012). *Clean Energy Technologies And Systems For A Sustainable World.*

Dewangan, S. K., Mohan, M., Kumar, V., Sharma, A., & Ahn, B. (2022). A comprehensive review of the prospects for future hydrogen storage in materials-application and outstanding issues. *International Journal of Energy Research, 46*(12), 16150–16177. Advance online publication. doi:10.1002/er.8322

DiLisi, G. A. (2017). The Hindenburg Disaster: Combining Physics and History in the Laboratory. *The Physics Teacher, 55*(5), 268–273. doi:10.1119/1.4981031

Dillman, K., & Heinonen, J. (2023). Towards a Safe Hydrogen Economy: An Absolute Climate Sustainability Assessment of Hydrogen Production. *Climate (Basel), 11*(1), 25. doi:10.3390/cli11010025

Dincer, I. (2020). Covid-19 coronavirus: Closing carbon age, but opening hydrogen age. *International Journal of Energy Research, 44*(8), 6093–6097. doi:10.1002/er.5569 PMID:34173427

Dincer, I., & Zamfirescu, C. (2012). Sustainable hydrogen production options and the role of IAHE. *International Journal of Hydrogen Energy, 37*(21), 16266–16286. doi:10.1016/j.ijhydene.2012.02.133

Ding, M., Flaig, R. W., Jiang, H.-L., & Yaghi, O. M. (2019). Carbon capture and conversion using metal–organic frameworks and MOF-based materials. *Chemical Society Reviews, 48*(10), 2783–2828. doi:10.1039/C8CS00829A PMID:31032507

Dinh, K. N., Liang, Q., Du, C. F., Zhao, J., Tok, A. I. Y., Mao, H., & Yan, Q. (2019). Nanostructured Metallic Transition Metal Carbides, Nitrides, Phosphides, And Borides For Energy Storage And Conversion. *Nano Today, 25*, 99–121. doi:10.1016/j.nantod.2019.02.008

Dones, R., Heck, T., & Hirschberg, S. (2004). Greenhouse gas emissions from energy systems, comparison and Overview. Encyclopedia of Energy, 77–95. doi:10.1016/B0-12-176480-X/00397-1

Dong, Y., Peng, W., Liu, Y., & Wang, Z. (2021). Photochemical origin of reactive radicals and halogenated organic substances in natural waters: A review. *Journal of Hazardous Materials, 401*, 123884. doi:10.1016/j.jhazmat.2020.123884 PMID:33113752

DSI. (2020). *Science and Innovation on South Africa's Platinum Valley project*. Department of Science and Innovation, SA Government.

Duan, Q. L., Zeng, Q., Jin, K. Q., Wang, Q. S., & Sun, J. H. (2022). Mechanism of self-ignition and flame propagation during high-pressure hydrogen release through a rectangular tube. *Process Safety and Environmental Protection, 164*, 283–290. doi:10.1016/j.psep.2022.06.017

Dudley, B. (2018). BP statistical review of world energy. *BP Statistical Review, 6*(2018).

Dufo-López, R., & Bernal-Agustín, J. L. (2020). *iHOGA PRO+ Software Version 2.5*. Electrical Engineering Department, University of Zaragoza. Available online: https://ihoga.unizar.es/

Durbin, D. J., & Malardier-Jugroot, C. (2013). Review of hydrogen storage techniques for on board vehicle applications. *International Journal of Hydrogen Energy*, *38*(34), 14595–14617. doi:10.1016/j.ijhydene.2013.07.058

Dwivedi, S. K., & Vishwakarma, M. (2018). Hydrogen embrittlement in different materials: A review. *International Journal of Hydrogen Energy*, *43*(46), 21603–21616. doi:10.1016/j.ijhydene.2018.09.201

Eberle, U., Felderhoff, M., & Schueth, F. (2009). Chemical and physical solutions for hydrogen storage. *Angewandte Chemie International Edition*, *48*(36), 6608–6630. doi:10.1002/anie.200806293 PMID:19598190

EDN. (2006). *The promise of more rugged and economical fuel cell stacks*. EDN. https://www.edn.com/the-promise-of-more-rugged-and-economical-fuel-cell-stacks/

Edwards, P. P., Kuznetsov, V. L., David, W. I., & Brandon, N. P. (2008). Hydrogen And Fuel Cells: Towards A Sustainable Energy Future. *Energy Policy*, *36*(12), 4356–4362. doi:10.1016/j.enpol.2008.09.036

Eiler, J. M. (2020). The Once and Future Hydrogen Economy. *Nature Journal*.

El-Emam, R. S., & Khamis, I. (2019). Advances in nuclear hydrogen production: Results from an IAEA International Collaborative Research Project. *International Journal of Hydrogen Energy*, *44*(35), 19080–19088. doi:10.1016/j.ijhydene.2018.04.012

Emadi, A., Williamson, S. S., & Khaligh, A. (2006). Power Electronics Intensive Solutions For Advanced Electric, Hybrid Electric, And Fuel Cell Vehicular Power Systems. *Ieee Transactions On Power Electronics*, *21*(3), 567–577. doi:10.1109/TPEL.2006.872378

Enescu, F. M., Birleanu, F. G., Raboaca, M. S., Raceanu, M., Bizon, N., & Thounthong, P. (2023). Electric Vehicle Charging Station Based on Photovoltaic Energy with or without the Support of a Fuel Cell–Electrolyzer Unit. *Energies*, *16*(2), 762. doi:10.3390/en16020762

Engelbrecht, J. (2012). Potential changes in the physical beneficiation processes that can improve the recovery grade or costs for the platinum group metals. *Proceedings of the Fifth International Platinum Conference: A Catalyst for Change*. Springer.

Engineering ToolBox. (2008). *Fossil and Alternative Fuels - Energy Content*. Available at: https://www.engineeringtoolbox.com/fossil-fuels-energy-content-d_1298.html

European Hydrogen and Fuel Cell Technology Platform. (2005), Strategic Research Agenda. fch. https://www.fch.europa.eu/sites/default/files/documents/hfp-sra004_v9-2004_sra-report-final_ 22jul2005.pdf

Evans, J. D., Bon, V., Senkovska, I., Lee, H.-C., & Kaskel, S. (2020). Four-dimensional metal-organic frameworks. *Nature Communications*, *11*(1), 2690. doi:10.103841467-020-16527-8 PMID:32483346

Falcone, P. M., Hiete, M., & Sapio, A. (2021). Hydrogen Economy And Sustainable Development Goals: Review And Policy Insights. *Current Opinion In Green And Sustainable Chemistry*, *31*, 100506. doi:10.1016/j.cogsc.2021.100506

Fan, L., Tu, Z., & Chan, S. H. (2021). Recent Development Of Hydrogen And Fuel Cell Technologies: A Review. *Energy Reports*, *7*, 8421–8446. doi:10.1016/j.egyr.2021.08.003

Farias, C. B. B., Barreiros, R. C. S., da Silva, M. F., Casazza, A. A., Converti, A., & Sarubbo, L. A. (2022). Use of hydrogen as fuel: A trend of the 21st century. *Energies*, *15*(1), 311. doi:10.3390/en15010311

Fayaz, H., Saidur, R., Razali, N., Anuar, F. S., Saleman, A. R., & Islam, M. R. (2012). An Overview Of Hydrogen As A Vehicle Fuel. *Renewable & Sustainable Energy Reviews*, *16*(8), 5511–5528. doi:10.1016/j.rser.2012.06.012

Faye, O., Szpunar, J., & Eduok, U. (2022). A critical review on the current technologies for the generation, storage, and transportation of hydrogen. *International Journal of Hydrogen Energy*, *47*(29), 13771–13802. doi:10.1016/j.ijhydene.2022.02.112

Felderhoff, M., & Bogdanovi, B. (2009). High temperature metal hydrides as heat storage materials for solar and related applications. *International Journal of Molecular Sciences*, *10*(1), 325–344. doi:10.3390/ijms10010325 PMID:19333448

Felseghi, R. A., & Papp, A. (2014). Considerations regarding hybrid systems of power generator from renewable energy sources. *Studia Universitatis Babes-Bolyai, Ambientum, 59.*

Felseghi, R. A., Şoimoşan, T. M., Filote, C., & Răboaca, M. S. (2019, b). Considerations Regarding the Green Retrofitting of Residential Buildings From Human Wellbeing Perspectives. In Retrofitting for Optimal Energy Performance (pp. 143-175). IGI Global.

Felseghi, R. A., Aşchilean, I., Cobîrzan, N., Bolboacă, A. M., & Raboaca, M. S. (2021). Optimal Synergy between Photovoltaic Panels and Hydrogen Fuel Cells for Green Power Supply of a Green Building—A Case Study. *Sustainability (Basel)*, *13*(11), 6304. doi:10.3390u13116304

Felseghi, R. A., Carcadea, E., Raboaca, M. S., Trufin, C. N., & Filote, C. (2019). a). Hydrogen fuel cell technology for the sustainable future of stationary applications. *Energies*, *12*(23), 4593. doi:10.3390/en12234593

Fernández, R. Á., Cilleruelo, F. B., & Martínez, I. V. (2016). A New Approach To Battery Powered Electric Vehicles: A Hydrogen Fuel-Cell-Based Range Extender System. *International Journal Of Hydrogen Energy*, *41*(8), 4808–4819. doi:10.1016/j.ijhydene.2016.01.035

Filote, C., Felseghi, R. A., Raboaca, M. S., & Aşchilean, I. (2020). Environmental impact assessment of green energy systems for power supply of electric vehicle charging station. *International Journal of Energy Research*, *44*(13), 10471–10494. doi:10.1002/er.5678

Flynn, R., Ricci, M., & Bellaby, P. (2012). Ambiguity, complexity and uncertainty surrounding the hazards of hydrogen and public views of emergent risks. *Journal of Risk Research*, *15*(4), 373–387. doi:10.1080/13669877.2011.634517

Foorginezhad, S., Mohseni-Dargah, M., Falahati, Z., Abbassi, R., Razmjou, A., & Asadnia, M. (2021). Sensing advancement towards safety assessment of hydrogen fuel cell vehicles. *Journal of Power Sources*, *489*. doi:ARTN 22945010.1016/j.jpowsour.2021.229450

Forrest, K., Mac Kinnon, M., Tarroja, B., & Samuelsen, S. (2020). Estimating The Technical Feasibility Of Fuel Cell And Battery Electric Vehicles For The Medium And Heavy Duty Sectors In California. *Applied Energy*, *276*, 115439. doi:10.1016/j.apenergy.2020.115439

Fredriksson, J., Weiefors, H., & Egardt, B. (2002). Powertrain Control For Active Damping Of Driveline Oscillations. *Vehicle System Dynamics*, *37*(5), 359–376. doi:10.1076/vesd.37.5.359.3527

Frenette, G., & Forthoffer, D. (2009). Economic & Commercial Viability Of Hydrogen Fuel Cell Vehicles From An Automotive Manufacturer Perspective. *International Journal Of Hydrogen Energy*, *34*(9), 3578–3588. doi:10.1016/j.ijhydene.2009.02.072

Froeling, H. A. J., Droge, M. T., Nane, G. F., & Van Wijk, A. J. M. (2021). Quantitative risk analysis of a hazardous jet fire event for hydrogen transport in natural gas transmission pipelines. *International Journal of Hydrogen Energy*, *46*(17), 10411–10422. doi:10.1016/j.ijhydene.2020.11.248

Furukawa, H., Cordova, K. E., O'Keeffe, M., & Yaghi, O. M. (2013). The chemistry and applications of metal-organic frameworks. *Science*, *341*(6149), 1230444. doi:10.1126cience.1230444 PMID:23990564

Fuster, B., Houssin-Agbomson, D., Jallais, S., Vyazmina, E., Guy, D. N., Bernard-Michel, G., ... Kinderen, J. D. (2017). Guidelines and recommendations for indoor use of fuel cells and hydrogen systems. *International Journal of Hydrogen Energy*, *42*(11), 7600–7607. doi:10.1016/j.ijhydene.2016.05.266

Gallagher, K. S. (2014). *The globalization of clean energy technology: Lessons from China*. MIT press. doi:10.7551/mitpress/9805.001.0001

Gao, Q. X., Wang, L. J., Peng, W., Zhang, P., & Chen, S. Z. (2022). Safety analysis of leakage in a nuclear hydrogen production system. *International Journal of Hydrogen Energy*, *47*(7), 4916–4931. doi:10.1016/j.ijhydene.2021.11.099

Gardiner, M., & Satyapal, S. (2009). *Energy Requirements for Hydrogen Gas Compression and Liquefaction as Related to Vehicle Storage Needs*. DOE Hydrogen and Fuel Cells Program Record. Available online:https://www.hydrogen.energy.gov/pdfs/9013_energy_requirements_for_hydrogen_gas_compression.pdf

Garland, N. L., Papageorgopoulos, D. C., & Stanford, J. M. (2012). Hydrogen and fuel cell technology: Progress, challenges, and Future Directions. *Energy Procedia*, *28*, 2–11. doi:10.1016/j.egypro.2012.08.034

Garrison, S. L., Hardy, B. J., Gorbounov, M. B., Tamburello, D. A., Corgnale, C., Bart, A., Mosher, D. A., & Anton, D. L. (2012). Optimization of internal heat exchangers for hydrogen storage tanks utilizing metal hydrides. *International Journal of Hydrogen Energy, 37*(3), 2850–2861. doi:10.1016/j.ijhydene.2011.07.044

Garside, M. (2023). *Production volume of platinum in South Africa from 2010 to 2022*. Statista. https://www.statista.com/statistics/1015227/platinum-production-south-africa/

Genovese, M., Schlüter, A., Scionti, E., Piraino, F., Corigliano, O., & Fragiacomo, P. (2023). Power-to-hydrogen and hydrogen-to-X energy systems for the industry of the future in Europe. *International Journal of Hydrogen Energy, 48*(44), 16545–16568. doi:10.1016/j.ijhydene.2023.01.194

Gerrard, M. B. (2011). *The law of clean energy: efficiency and renewables.*

Gharari, R., Kazeminejad, H., Kojouri, N. M., & Hedayat, A. (2018). A review on hydrogen generation, explosion, and mitigation during severe accidents in light water nuclear reactors. *International Journal of Hydrogen Energy, 43*(4), 1939–1965. doi:10.1016/j.ijhydene.2017.11.174

Gökçek, M., & Kale, C. (2018). Optimal design of a hydrogen refuelling station (HRFS) powered by hybrid power system. *Energy Conversion and Management, 161*, 215–224. doi:10.1016/j.enconman.2018.02.007

Gomes, S., Renaudin, G., Hagemann, H., Yvon, K., Sulic, M. P., & Jensen, C. M. (2005). Effects of milling, doping and cycling of NaAlH4 studied by vibrational spectroscopy and X-ray diffraction. *Journal of Alloys and Compounds, 390*(1-2), 305–313. doi:10.1016/j.jallcom.2004.08.036

Goor, M., Menkin, S., & Peled, E. (2018). High power direct methanol fuel cell for mobility and portable applications. *International Journal of Hydrogen Energy, 44*(5), 3138–3143. doi:10.1016/j.ijhydene.2018.12.019

Gopal, M. R., & Murthy, S. S. (1995). Performance of a metal hydride cooling system. *International Journal of Refrigeration, 18*(6), 413–420. doi:10.1016/0140-7007(95)98164-G

Gosling, S. N., & Arnell, N. W. (2016). A global assessment of the impact of climate change on water scarcity. *Climatic Change, 134*(3), 371–385. doi:10.100710584-013-0853-x

Grimm, A., de Jong, W. A., & Kramer, G. J. (2020). Renewable hydrogen production: A techno-economic comparison of photoelectrochemical cells and photovoltaic-electrolysis. *International Journal of Hydrogen Energy, 45*(43), 22545–22555. doi:10.1016/j.ijhydene.2020.06.092

Groth, K. M., & Hecht, E. S. (2017). HyRAM: A methodology and toolkit for quantitative risk assessment of hydrogen systems. *International Journal of Hydrogen Energy, 42*(11), 7485–7493. doi:10.1016/j.ijhydene.2016.07.002

Guler, M. G., Gecici, E., & Erdogan, A. (2021). Design of a future hydrogen supply chain: A multi period model for Turkey. *International Journal of Hydrogen Energy, 46*(30), 16279–16298. doi:10.1016/j.ijhydene.2020.09.018

Guoxing, C., Tu, X., Homm, G., & Weidenkaff, A. (2022). Plasma Pyrolysis For A Sustainable Hydrogen Economy. *Nature Reviews. Materials*.

Gye, H. R., Seo, S. K., Bach, Q. V., Ha, D., & Lee, C. J. (2019). Quantitative risk assessment of an urban hydrogen refueling station. *International Journal of Hydrogen Energy*, *44*(2), 1288–1298. doi:10.1016/j.ijhydene.2018.11.035

H2Tools.org. (2022). *Lessons Learned*. Retrieved September 1, 2022, from https://h2tools.org/lessons?search_api_fulltext=

Hadef, H., Negrou, B., Ayuso, T. G., Djebabra, M., & Ramadan, M. (2020). Preliminary hazard identification for risk assessment on a complex system for hydrogen production. *International Journal of Hydrogen Energy*, *45*(20), 11855–11865. doi:10.1016/j.ijhydene.2019.10.162

Hagen, M., & Azevedo, A. (2022). Climate Changes Consequences from Sun-Earth Connections and Anthropogenic Relationships. *Nature and Science*, *14*(2), 24–41.

Hankin, A., & Shah, N. (2017). Process exploration and assessment for the production of methanol and dimethyl ether from carbon dioxide and water. *Sustainable Energy & Fuels*, *1*(7), 1541–1556. doi:10.1039/C7SE00206H

Han, S. S., & Goddard, W. A. (2007a). Lithium-Doped Metal-Organic Frameworks for Reversible H2 Storage at Ambient Temperature. *Journal of the American Chemical Society*, *129*(27), 8422–8423. doi:10.1021/ja072599+ PMID:17569539

Hansen, O. R. (2020). Hydrogen infrastructure-Efficient risk assessment and design optimization approach to ensure safe and practical solutions. *Process Safety and Environmental Protection*, *143*, 164–176. doi:10.1016/j.psep.2020.06.028

Hao, X., Zhou, Y., Wang, H., & Ouyang, M. (2020). Plug-In Electric Vehicles In China And The Usa: A Technology And Market Comparison. *Mitigation And Adaptation Strategies For Global Change*, *25*(3), 329–353. doi:10.100711027-019-09907-z

Harries, B. D. N., Paskevicius, M., Sheppard, D. A., Edward, T., Price, C., & Buckley, C. E. (2012). Concentrating solar thermal heat storage using metal hydrides. *Proceedings of the IEEE*, *100*(2), 539–549. doi:10.1109/JPROC.2011.2158509

Hart, D. (2000). Sustainable energy conversion: Fuel cells — the competitive option? *Journal of Power Sources*, *86*(1-2), 23–27. doi:10.1016/S0378-7753(99)00455-3

Hassan, I. A., Ramadan, H. S., Saleh, M. A., & Hissel, D. (2021). Hydrogen storage technologies for stationary and mobile applications: Review, analysis and perspectives. *Renewable & Sustainable Energy Reviews, 149*. doi:ARTN 11131110.1016/j.rser.2021.111311

He, R., Tian, X., Ni, Y., & Xu, Y. (2017). Mode Transition Coordination Control For Parallel Hybrid Electric Vehicle Based On Switched System. *Advances In Mechanical Engineering*, *9*(8), 1687814017715564. doi:10.1177/1687814017715564

Hernandez-Ochoa, I. M., Asseng, S., Kassie, B. T., Xiong, W., Robertson, R., Pequeno, D. N. L., & Hoogenboom, G. (2018). Climate change impact on Mexico wheat production. *Agricultural and Forest Meteorology*, *263*, 373–387. doi:10.1016/j.agrformet.2018.09.008

Hibbeler, R. C. (2018). *Mechanics of materials*. Pearson Education Limited.

Hirose, K. (2012). 2011 Fukushima Dai-ichi nuclear power plant accident: Summary of regional radioactive deposition monitoring results. *Journal of Environmental Radioactivity*, *111*, 13–17. doi:10.1016/j.jenvrad.2011.09.003 PMID:22119330

Hirscher, M., Becher, M., Haluska, M., Quintel, A., Skakalova, V., Choi, Y. M., Dettlaff-Weglikowska, U., Roth, S., Stepanek, I., Bernier, P., Leonhardt, A., & Fink, J. (2002). Hydrogen storage in carbon nanostructures. *Journal of Alloys and Compounds*, *330*, 654–658. doi:10.1016/S0925-8388(01)01643-7

Hirscher, M., & Panella, B. (2007). Hydrogen storage in metal–organic frameworks. *Scripta Materialia*, *56*(10), 809–812. doi:10.1016/j.scriptamat.2007.01.005

Hoffert, M. I., Caldeira, K., Benford, G., Criswell, D. R., Green, C., Herzog, H., & Wigley, T. M. (2002). Advanced Technology Paths To Global Climate Stability: Energy For A Greenhouse Planet. *Science*, *298*(5595), 981–987. doi:10.1126cience.1072357 PMID:12411695

Holdcroft, S. (2014). Fuel cell catalyst layers: A polymer science perspective. *Chemistry of Materials*, *26*(1), 381–393. doi:10.1021/cm401445h

Holladay, J. D., Hu, J., King, D. L., & Wang, Y. (2009). An overview of hydrogen production technologies. *Catalysis Today*, *139*(4), 244–260. doi:10.1016/j.cattod.2008.08.039

Hosseini, S. E., & Wahid, M. A. (2020). Hydrogen from solar energy, a clean energy carrier from a sustainable source of energy. *International Journal of Energy Research*, *44*(6), 4110–4131. doi:10.1002/er.4930

Huang, Z. G., Guo, Z. P., Calka, A., Wexler, D., Lukey, C., & Liu, H. K. (2006). Effects of iron oxide (Fe2O3, Fe3O4) on hydrogen storage properties of Mg-based composites. *Journal of Alloys and Compounds*, *422*(1-2), 299–304. doi:10.1016/j.jallcom.2005.11.074

Hu, J., Chen, J., Sundararaman, S., Chandrashekhara, K., & Chernicoff, W. (2008). Analysis of composite hydrogen storage cylinders subjected to localized flame impingements. *International Journal of Hydrogen Energy*, *33*(11), 2738–2746. doi:10.1016/j.ijhydene.2008.03.012

Huston, E. L., & Sandrock, G. D. (1980). Engineering properties of metal hydrides. *Journal of the Less Common Metals*, *74*(2), 435–443. doi:10.1016/0022-5088(80)90182-4

Hwang, H. T., & Varma, A. (2014). Hydrogen storage for fuel cell vehicles. *Current Opinion in Chemical Engineering*, *5*, 42–48. doi:10.1016/j.coche.2014.04.004

HySA_Systems. (2014). *Hydrogen fuelled vehicles*. HySA. http://www.hysasystems.com/index.php/programmes/hydrogen-fuel-cell-vehicles

IEA. (2019). *The Future of Hydrogen – Analysis*. IEA.

İnci, M., Büyük, M., Demir, M. H., & İlbey, G. (2021). A Review And Research On Fuel Cell Electric Vehicles: Topologies, Power Electronic Converters, Energy Management Methods, Technical Challenges, Marketing And Future Aspects. *Renewable & Sustainable Energy Reviews*, *137*, 110648. doi:10.1016/j.rser.2020.110648

Inomata, A., Aoki, H., & Miura, T. (1998). Measurement and modelling of hydriding and dehydriding kinetics. *Journal of Alloys and Compounds*, *278*(1-2), 103–109. doi:10.1016/S0925-8388(98)00008-5

Iordache, I. & Stefănescu, I. (2010). Tehnici Avansate De Tratare a Apei ¸Si Separare a Hidrogenului. Editura AGIR: Bucureşti, Romania.

Irena, I. E. A. (2011). Prospects for the African power sector. *Scenarios and Strategies for Africa Project*.

Ishaq, H., & Dincer, I. (2019). A comparative evaluation of three CuCl cycles for hydrogen production. *International Journal of Hydrogen Energy*, *44*(16), 7958–7968. doi:10.1016/j.ijhydene.2019.01.249

Ishaq, H., & Dincer, I. (2020). A comprehensive study on using new hydrogen-natural gas and ammonia-natural gas blends for better performance. *Journal of Natural Gas Science and Engineering*, *81*, 103362. doi:10.1016/j.jngse.2020.103362

Ishaq, H., Dincer, I., & Crawford, C. (2022). A review on hydrogen production and utilization: Challenges and opportunities. *International Journal of Hydrogen Energy*, *47*(62), 26238–26264. doi:10.1016/j.ijhydene.2021.11.149

Iulianelli, A., & Basile, A. (Eds.). (2014). *Advances in hydrogen production, storage and distribution*. Elsevier.

Jacobsen, R. T., Penoncello, S. G., & Lemmon, E. W. (1997). Thermodynamic Properties of Cryogenic Fluids. Springer. https://doi.org/https://doi.org/10.1007/978-1-4899-1798-0_5

Jain, I. P., Jain, P., & Jain, A. (2010). Novel hydrogen storage materials : A review of lightweight complex hydrides. *Journal of Alloys and Compounds*, *503*(2), 303–339. doi:10.1016/j.jallcom.2010.04.250

Jaiswal, A., Chakraborty, B., & Sahu, S. (2022). High capacity reversible hydrogen storage in Si substituted and Li decorated C20 fullerene: Acumen from density functional theory simulations. *International Journal of Energy Research*, *46*(14), 19521–19537. doi:10.1002/er.8524

Jaiswal, A., Sahoo, R. K., Ray, S. S., & Sahu, S. (2022). Alkali metals decorated silicon clusters (SinMn, n= 6, 10; M= Li, Na) as potential hydrogen storage materials: A DFT study. *International Journal of Hydrogen Energy*, *47*(3), 1775–1789. doi:10.1016/j.ijhydene.2021.10.228

Japan's Hydrogen Industrial Strategy. (n.d.). Center for Strategic and International Studies. https://www.csis.org/analysis/japans-hydrogen-industrial-strategy

Jena, P. (2011). Materials for hydrogen storage: Past, present, and future. *The Journal of Physical Chemistry Letters*, *2*(3), 206–211. doi:10.1021/jz1015372

Jeon, J., Choi, W., & Kim, S. J. (2019). A flammability limit model for hydrogen-air-diluent mixtures based on heat transfer characteristics in flame propagation. *Nuclear Engineering and Technology*, *51*(7), 1749–1757. doi:10.1016/j.net.2019.05.005

Jones, E. R., van Vliet, M. T. H., Qadir, M., & Bierkens, M. F. P. (2021). Country-level and gridded estimates of wastewater production, collection, treatment and reuse. *Earth System Science Data*, *13*(2), 237–254. doi:10.5194/essd-13-237-2021

Jones, J., Genovese, A., & Tob-Ogu, A. (2020). Hydrogen vehicles in urban logistics: A total cost of ownership analysis and some policy implications. *Renewable & Sustainable Energy Reviews*, *119*, 109595. doi:10.1016/j.rser.2019.109595

Ju, H., Meng, H., & Wang, C.-Y. (2005). A single-phase, non-isothermal model for PEM fuel cells. *International Journal of Heat and Mass Transfer*, *48*(7), 1303–1315. doi:10.1016/j.ijheatmasstransfer.2004.10.004

Kabir, E., Kumar, P., Kumar, S., Adelodun, A. A., & Kim, K.-H. (2018). Solar energy: Potential and future prospects. *Renewable & Sustainable Energy Reviews*, *82*, 894–900. doi:10.1016/j.rser.2017.09.094

Kachroo, H., Chaurasia, A. K., Chaurasia, S. K., & Yadav, V. K. (2022). Sustainable Clean Energy Production from the Bio-electrochemical Process Using Cathode as Nanocatalyst. Handbook of Green and Sustainable Nanotechnology, 1–30. doi:10.1007/978-3-030-69023-6_58-1

Kakodkar, R., He, G., Demirhan, C. D., Arbabzadeh, M., Baratsas, S. G., Avraamidou, S., & Pistikopoulos, E. N. (2022). A Review Of Analytical And Optimization Methodologies For Transitions In Multi-Scale Energy Systems. *Renewable & Sustainable Energy Reviews*, *160*, 112277. doi:10.1016/j.rser.2022.112277

Kakoulaki, G., Kougias, I., Taylor, N., Dolci, F., Moya, J., & Jäger-Waldau, A. (2021). Green hydrogen in Europe–A regional assessment: Substituting existing production with electrolysis powered by renewables. *Energy Conversion and Management*, *228*, 113649. doi:10.1016/j.enconman.2020.113649

Kalamaras, C. M., & Efstathiou, A. M. (2013). Hydrogen Production Technologies: Current State And Future Developments. In *Conference Papers In Science*. Hindawi.

Kam, P. M., Aznar-Siguan, G., Schewe, J., Milano, L., Ginnetti, J., Willner, S., McCaughey, J. W., & Bresch, D. N. (2021). Global warming and population change both heighten future risk of human displacement due to river floods. *Environmental Research Letters*, *16*(4), 044026. doi:10.1088/1748-9326/abd26c

Kandidayeni, M., Macias, A., Boulon, L., & Trovao, J. P. F. (2020). Online Modeling of a Fuel Cell System for an Energy Management Strategy Design. *Energies*, *13*(14). doi:ARTN 371310.3390/en13143713

Kandlikar, S. G., & Lu, Z. (2009). Thermal management issues in a PEMFC stack–A brief review of current status. *Applied Thermal Engineering*, *29*(7), 1276–1280. doi:10.1016/j.applthermaleng.2008.05.009

Karaca, A. E., & Dincer, I. (2023). Development and evaluation of a solar based integrated hydrogen energy system for mobile applications. *Energy Conversion and Management*, *280*, 116808. doi:10.1016/j.enconman.2023.116808

Karlsson, M., Alfredsson, E., & Westling, N. (2020). Climate Policy Co-Benefits: A Review. *Climate Policy*, *20*(3), 292–316. doi:10.1080/14693062.2020.1724070

Khalil, Y. F. (2013). Experimental determination of dust cloud deflagration parameters of selected hydrogen storage materials: Complex metal hydrides, chemical hydrides, and adsorbents. *Journal of Loss Prevention in the Process Industries*, *26*(1), 96–103. doi:10.1016/j.jlp.2012.09.010

Kiani, R., Mirzaei, F., Ghanbari, F., Feizi, R., & Mehdipour, F. (2020). Real textile wastewater treatment by a sulfate radicals-Advanced Oxidation Process: Peroxydisulfate decomposition using copper oxide (CuO) supported onto activated carbon. *Journal of Water Process Engineering*, *38*, 101623. doi:10.1016/j.jwpe.2020.101623

Kikukawa, S., Mitsuhashi, H., & Miyake, A. (2009). Risk assessment for liquid hydrogen fueling stations. *International Journal of Hydrogen Energy*, *34*(2), 1135–1141. doi:10.1016/j.ijhydene.2008.10.093

Kim, K. S., Kang, J. H., & Kim, S. J. (2021). Carbon effect on hydrogen diffusivity and embrittlement in austenitic stainless steels. *Corrosion Science*, *180*. doi:ARTN 10922610.1016/j.corsci.2020.109226

Kirubakaran, A., Jain, S., & Nema, R. K. (2009). A review on fuel cell technologies and power electronic interface. *Renewable & Sustainable Energy Reviews*, *13*(9), 2430–2440. . doi:10.1016/j.rser.2009.04.004

Kobayashi, H., Hayakawa, A., Somarathne, K. D. K. A., & Okafor, E. C. (2019). Science and technology of ammonia combustion. *Proceedings of the Combustion Institute*, *37*(1), 109–133. doi:10.1016/j.proci.2018.09.029

Kondo, S., Takizawa, K., Takahashi, A., & Tokuhashi, K. (2011). On the temperature dependence of flammability limits of gases. *Journal of Hazardous Materials*, *187*(1-3), 585–590. doi:10.1016/j.jhazmat.2011.01.037 PMID:21288637

Kong, D., Eckhoff, R. K., & Alfert, F. (1995). Auto-Ignition of Ch4 Air, C3h8 Air, Ch4 C3h8 Air and Ch4 Co2 Air Using an Ignition Bomb. *Journal of Hazardous Materials*, *40*(1), 69–84. doi:10.1016/0304-3894(94)00082-R

Koscher, G. A., & Kordesch, K. (2003). *Handbook of fuel cells- fundamentals, technology and applications* (Vol. 4). John Wiley and Sons.

Kovac, A., Paranos, M., & Marcius, D. (2021). Hydrogen in energy transition: A review. *International Journal of Hydrogen Energy*, *46*(16), 10016–10035. doi:10.1016/j.ijhydene.2020.11.256

Kovats, R. S., & Haines, A. (2005). Global Climate Change And Health: Recent Findings And Future Steps. *Canadian Medical Association Journal*, *172*(4), 501–502. doi:10.1503/cmaj.050020 PMID:15710942

Krainz, G., Hödl, P., & Hofmeister, F. (2004). Automotive Production of Liquid Hydrogen Storage Systems. *The International German Hydrogen Energy Congress.*

Krasnov, P. O., Ding, F., Singh, A. K., & Yakobson, B. I. (2007). Clustering of Sc on SWNT and reduction of hydrogen uptake: Ab-initio all-electron calculations. *The Journal of Physical Chemistry C*, *111*(49), 17977–17980. doi:10.1021/jp077264t

Kubas, G. J. (2001). Metal–dihydrogen and σ-bond coordination: The consummate extension of the Dewar–Chatt–Duncanson model for metal–olefin π bonding. *Journal of Organometallic Chemistry*, *635*(1-2), 37–68. doi:10.1016/S0022-328X(01)01066-X

Kumar Gupta, S., Kumari, S., Reddy, K., & Bux, F. (2013). Trends in biohydrogen production: Major challenges and state-of-the-art developments. *Environmental Technology*, *34*(13–14), 1653–1670. doi:10.1080/09593330.2013.822022 PMID:24350426

Kumar, E. M., Sinthika, S., & Thapa, R. (2015). First principles guide to tune h-BN nanostructures as superior light-element-based hydrogen storage materials: Role of the bond exchange spillover mechanism. *Journal of Materials Chemistry. A, Materials for Energy and Sustainability*, *3*(1), 304–313. doi:10.1039/C4TA04706K

Kuvvetli, Y. (2020). Multi-objective and multi-period hydrogen refueling station location problem. *International Journal of Hydrogen Energy*, *45*(55), 30845–30858. doi:10.1016/j.ijhydene.2020.08.126

Lagioia, G., Spinelli, M. P., & Amicarelli, V. (2023). Blue and green hydrogen energy to meet European Union decarbonisation objectives. An overview of perspectives and the current state of affairs. *International Journal of Hydrogen Energy*, *48*(4), 1304–1322. doi:10.1016/j.ijhydene.2022.10.044

Lamy, C., & Millet, P. (2020). A critical review on the definitions used to calculate the energy efficiency coefficients of water electrolysis cells working under near ambient temperature conditions. In *Journal of Power Sources* (Vol. 447). Elsevier B.V. doi:10.1016/j.jpowsour.2019.227350

Lan, J., Cao, D., & Wang, W. (2009). Li12Si60H60 fullerene composite: A promising hydrogen storage medium. *ACS Nano*, *3*(10), 3294–3300. doi:10.1021/nn900842j PMID:19761195

Latypov, R. M., Heinonen, J. S., & Chistyakova, S. Y. (2022). Magmatic erosion of high-temperature-melting cumulates in the Bushveld Complex by chemical dissolution. *Geosystems and Geoenvironment*, *1*(3), 100077.

Le Quéré, C., Jackson, R. B., Jones, M. W., Smith, A. J., Abernethy, S., Andrew, R. M., De-Gol, A. J., Willis, D. R., Shan, Y., Canadell, J. G., Friedlingstein, P., Creutzig, F., & Peters, G. P. (2020). Temporary reduction in daily global CO2 emissions during the COVID-19 forced confinement. *Nature Climate Change*, *10*(7), 647–653. doi:10.103841558-020-0797-x

Lebrouhi, B. E., Djoupo, J. J., Lamrani, B., Benabdelaziz, K., & Kousksou, T. (2022). Global Hydrogen Development-A Technological And Geopolitical Overview. *International Journal Of Hydrogen Energy, 47*(11), 7016–7048. doi:10.1016/j.ijhydene.2021.12.076

Lee, J. Y., Yu, D. M., Kim, T. H., Yoon, S. J., & Hong, Y. T. (2015). Multi-Block Copolymers Based On Poly (P-Phenylene) S With Excellent Durability And Fuel Cell Performance. *Journal Of Membrane Science, 492*, 209–219. doi:10.1016/j.memsci.2015.04.013

Leng, H., Ichikawa, T., & Fujii, H. (2006). Hydrogen Storage Properties of Li - Mg - N - H Systems with Different Ratios of. *Journal of Physical Chemistry, 110*(26), 12964–12968. doi:10.1021/jp061120h PMID:16805600

Letcher, T. M. (2019). Why do we have global warming? In *Managing global warming* (pp. 3–15). Academic Press. doi:10.1016/B978-0-12-814104-5.00001-6

Ley, M. B., Jepsen, L. H., Lee, Y., Cho, Y. W., Von Colbe, M. B., Dornheim, M., Rokni, M., Jensen, J. O., Sloth, M., Filinchuk, Y., Jørgensen, J. E., Besenbacher, F., & Jensen, T. R. (2014). Complex hydrides for hydrogen storage – new perspectives. *Materials Today, 17*(3), 122–128. doi:10.1016/j.mattod.2014.02.013

Ley, M. B., Meggouh, M., Moury, R., Peinecke, K., & Felderhoff, M. (2015). Development of hydrogen storage tank systems based on complex metal hydrides. *Materials (Basel), 8*(9), 5891–5921. doi:10.3390/ma8095280 PMID:28793541

Lhotka, O., Kyselý, J., & Farda, A. (2018). Climate change scenarios of heat waves in Central Europe and their uncertainties. *Theoretical and Applied Climatology, 131*(3), 1043–1054. doi:10.100700704-016-2031-3

Li, J., Cheng, S., Zhao, Q., Long, P., & Dong, J. (2009). Synthesis and hydrogen-storage behavior of metal–organic framework MOF-5. *International Journal of Hydrogen Energy, 34*(3), 1377-1382.

Li, S., & Zhang, C. (2009, March). Study On Battery Management System And Lithium-Ion Battery. In 2009 *International Conference On Computer And Automation Engineering* (Pp. 218-222). IEEE.

Liang, Y., Pan, X. M., Zhang, C. M., Xie, B., & Liu, S. J. (2019). The simulation and analysis of leakage and explosion at a renewable hydrogen refuelling station. *International Journal of Hydrogen Energy, 44*(40), 22608–22619. doi:10.1016/j.ijhydene.2019.05.140

Li, H., Cao, X. W., Liu, Y., Shao, Y. B., Nan, Z. L., Teng, L., Peng, W., & Bian, J. (2022). Safety of hydrogen storage and transportation: An overview on mechanisms, techniques, and challenges. *Energy Reports, 8*, 6258–6269. doi:10.1016/j.egyr.2022.04.067

Linssen, J., & Hake, J. F. (2016). Hydrogen research, development, demonstration, and market deployment activities. *Hydrogen Science and Engineering: Materials, Processes, Systems and Technology*, 57-84.

Li, Q., He, R., Gao, J.-A., Jensen, J. O., & Bjerrum, N. J. (2003). The CO poisoning effect in PEMFCs operational at temperatures up to 200 C. *Journal of the Electrochemical Society, 150*(12), A1599. doi:10.1149/1.1619984

Liu, F., Sun, Z. N., Ding, M., & Bian, H. Z. (2021). Research progress of hydrogen behaviors in nuclear power plant containment under severe accident conditions. *International Journal of Hydrogen Energy, 46*(73), 36477–36502. doi:10.1016/j.ijhydene.2021.08.151

Liu, F., Zhao, F., Liu, Z., & Hao, H. (2018). The Impact Of Fuel Cell Vehicle Deployment On Road Transport Greenhouse Gas Emissions: The China Case. *International Journal Of Hydrogen Energy, 43*(50), 22604–22621. doi:10.1016/j.ijhydene.2018.10.088

Liu, W., Sun, L., Li, Z., Fujii, M., Geng, Y., Dong, L., & Fujita, T. (2020). Trends and future challenges in hydrogen production and storage research. *Environmental Science and Pollution Research International, 27*(25), 31092–31104. doi:10.100711356-020-09470-0 PMID:32529621

Liu, Y. L., Liu, Z., Wei, J. J., Lan, Y. Q., Yang, S. Y., & Jin, T. (2021). Evaluation and prediction of the safe distance in liquid hydrogen spill accident. *Process Safety and Environmental Protection, 146*, 1–8. doi:10.1016/j.psep.2020.08.037

Li, Y., Sun, Y., Qin, Y., Zhang, W., Wang, L., Luo, M., Yang, H., & Guo, S. (2020). Recent advances on water-splitting electrocatalysis mediated by noble-metal-based nanostructured materials. *Advanced Energy Materials, 10*(11), 1903120. doi:10.1002/aenm.201903120

Li, Y., & Taghizadeh-Hesary, F. (2022). The economic feasibility of green hydrogen and fuel cell electric vehicles for road transport in China. *Energy Policy, 160*, 112703. doi:10.1016/j.enpol.2021.112703

Li, Y., Yang, F. H., & Yang, R. T. (2007). Kinetics and mechanistic model for hydrogen spillover on bridged metal– organic frameworks. *The Journal of Physical Chemistry C, 111*(8), 3405–3411. doi:10.1021/jp065367q

Li, Y., & Yang, R. T. (2006a). Hydrogen storage in metal– organic frameworks by bridged hydrogen spillover. *Journal of the American Chemical Society, 128*(25), 8136–8137. doi:10.1021/ja061681m PMID:16787068

Li, Y., & Yang, R. T. (2006b). Significantly enhanced hydrogen storage in metal– organic frameworks via spillover. *Journal of the American Chemical Society, 128*(3), 726–727. doi:10.1021/ja056831s PMID:16417355

Lochan, R. C., & Head-Gordon, M. (2006). Computational studies of molecular hydrogen binding affinities: The role of dispersion forces, electrostatics, and orbital interactions. *Physical Chemistry Chemical Physics, 8*(12), 1357–1370. doi:10.1039/b515409j PMID:16633617

Longden, T., Beck, F. J., Jotzo, F., Andrews, R., & Prasad, M. (2022). 'Clean'hydrogen?– Comparing the emissions and costs of fossil fuel versus renewable electricity based hydrogen. *Applied Energy, 306*, 118145. doi:10.1016/j.apenergy.2021.118145

Lu, L., & Ren, Z. J. (2016). Microbial electrolysis cells for waste biorefinery: A state of the art review. In Bioresource Technology (Vol. 215, pp. 254–264). doi:10.1016/j.biortech.2016.03.034

Lu, H., Guan, Y., He, L., Adhikari, H., Pellikka, P., Heiskanen, J., & Maeda, E. (2020). Patch aggregation trends of the global climate landscape under future global warming scenario. *International Journal of Climatology*, *40*(5), 2674–2685. doi:10.1002/joc.6358

Luo, X., Wang, J., Dooner, M., & Clarke, J. (2015). Overview of current development in electrical energy storage technologies and the application potential in power system operation. *Applied Energy*, *137*, 511–536. doi:10.1016/j.apenergy.2014.09.081

Lyu, J., Kudiiarov, V., & Lider, A. (2020). An overview of the recent progress in modifications of carbon nanotubes for hydrogen adsorption. *Nanomaterials (Basel, Switzerland)*, *10*(2), 255. doi:10.3390/nano10020255 PMID:32024092

Maggio, G., Nicita, A., & Squadrito, G. (2019). How the hydrogen production from RES could change energy and fuel markets: A review of recent literature. *International Journal of Hydrogen Energy*, *44*(23), 11371–11384. doi:10.1016/j.ijhydene.2019.03.121

Mahfuz, M. H., Kamyar, A., Afshar, O., Sarraf, M., Anisur, M. R., Kibria, M. A., Saidur, R., & Metselaar, I. H. S. C. (2014). Exergetic analysis of a solar thermal power system with PCM storage. *Energy Conversion and Management*, *78*, 486–492. doi:10.1016/j.enconman.2013.11.016

Makepeace, J. W., He, T., Weidenthaler, C., Jensen, T. R., Chang, F., Vegge, T., & David, W. I. (2019). Reversible ammonia-based and liquid organic hydrogen carriers for high-density hydrogen storage: Recent progress. *international journal of hydrogen energy*, *44*(15), 7746-7767.

Mandal, S., & Jana, A. K. (2020). Simulating reactive distillation of HIx (HI–H2O–I2) system in Sulphur-Iodine cycle for hydrogen production. *Nuclear Engineering and Technology*, *52*(2), 279–286. doi:10.1016/j.net.2019.07.033

Mantell, C. L. (1982). *Batteries And Energy Systems*.

Marchi, C. S., Hecht, E. S., Ekoto, I. W., Groth, K. M., LaFleur, C., Somerday, B. P., ... James, C. W. (2017). Overview of the DOE hydrogen safety, codes and standards program, part 3: Advances in research and development to enhance the scientific basis for hydrogen regulations, codes and standards. *International Journal of Hydrogen Energy*, *42*(11), 7263–7274. doi:10.1016/j.ijhydene.2016.07.014

Martin, N. J., & Rice, J. L. (2014). Influencing clean energy laws: An analysis of business stakeholder engagement. *Business Strategy and the Environment*, *23*(7), 447–460. doi:10.1002/bse.1795

Mastoi, M. S., Zhuang, S., Munir, H. M., Haris, M., Hassan, M., Usman, M., Bukhari, S. S. H., & Ro, J. S. (2022). An in-depth analysis of electric vehicle charging station infrastructure, policy implications, and future trends. *Energy Reports*, *8*, 11504–11529. doi:10.1016/j.egyr.2022.09.011

McCarty, R. D., Hord, J., & Roder, H. M. (1981). *Selected properties of hydrogen (engineering design data)* (Vol. 168). US Department of Commerce, National Bureau of Standards. doi:10.6028/NBS.MONO.168

Mccullough, F. P., Levine, C. A., & Snelgrove, R. V. (1989). Secondary Battery. *Noûs (Detroit, Mich.)*, 4830938.

McMichael, A. J., Campbell-Lendrum, D. H., Corvalán, C. F., Ebi, K. L., Githeko, A., Scheraga, J. D., & Woodward, A. (2003). *Climate change and human health: risks and responses*. World Health Organization.

McWhorter, S., Read, C., Ordaz, G., & Stetson, N. (2011). Materials-based hydrogen storage: Attributes for near-term, early market PEM fuel cells. *Current Opinion in Solid State and Materials Science*, *15*(2), 29–38. doi:10.1016/j.cossms.2011.02.001

Medeiros, C. P., Alencar, M. H., & de Almeida, A. T. (2016). Hydrogen pipelines: Enhancing information visualization and statistical tests for global sensitivity analysis when evaluating multidimensional risks to support decision-making. *International Journal of Hydrogen Energy*, *41*(47), 22192–22205. doi:10.1016/j.ijhydene.2016.09.113

Mehrpooya, M., & Habibi, R. (2020). A review on hydrogen production thermochemical water-splitting cycles. *Journal of Cleaner Production*, *275*, 123836. doi:10.1016/j.jclepro.2020.123836

Mekhilef, S., Saidur, R., & Safari, A. (2012). Comparative study of different fuel cell technologies. *Renewable & Sustainable Energy Reviews*, *16*(1), 981–989. doi:10.1016/j.rser.2011.09.020

Mellouli, S., Askri, F., Dhaou, H., Jemni, A., & Ben Nasrallah, S. (2009). Numerical study of heat exchanger effects on charge / discharge times of metal – hydrogen storage vessel. *International Journal of Hydrogen Energy*, *34*(7), 3005–3017. doi:10.1016/j.ijhydene.2008.12.099

Meng, X., Gu, A., Wu, X., Zhou, L., Zhou, J., Liu, B., & Mao, Z. (2021). Status quo of China hydrogen strategy in the field of Transportation and International Comparisons. *International Journal of Hydrogen Energy*, *46*(57), 28887–28899. doi:10.1016/j.ijhydene.2020.11.049

Messaoudani, Z. L., Rigas, F., Hamid, M. D. B., & Hassan, C. R. C. (2016). Hazards, safety and knowledge gaps on hydrogen transmission via natural gas grid: A critical review. *International Journal of Hydrogen Energy*, *41*(39), 17511–17525. doi:10.1016/j.ijhydene.2016.07.171

Midilli, A. D. N. A. N., Ay, M., Dincer, I., & Rosen, M. A. (2005). a). On hydrogen and hydrogen energy strategies II: Future projections affecting global stability and unrest. *Renewable & Sustainable Energy Reviews*, *9*(3), 273–287. doi:10.1016/j.rser.2004.05.002

Midilli, A., Ay, M., Dincer, I., & Rosen, M. A. (2005). b). On hydrogen and hydrogen energy strategies: I: current status and needs. *Renewable & Sustainable Energy Reviews*, *9*(3), 255–271. doi:10.1016/j.rser.2004.05.003

Miller, E. L., Thompson, S. T., Randolph, K., Hulvey, Z., Rustagi, N., & Satyapal, S. (2020). Us Department Of Energy Hydrogen And Fuel Cell Technologies Perspectives. *Mrs Bulletin*, *45*(1), 57–64. doi:10.1557/mrs.2019.312

Milner-Elkharouf, L., Khzouz, M., & Steinberger-Wilckens, R. (2020). Catalyst development for indirect internal reforming (IIR) of methane by partial oxidation. *International Journal of Hydrogen Energy*, *45*(8), 5285–5296. doi:10.1016/j.ijhydene.2019.05.026

Ministry of Foreign Affairs of Japan Prime minister Suga's Attendance at the Leaders Summit on Climate. (2021). https://www.mofa.go.jp/ic/ch/page6e_000236.html

Mishra, S., Verma, S., Chowdhury, S., Gaur, A., Mohapatra, S., Dwivedi, G., & Verma, P. (2021). A Comprehensive Review On Developments In Electric Vehicle Charging Station Infrastructure And Present Scenario Of India. *Sustainability (Basel)*, *13*(4), 2396. doi:10.3390u13042396

Mneimneh, F., Ghazzawi, H., Abu Hejjeh, M., Manganelli, M., & Ramakrishna, S. (2023). Roadmap to achieving sustainable development via green hydrogen. *Energies*, *16*(3), 1368. doi:10.3390/en16031368

Mohammadfam, I., & Zarei, E. (2015). Safety risk modeling and major accidents analysis of hydrogen and natural gas releases: A comprehensive risk analysis framework. *International Journal of Hydrogen Energy*, *40*(39), 13653–13663. doi:10.1016/j.ijhydene.2015.07.117

Molnarne, M., & Schroeder, V. (2019). Hazardous properties of hydrogen and hydrogen containing fuel gases. *Process Safety and Environmental Protection*, *130*, 1–5. doi:10.1016/j.psep.2019.07.012

Momirlan, M., & Veziroglu, T. N. (2002). Current status of hydrogen energy. *Renewable & Sustainable Energy Reviews*, *6*(1-2), 141–179. doi:10.1016/S1364-0321(02)00004-7

Moradi, R., & Groth, K. M. (2019). Hydrogen storage and delivery: Review of the state of the art technologies and risk and reliability analysis. *International Journal of Hydrogen Energy*, *44*(23), 12254–12269. doi:10.1016/j.ijhydene.2019.03.041

Mormann, F. (2015). Clean Energy Federalism. *Florida Law Review*, *67*, 1621.

Msimanga, B., & Sebitosi, A. B. (2014). South Africa's non-policy driven options for renewable energy development. *Renewable Energy*, *69*, 420–427. doi:10.1016/j.renene.2014.03.041

Munjewar, S. S., Thombre, S. B., & Mallick, R. K. (2017). Approaches to overcome the barrier issues of passive direct methanol fuel cell – Review. *Renewable & Sustainable Energy Reviews*, *67*, 1087–1104. doi:10.1016/j.rser.2016.09.002

Murthy, S. S. (2012). Heat and Mass Transfer in Solid State Hydrogen Storage : A Review. *Journal of Heat Transfer*, *134*(3), 1–11. doi:10.1115/1.4005156

Muthukumar, M., Rengarajan, N., Velliyangiri, B., Omprakas, M. A., Rohit, C. B., & Raja, U. K. (2021). The Development Of Fuel Cell Electric Vehicles–A Review. *Materials Today: Proceedings*, *45*, 1181–1187. doi:10.1016/j.matpr.2020.03.679

Najjar, Y. S. (2013). Hydrogen Safety: The Road Toward Green Technology. *International Journal Of Hydrogen Energy*, *38*(25), 10716–10728. doi:10.1016/j.ijhydene.2013.05.126

Ng, H. D., & Lee, J. H. S. (2008). Comments on explosion problems for hydrogen safety. *Journal of Loss Prevention in the Process Industries*, *21*(2), 136–146. doi:10.1016/j.jlp.2007.06.001

Ngqalakwezi, A., & Nkazi, D. B. (2020). Hydrogen Storage: Materials, Kinetics and Thermodynamics. In Advanced Applications of Hydrogen and Engineering Systems in the Automotive Industry. IntechOpen.

Ngqalakwezi, A., Nkazi, D. B., Baloyi, S. J., & Ntho, T. A. (2020). Hydrogen: An Environmental Remediation. In Global Issues and Innovative Solutions in Healthcare, Culture, and the Environment (pp. 168–194). IGI Global.

Nicoletti, G., Arcuri, N., Nicoletti, G., & Bruno, R. (2015). A Technical And Environmental Comparison Between Hydrogen And Some Fossil Fuels. *Energy Conversion And Management*, *89*, 205–213. doi:10.1016/j.enconman.2014.09.057

Nikolaidis, P., & Poullikkas, A. (2017). A Comparative Overview Of Hydrogen Production Processes. *Renewable & Sustainable Energy Reviews*, *67*, 597–611. doi:10.1016/j.rser.2016.09.044

Ni, M. (2018). New developments and challenges of solid oxide fuel cell (SOFC)-based technologies. *International Journal of Energy Research*, *42*(15), 1–6. doi:10.1002/er.4213

Nimir, W., Al-Othman, A., Tawalbeh, M., Al Makky, A., Ali, A., Karimi-Maleh, H., Karimi, F., & Karaman, C. (2023). Approaches towards the development of heteropolyacid-based high temperature membranes for PEM fuel cells. *International Journal of Hydrogen Energy*, *48*(17), 6638–6656. doi:10.1016/j.ijhydene.2021.11.174

Niu, J., Rao, B. K., & Jena, P. (1992). Binding of hydrogen molecules by a transition-metal ion. *Physical Review Letters*, *68*(15), 2277–2280. doi:10.1103/PhysRevLett.68.2277 PMID:10045354

Noaa, N. (2019). *State of the climate: Global climate report for annual 2018.* www. ncdc. noaa. gov/sotc/global/201813

Noor, M., Khan, D., Khan, A., & Rasheed, N. (2023). The impact of renewable and non-renewable energy on sustainable development in South Asia. *Environment, Development and Sustainability*, •••, 1–18. doi:10.100710668-023-03210-3

Notteboom, T., & Haralambides, H. (2023). Seaports as green hydrogen hubs: Advances, opportunities and challenges in Europe. *Maritime Economics & Logistics*, *25*(1), 1–27. doi:10.105741278-023-00253-1

Nuttall, W. J., & Bakenne, A. T. (2020). Towards a Hydrogen Economy. In *Fossil Fuel Hydrogen* (pp. 43–52). Springer. doi:10.1007/978-3-030-30908-4_4

O'Connor, C., & Alexandrova, T. (2021). The Geological Occurrence, Mineralogy, and Processing by Flotation of Platinum Group Minerals (PGMs) in South Africa and Russia. *Minerals (Basel)*, *11*(1), 54. doi:10.3390/min11010054

O'hayre, R., Cha, S.-W., Colella, W., & Prinz, F. B. (2016). *Fuel cell fundamentals*. John Wiley & Sons. doi:10.1002/9781119191766

Oberoi, A. S. (2015). *Reversible electrochemical storage of hydrogen in activated carbons from Victorian brown coal and other precursors*. [PhD thesis, RMIT University Australia]

Oetari, P. S., Hadi, S. P., & Huboyo, H. S. (2019). Trace elements in fine and coarse particles emitted from coal-fired power plants with different air pollution control systems. *Journal of Environmental Management*, *250*, 109497. doi:10.1016/j.jenvman.2019.109497 PMID:31491717

Offer, G. J., Howey, D., Contestabile, M., Clague, R., & Brandon, N. P. (2010). Comparative Analysis Of Battery Electric, Hydrogen Fuel Cell And Hybrid Vehicles In A Future Sustainable Road Transport System. *Energy Policy*, *38*(1), 24–29. doi:10.1016/j.enpol.2009.08.040

Ohno, T. (2019). *Toyota Production System: Beyond Large-Scale Production*. Productivity Press..

Okolie, J. A., Patra, B. R., Mukherjee, A., Nanda, S., Dalai, A. K., & Kozinski, J. A. (2021). Futuristic applications of hydrogen in energy, biorefining, aerospace, pharmaceuticals and metallurgy. *International Journal of Hydrogen Energy*, *46*(13), 8885–8905. doi:10.1016/j.ijhydene.2021.01.014

Olympics Demonstrate Hydrogen Applications. (n.d.). Retrieved March 26, 2023, from https://fuelcellsworks.com/news/olympics-demonstrate-hydrogen-applications/

Öner, E., Öztürk, A., & Yurtcan, A. B. (2020). Utilization Of The Graphene Aerogel As Pem Fuel Cell Catalyst Support: Effect Of Polypyrrole (Ppy) And Polydimethylsiloxane (Pdms) Addition. *International Journal Of Hydrogen Energy*, *45*(60), 34818–34836. doi:10.1016/j.ijhydene.2020.05.053

Online-Content. (2023). *PGM Markets*. Platinum Group Metals.

Orimo, S., Nakamori, Y., Eliseo, J. R., Züttel, A., & Jensen, C. M. (2007). Complex hydrides for hydrogen storage. *Chemical Reviews*, *107*(10), 4111–4132. doi:10.1021/cr0501846 PMID:17848101

Oruc, O., & Dincer, I. (2021). Assessing the potential of thermochemical water splitting cycles: A bridge towards for clean and sustainable hydrogen generation. *Fuel*, *286*, 119325. doi:10.1016/j.fuel.2020.119325

Otsubo, H., Mizuno, M., Negishi, Y., & Ogami, N. (2007). *High-pressure hydrogen tank for FCHV* (No. 2007-01-2009). SAE Technical Paper.

Özcan, O., & Akın, A. N. (2019). Thermodynamic analysis of methanol steam reforming to produce hydrogen for HT-PEMFC: An optimization study. *International Journal of Hydrogen Energy*, *44*(27), 14117–14126. doi:10.1016/j.ijhydene.2018.12.211

Öztürk, A., Akay, R. G., Erkan, S., & Bayrakçeken Yurtcan, A. (2020). Introduction to fuel cells. In Direct Liquid Fuel Cells Fundamentals, Advances and Future (pp. 1-47). Academic Press.

Palo, D. R., Dagle, R. A., & Holladay, J. D. (2007). Methanol steam reforming for hydrogen production. *Chemical Reviews*, *107*(10), 3992–4021. doi:10.1021/cr050198b PMID:17845061

Palys, M. J., & Daoutidis, P. (2020). Using hydrogen and ammonia for renewable energy storage: A geographically comprehensive techno-economic study. *Computers & Chemical Engineering*, *136*, 106785. doi:10.1016/j.compchemeng.2020.106785

Pan, S.-Y., Snyder, S. W., Packman, A. I., Lin, Y. J., & Chiang, P.-C. (2018). Cooling water use in thermoelectric power generation and its associated challenges for addressing water-energy nexus. *Water-Energy Nexus*, *1*(1), 26–41. doi:10.1016/j.wen.2018.04.002

Pan, X. H., Yan, W. Y., Jiang, Y. M., Wang, Z. L., Hua, M., Wang, Q. Y., & Jiang, J. C. (2019). Experimental Investigation of the Self-Ignition and Jet Flame of Hydrogen Jets Released under Different Conditions. *ACS Omega*, *4*(7), 12004–12011. doi:10.1021/acsomega.9b01214 PMID:31460312

Pareek, A., Dom, R., Gupta, J., Chandran, J., Adepu, V., & Borse, P. H. (2020). Insights into renewable hydrogen energy: Recent advances and prospects. *Materials Science for Energy Technologies*, *3*, 319–327. doi:10.1016/j.mset.2019.12.002

Park, C., Lim, S., Shin, J., & Lee, C. Y. (2022). How much hydrogen should be supplied in the transportation market? Focusing on hydrogen fuel cell vehicle demand in South Korea: Hydrogen demand and fuel cell vehicles in South Korea. *Technological Forecasting and Social Change*, *181*, 121750. doi:10.1016/j.techfore.2022.121750

Park, S., Lee, J.-W., & Popov, B. N. (2012). A review of gas diffusion layer in PEM fuel cells: Materials and designs. *International Journal of Hydrogen Energy*, *37*(7), 5850–5865. doi:10.1016/j.ijhydene.2011.12.148

Paster, M. D., Ahluwalia, R. K., Berry, G., Elgowainy, A., Lasher, S., McKenney, K., & Gardiner, M. (2011). Hydrogen storage technology options for fuel cell vehicles: Well-to-wheel costs, energy efficiencies, and greenhouse gas emissions. *International Journal of Hydrogen Energy*, *36*(22), 14534–14551. doi:10.1016/j.ijhydene.2011.07.056

Patel, P., Maru, H. C., Borglum, B., Stokes, R. A., Petri, R. J., Sishtla, C., Krist, K., Armstrong, T., & Virkar, A. (2004). *Thermally integrated power systems, high power density SOFC generator. Proceedings of the Fuel Cell Seminar*, San Antonio, TX.

Patil, S. D., & Gopal, M. R. (2013). Entropy generation and its relation to heating and cooling requirements of a metal hydride hydrogen storage reactor. *International Journal of Hydrogen Energy*, *38*(31), 13687–13693. doi:10.1016/j.ijhydene.2013.08.018

Pehnt, M. (2003). Assessing future energy and transport systems: The case of Fuel Cells. *The International Journal of Life Cycle Assessment*, *8*(5), 283–289. doi:10.1007/BF02978920

Pesonen, O., & Alakunnas, T. (2017). *Energy Storage: a missing piece of the puzzle for the self-sufficient living*. UAS publication series B: research reports and compilations.

Petitpas, G., & Aceves, S. M. (2012). Modeling of sudden hydrogen expansion from cryogenic pressure vessel failure. *International Journal of Hydrogen Energy*, *38*(19), 8190–8198. doi:10.1016/j.ijhydene.2012.03.166

Pezhmani, Y., Oskouei, M. Z., Rezaei, N., & Mehrjerdi, H. (2023). Optimal operation of active distribution networks hosting hybrid hydrogen-electricity refuelling stations considering water demand under a stochastic-IGDT approach. *IET Generation, Transmission & Distribution*, *17*(7), 1461–1476. doi:10.1049/gtd2.12752

Pingkuo, L., & Xue, H. (2022). Comparative analysis on similarities and differences of hydrogen energy development in the World's top 4 largest economies: A novel framework. *International Journal of Hydrogen Energy*, *47*(16), 9485–9503. doi:10.1016/j.ijhydene.2022.01.038

Pollet, B. G., Pasupathi, S., Swart, G., Mouton, K., Lototskyy, M., Williams, M., Bujlo, P., Ji, S., Bladergroen, B. J., & Linkov, V. (2014a). Hydrogen South Africa (HySA) Systems Competence Centre: Mission, objectives, technological achievements and breakthroughs. *International Journal of Hydrogen Energy*, *39*(8), 3577–3596. doi:10.1016/j.ijhydene.2013.11.116

Pollet, B. G., Staffell, I., & Adamson, K.-A. (2015). Current energy landscape in the Republic of South Africa. *International Journal of Hydrogen Energy*, *40*(46), 16685–16701. doi:10.1016/j.ijhydene.2015.09.141

Posso, F., Pulido, A., & Acevedo-Páez, J. C. (2023). Towards the hydrogen economy: Estimation of green hydrogen production potential and the impact of its uses in Ecuador as a case study. *International Journal of Hydrogen Energy*, *48*(32), 11922–11942. doi:10.1016/j.ijhydene.2022.05.128

Prabhukhot, P. R., Wagh, M. M., & Gangal, A. C. (2016). A Review on Solid State Hydrogen Storage Material. *Advances in Energy and Power*, *4*(2), 11–22. doi:10.13189/aep.2016.040202

Punekar, N. S. (2018). Hallmarks of an Enzyme Catalyst. In *ENZYMES: Catalysis, Kinetics and Mechanisms* (pp. 43–51). Springer. doi:10.1007/978-981-13-0785-0_5

Puszkiel, J., Garroni, S., Milanese, C., Gennari, F., Klassen, T., Dornheim, M., & Pistidda, C. (2017). Tetrahydroborates: Development and potential as hydrogen storage medium. *Inorganics*, *5*(4), 74. doi:10.3390/inorganics5040074

Qian, J. Y., Li, X. J., Gao, Z. X., & Jin, Z. J. (2020). A numerical study of unintended hydrogen release in a hydrogen refueling station. *International Journal of Hydrogen Energy*, *45*(38), 20142–20152. doi:10.1016/j.ijhydene.2020.05.063

Qin, N., Brooker, P., & Raissi, A. (2017). *Fuel Cell Vehicle Technologies, Infrastructure And Requirements* (No. Evtc Report Number: Fsec-Cr-2059-17). University Of Central Florida. Electric Vehicle Transportation Center (Evtc)

Qin, N., Raissi, A., & Brooker, P. (2014). Analysis Of Fuel Cell Vehicle Developments. The Florida Solar Energy Center (Fsec): Cocoa, Fl, Usa.

Rady, A. C., Giddey, S., Badwal, S. P. S., Ladewig, B. P., & Bhattacharya, S. (2012). Review of Fuels for Direct Carbon Fuel Cells. *Energy & Fuels*, *26*(3), 1471–1488. doi:10.1021/ef201694y

Raju, M., & Kumar, S. (2012). Optimization of heat exchanger designs in metal hydride based hydrogen storage systems. *International Journal of Hydrogen Energy*, *37*(3), 2767–2778. doi:10.1016/j.ijhydene.2011.06.120

Ram Gopal, M., & Murthy, S. S. (1993). Parametric studies on heat and mass transfer in metal hydride beds. *Chemical Engineering and Processing*, *32*(4), 217–223. doi:10.1016/0255-2701(93)80003-Y

Rani, M., Shanker, U., & Chaurasia, A. K. (2017). Catalytic potential of laccase immobilized on transition metal oxides nanomaterials: Degradation of alizarin red S dye. *Journal of Environmental Chemical Engineering*, *5*(3), 2730–2739. Advance online publication. doi:10.1016/j.jece.2017.05.026

Ravi, S. S., & Aziz, M. (2022). Clean hydrogen for mobility–Quo vadis? *International Journal of Hydrogen Energy*, *47*(47), 20632–20661. doi:10.1016/j.ijhydene.2022.04.158

Razi, F., & Dincer, I. (2022). Challenges, opportunities and future directions in hydrogen sector development in Canada. *International Journal of Hydrogen Energy*, *47*(15), 9083–9102. doi:10.1016/j.ijhydene.2022.01.014

Rei, M.-H., Kao, Y.-L., Yeh, G.-T., & Chen, S.-C. (2020). The Influence of the Hydrogen Supply Modes on a Hydrogen Refueling Station. *Modern Economy*, *11*(4), 867–880. doi:10.4236/me.2020.114065

Ren, J., Gao, S., Liang, H., Tan, S., & Dong, L. (2023). The role of hydrogen energy: Strengths, weaknesses, opportunities, and threats. In *Hydrogen Economy* (pp. 3–43). Academic Press. doi:10.1016/B978-0-323-99514-6.00014-5

Ren, J., Musyoka, N. M., Langmi, H. W., Mathe, M., & Liao, S. (2017). Current research trends and perspectives on materials-based hydrogen storage solutions: A critical review. *International Journal of Hydrogen Energy*, *42*(1), 289–311. doi:10.1016/j.ijhydene.2016.11.195

Ren, L., Zhou, S., & Ou, X. (2020). Life-Cycle Energy Consumption And Greenhouse-Gas Emissions Of Hydrogen Supply Chains For Fuel-Cell Vehicles In China. *Energy*, *209*, 118482. doi:10.1016/j.energy.2020.118482

Ren, X., Dong, L., Xu, D., & Hu, B. (2020). Challenges towards hydrogen economy in China. *International Journal of Hydrogen Energy*, *45*(59), 34326–34345. doi:10.1016/j.ijhydene.2020.01.163

Ren, Z., Zhang, D., & Wang, Z. (2012). Stacks With Tin/Titanium As The Bipolar Plate For Pemfcs. *Energy*, *48*(1), 577–581. doi:10.1016/j.energy.2012.10.020

Rivard, E., Trudeau, M., & Zaghib, K. (2019). Hydrogen Storage for Mobility: A Review. *Materials, 12*(12). doi:ARTN 197310.3390/ma12121973

Rivard, E., Trudeau, M., & Zaghib, K. (2019). Hydrogen storage for mobility: A review. *Materials (Basel)*, *12*(12), 1973. doi:10.3390/ma12121973 PMID:31248099

Rodriguez-Prieto, A., Camacho, A. M., Mendoza, C., Kickhofel, J., & Lomonaco, G. (2021). Evolution of Standardized Specifications on Materials, Manufacturing and In-Service Inspection of Nuclear Reactor Vessels. *Sustainability, 13*(19). doi:ARTN 1051010.3390/su131910510

Rojas, J. M. L., Lopez, R. D., & Navarro, J. A. D. (2022). *Genetic Optimization Techniques for Sizing and Management of Modern Power Systems*. Elsevier.

Rosenzweig, C., & Parry, M. L. (1994). Potential impact of climate change on world food supply. *Nature*, *367*(6459), 133–138. doi:10.1038/367133a0

Rosi, N. L., Eckert, J., Eddaoudi, M., Vodak, D. T., Kim, J., O'Keeffe, M., & Yaghi, O. M. (2003). Hydrogen storage in microporous metal-organic frameworks. *Science*, *300*(5622), 1127–1129. doi:10.1126cience.1083440 PMID:12750515

Ross, D. K. (2006). Hydrogen Storage: The Major Technological Barrier To The Development Of Hydrogen Fuel Cell Cars. *Vacuum*, *80*(10), 1084–1089. doi:10.1016/j.vacuum.2006.03.030

Rowsell, J. L. C., & Yaghi, O. M. (2005). Strategies for Hydrogen Storage in Metal-Organic Frameworks. *Angewandte Chemie International Edition*, *44*(30), 4670–4679. doi:10.1002/anie.200462786 PMID:16028207

Rozendal, R. A., Hamelers, H. V. M., Euverink, G. J. W., Metz, S. J., & Buisman, C. J. N. (2006). Principle and perspectives of hydrogen production through biocatalyzed electrolysis. *International Journal of Hydrogen Energy*, *31*(12), 1632–1640. doi:10.1016/j.ijhydene.2005.12.006

Rusman, N. A. A., & Dahari, M. (2016). A Review On The Current Progress Of Metal Hydrides Material For Solid-State Hydrogen Storage Applications. *International Journal Of Hydrogen Energy*, *41*(28), 12108–12126. doi:10.1016/j.ijhydene.2016.05.244

Safari, F., & Dincer, I. (2020). A review and comparative evaluation of thermochemical water splitting cycles for hydrogen production. *Energy Conversion and Management*, *205*, 112182. doi:10.1016/j.enconman.2019.112182

Sakamoto, J., Misono, H., Nakayama, J., Kasai, N., Shibutani, T., & Miyake, A. (2018). Evaluation of Safety Measures of a Hydrogen Fueling Station Using Physical Modeling. *Sustainability, 10*(11). doi:ARTN 384610.3390/su10113846

Sakintuna, B., Lamari-darkrim, F., & Hirscher, M. (2007). Metal hydride materials for solid hydrogen storage : A review. *International Journal of Hydrogen Energy*, *32*(9), 1121–1140. doi:10.1016/j.ijhydene.2006.11.022

Salehi, A., Mousavi, S. M., Fasihfar, A., & Ravanbakhsh, M. (2019). Energy, exergy, and environmental (3E) assessments of an integrated molten carbonate fuel cell (MCFC), Stirling engine and organic Rankine cycle (ORC) cogeneration system fed by a biomass-fueled gasifier. *International Journal of Hydrogen Energy*, *44*(59), 31488–31505. doi:10.1016/j.ijhydene.2019.10.038

Sandí, G. (2004). Hydrogen storage and its limitations. *The Electrochemical Society Interface*, *13*(3), 40–44. doi:10.1149/2.F06043IF

Sandrock, G. (1999). A panoramic overview of hydrogen storage alloys from a gas reaction point of view. *Journal of Alloys and Compounds*, *293*, 877–888. doi:10.1016/S0925-8388(99)00384-9

Sandrock, G., Gross, K., Thomas, G., Jensen, C., Meeker, D., & Takara, S. (2002). Engineering considerations in the use of catalyzed sodium alanates for hydrogen storage. *Journal of Alloys and Compounds*, *330*, 696–701. doi:10.1016/S0925-8388(01)01505-5

Santhanam, K. S. V., Press, R. J., Miri, M. J., Bailey, A. V., & Takacs, G. A. (2017). *Introduction to hydrogen technology*. John Wiley & Sons.

Saritas, O., Meissner, D., & Sokolov, A. (2019). A Transition Management Roadmap For Fuel Cell Electric Vehicles (Fcevs). *Journal Of The Knowledge Economy*, *10*(3), 1183–1203. doi:10.100713132-018-0523-3

Sarker, A. K., Azad, A. K., Rasul, M. G., & Doppalapudi, A. T. (2023). Prospect of Green Hydrogen Generation from Hybrid Renewable Energy Sources: A Review. *Energies*, *16*(3), 1556. doi:10.3390/en16031556

Sartbaeva, A., Kuznetsov, V. L., Wells, S. A., & Edwards, P. P. (2008). Hydrogen Nexus In A Sustainable Energy Future. *Energy & Environmental Science*, *1*(1), 79–85. doi:10.1039/b810104n

Sazali, N. (2020). Emerging technologies by hydrogen: A review. *International Journal of Hydrogen Energy*, *45*(38), 18753–18771. doi:10.1016/j.ijhydene.2020.05.021

Scheller, F., Wald, S., Kondziella, H., Gunkel, P. A., Bruckner, T., & Keles, D. (2023). Future role and economic benefits of hydrogen and synthetic energy carriers in Germany: A review of long-term energy scenarios. *Sustainable Energy Technologies and Assessments*, *56*, 103037. doi:10.1016/j.seta.2023.103037

Scherer, G. G. (1990). Polymer membranes for fuel cells. *Berichte der Bunsengesellschaft fur Physikalische Chemie*, *94*(9), 1008–1014. doi:10.1002/bbpc.19900940926

Schlapbach, L., & Züttel, A. (2001). Hydrogen-storage materials for mobile applications. *Nature*, *414*(6861), 353–358. doi:10.1038/35104634 PMID:11713542

Schneemann, A., White, J. L., Kang, S., Jeong, S., Wan, L. F., Cho, E. S., Heo, T. W., Prendergast, D., Urban, J. J., Wood, B. C., Allendorf, M. D., & Stavila, V. (2018). Nanostructured metal hydrides for hydrogen storage. *Chemical Reviews*, *118*(22), 10775–10839. doi:10.1021/acs.chemrev.8b00313 PMID:30277071

Schouwstra, R. P., Kinloch, E. D., & Lee, C. A. (2000). A short geological review of the Bushveld Complex. *Platinum Metals Review*, *44*(1), 33–39.

Schüth, F., Bogdanović, B., & Felderhoff, M. (2004). Light metal hydrides and complex hydrides for hydrogen storage. *Chemical Communications*, (20), 2249–2258. doi:10.1039/B406522K PMID:15489969

Schwarze, K., Posdziech, O., Kroop, S., Lapeña-Rey, N., & Mermelstein, J. (2017). Green Industrial Hydrogen Via Reversible High-Temperature Electrolysis. *Ecs Transactions*, *78*(1), 2943–2952. doi:10.1149/07801.2943ecst

Scientific consensus on climate change. (2022, October 15). In *Wikipedia*. https://en.wikipedia.org/wiki/Scientific_consensus_on_climate_change

Shadman, M., Roldan-Carvajal, M., Pierart, F. G., Haim, P. A., Alonso, R., Silva, C., Osorio, A. F., Almonacid, N., Carreras, G., Maali Amiri, M., Arango-Aramburo, S., Rosas, M. A., Pelissero, M., Tula, R., Estefen, S. F., Pastor, M. L., & Saavedra, O. R. (2023). A review of offshore renewable energy in South America: Current status and future perspectives. *Sustainability (Basel)*, *15*(2), 1740. doi:10.3390u15021740

Shamardina, O., Chertovich, A., Kulikovsky, A. A., & Khokhlov, A. R. (2010). A simple model of a high temperature PEM fuel cell. *International Journal of Hydrogen Energy*, *35*(18), 9954–9962. doi:10.1016/j.ijhydene.2009.11.012

Shankar, R., Pathak, N., Chaurasia, A. K., Mondal, P., & Chand, S. (2017). Energy production through microbial fuel cells. *Sustainable Utilization of Natural Resources*, 354–380. doi:10.1201/9781315153292

Shao, Y., Yin, G., Wang, Z., & Gao, Y. (2007). Proton exchange membrane fuel cell from low temperature to high temperature: Material challenges. *Journal of Power Sources*, *167*(2), 235–242. doi:10.1016/j.jpowsour.2007.02.065

Shen, H., Li, H., Yang, Z., & Li, C. (2022). *Magic of hydrogen spillover: Understanding and application*. Green Energy & Environment.

Sierzchula, W., Bakker, S., Maat, K., & Van Wee, B. (2012). Technological Diversity Of Emerging Eco-Innovations: A Case Study Of The Automobile Industry. *Journal Of Cleaner Production*, *37*, 211–220. doi:10.1016/j.jclepro.2012.07.011

Sillero, L., Solera, R., & Perez, M. (2022). Anaerobic co-digestion of sewage sludge, wine vinasse and poultry manure for bio-hydrogen production. *International Journal of Hydrogen Energy*, *47*(6), 3667–3678. doi:10.1016/j.ijhydene.2021.11.032

Sinclair, K. (2001). *Status Of Avian Research At The National Renewable Energy Laboratory (No. Nrel/Cp-500-30546). National Renewable Energy Lab.(Nrel), Golden*. United States.

Singh Thakur, L., Parmar, H., Kumar Varma, A., Kumar Chaurasia, A., & Mondal, P. (2022). Removal of manganese from synthetic wastewater by Vetiveria zizanioides. *Materials Today: Proceedings*, 8–11. doi:10.1016/j.matpr.2022.08.395

Skob, Y. A., Ugryumov, M. L., & Granovskiy, E. A. (2021). Numerical assessment of hydrogen explosion consequences in a mine tunnel. *International Journal of Hydrogen Energy*, *46*(23), 12361–12371. doi:10.1016/j.ijhydene.2020.09.067

Slowik, P., Pavlenko, N., & Lutsey, N. (2016). Assessment Of Next-Generation Electric Vehicle Technologies. *International Council On Clean Transportation*, 1225.

Smalley, R. E. (2005). Future global energy prosperity: The terawatt challenge. *MRS Bulletin*, *30*(6), 412–417. doi:10.1557/mrs2005.124

Smith, B. J., Graziano, D. J., Riddle, M. E., Liu, D. J., Sun, P., Iloeje, C., & Diamond, D. (2022). Platinum Group Metal Catalysts-Supply Chain Deep Dive Assessment. *Usdoe Office Of Policy* (Po).

Song, Y., Zhang, C., Ling, C. Y., Han, M., Yong, R. Y., Sun, D., & Chen, J. (2020). Review On Current Research Of Materials, Fabrication And Application For Bipolar Plate In Proton Exchange Membrane Fuel Cell. *International Journal Of Hydrogen Energy*, *45*(54), 29832–29847. doi:10.1016/j.ijhydene.2019.07.231

Souahlia, A., Dhaou, H., Askri, F., Mellouli, S., Jemni, A., & Ben Nasrallah, S. (2011). Experimental study and characterization of metal hydride containers. *International Journal of Hydrogen Energy*, *36*(8), 4952–4957. doi:10.1016/j.ijhydene.2011.01.074

Sovacool, B. K. (2021). Clean, low-carbon but corrupt? Examining corruption risks and solutions for the renewable energy sector in Mexico, Malaysia, Kenya and South Africa. *Energy Strategy Reviews*, *38*, 100723. doi:10.1016/j.esr.2021.100723

Srinivasan, S., Diego, E., Goswami, Y., & Stefanakos, E. (2008). Effects of catalysts doping on the thermal decomposition behavior of Zn(BH4)2. *International Journal of Hydrogen Energy*, *33*(9), 2268–2272. doi:10.1016/j.ijhydene.2008.02.062

Staffell, I., Scamman, D., Abad, A. V., Balcombe, P., Dodds, P. E., Ekins, P., Shah, N., & Ward, K. R. (2018). The role of hydrogen and fuel cells in the global energy system. doi:10.31224/osf.io/rzm4gosf.io/rzm4g

Staffell, I., Scamman, D., Velazquez Abad, A., Balcombe, P., Dodds, P. E., Ekins, P., Shah, N., & Ward, K. R. (2019). The role of hydrogen and fuel cells in the global energy system. *Energy & Environmental Science*, *12*(2), 463–491. doi:10.1039/C8EE01157E

Stassen, I., Burtch, N., Talin, A., Falcaro, P., Allendorf, M., & Ameloot, R. (2017). An updated roadmap for the integration of metal–organic frameworks with electronic devices and chemical sensors. *Chemical Society Reviews*, *46*(11), 3185–3241. doi:10.1039/C7CS00122C PMID:28452388

Steen, M. L., Jordan, A. C., & Fisher, E. R. (2002). Hydrophilic modification of polymeric membranes by low temperature H2O plasma treatment. *Journal of Membrane Science*, *204*(1–2), 341–357. doi:10.1016/S0376-7388(02)00061-3

Stennikov, V. (2023, January). Transformation challenges faced by energy systems. In. AIP Conference Proceedings: Vol. 2552. *No. 1* (p. 080030). AIP Publishing LLC.

Story, G. C. (2000). Hydride bed/fuelcell proect. *Proceedings of the 2000 Hydrogen Program Review*.

Sudjoko, C., Sasongko, N. A., Utami, I., & Maghfuri, A. (2021, November). Utilization Of Electric Vehicles As An Energy Alternative To Reduce Carbon Emissions. In I*op Conference Series: Earth And Environmental Science* (Vol. 926, No. 1, P. 012094). Iop Publishing.

Suh, M. P., Park, H. J., Prasad, T. K., & Lim, D.-W. (2012). Hydrogen Storage in Metal–Organic Frameworks. *Chemical Reviews*, *112*(2), 782–835. doi:10.1021/cr200274s PMID:22191516

Sun, R., Pu, L., Yu, H., Dai, M., & Li, Y. (2022). Modeling the diffusion of flammable hydrogen cloud under different liquid hydrogen leakage conditions in a hydrogen refueling station. *International Journal of Hydrogen Energy*, *47*(61), 25849–25863. doi:10.1016/j.ijhydene.2022.05.303

Suzuki, S., Shoji, N., & Tsuru, T. (2021). Performance Evaluation Of Water Vapor Permeation Through Perfluorosulfonic Acid Capillary Membranes. *Separation And Purification Technology*, *266*, 118508. doi:10.1016/j.seppur.2021.118508

Suzuki, T., Shiota, K., Izato, Y., Komori, M., Sato, K., Takai, Y., Ninomiya, T., & Miyake, A. (2021). Quantitative risk assessment using a Japanese hydrogen refueling station model. *International Journal of Hydrogen Energy*, *46*(11), 8329–8343. doi:10.1016/j.ijhydene.2020.12.035

Tabtimsai, C., Ruangpornvisuti, V., Tontapha, S., & Wanno, B. (2018). A DFT investigation on group 8B transition metal-doped silicon carbide nanotubes for hydrogen storage application. *Applied Surface Science*, *439*, 494–505. doi:10.1016/j.apsusc.2017.12.255

Tawfik, H., Hung, Y., & Mahajan, D. (2007). Metal Bipolar Plates For Pem Fuel Cell—A Review. *Journal Of Power Sources*, *163*(2), 755–767. doi:10.1016/j.jpowsour.2006.09.088

Technology, D. of S. and. (2007). Innovation towards a knowledge-based economy: Ten-year plan for South Africa (2008-2018). Department of Science and Technology Pretoria.

Terlouw, T., Bauer, C., Mckenna, R., & Mazzotti, M. (2022). Large-Scale Hydrogen Production Via Water Electrolysis: A Techno-Economic And Environmental Assessment. *Energy & Environmental Science*, *15*(9), 3583–3602. doi:10.1039/D2EE01023B

Thomas, C. E. (2009). Fuel Cell And Battery Electric Vehicles Compared. *International Journal Of Hydrogen Energy*, *34*(15), 6005–6020. doi:10.1016/j.ijhydene.2009.06.003

Thompson, S. T., & Papageorgopoulos, D. (2019). Platinum Group Metal-Free Catalysts Boost Cost Competitiveness Of Fuel Cell Vehicles. *Nature Catalysis*, *2*(7), 558–561. doi:10.103841929-019-0291-x

Tian, Z., Lv, H., Zhou, W., Zhang, C. M., & He, P. F. (2022). Review on equipment configuration and operation process optimization of hydrogen refueling station. *International Journal of Hydrogen Energy*, *47*(5), 3033–3053. doi:10.1016/j.ijhydene.2021.10.238

Timmerhaus, K. D., & Flynn, T. M. (1989). Cryogenic Process Engineering (K. Mendelssohn, C. Rizzuto, & Klaus, Eds.). Plenum Press.

Tiwari, A. (2021). The emerging global trends in hydrogen energy research for achieving the net zero goals. *Advanced Materials Letters*, *12*(10), 1–5. doi:10.5185/aml.2021.15697

Trapp, C. T., Kanbach, D. K., & Kraus, S. (2022). Sector coupling and business models towards sustainability: The case of the hydrogen vehicle industry. *Sustainable Technology and Entrepreneurship*, *1*(2), 100014. doi:10.1016/j.stae.2022.100014

Trivedi & Bandyopadhyay. (2015). Hydrogen storage in small size MgnCo clusters: A density functional study. *International Journal of Hydrogen Energy, 40*(37), 12727-12735.

Tsao, C.-S., Yu, M.-S., Wang, C.-Y., Liao, P.-Y., Chen, H.-L., Jeng, U.-S., Tzeng, Y.-R., Chung, T.-Y., & Wu, H.-C. (2009). Nanostructure and hydrogen spillover of bridged metal-organic frameworks. *Journal of the American Chemical Society, 131*(4), 1404–1406. doi:10.1021/ja802741b PMID:19140765

Tseng, P., Lee, J., & Friley, P. (2005). A hydrogen economy: Opportunities and challenges. *Energy, 30*(14), 2703–2720. doi:10.1016/j.energy.2004.07.015

Turoń, K. (2020). Hydrogen-powered vehicles in urban transport systems–current state and development. *Transportation Research Procedia, 45*, 835–841. doi:10.1016/j.trpro.2020.02.086

Tzimas, E., Filiou, C., Peteves, S., & Veyret, J. (2003). *Hydrogen Storage: State-of-the-Art and Future Perspective.* EUR 20995 EN. cat. No. LD-NA-20995-EN-C.

Tzimas, E., Filiou, C., Peteves, S. D., & Veyret, J. (2003). *Hydrogen storage : state of the art and future perspective.* Institute for Energy, Directorate General Joint Research Centre.

U.S. Department of Energy's Energy Efficiency and Renewable Energy Website. (n.d.). https://www1.eere.energy.gov/hydrogenandfuelcells/storage/current_technology.html

United States Environmental Protection Agency. (2017). *Inventory of U.S. Greenhouse Gas Emissions and Sinks: 1990–2015.* Available online: https://www.epa.gov/sites/production/files/2017-02/documents/2017_ complete_report.pdf

Urunkar, R. U., & Patil, S. D. (2021b). Numerical study on performance evaluation of multi-tubular sodium alanate hydride reactor by enhancing heat and mass transfer characteristics using nanofluids. *Proceedings of the 26th National and 4th International ISHMT-ASTFE Heat and Mass Transfer Conference December 17-20, 2021, IIT Madras, Chennai-600036, Tamil Nadu, India*, 2613–2618. 10.1615/IHMTC-2021.3950

Urunkar, R. U., & Patil, S. D. (2021a). Enhancement of heat and mass transfer characteristics of metal hydride reactor for hydrogen storage using various nanofluids. *International Journal of Hydrogen Energy, 46*(37), 19486–19497. doi:10.1016/j.ijhydene.2021.03.090

Valente, A., Iribarren, D., & Dufour, J. (2017). Life cycle assessment of hydrogen energy systems: A review of methodological choices. *The International Journal of Life Cycle Assessment, 22*(3), 346–363. doi:10.100711367-016-1156-z

Van de Krol, R., & Grätzel, M. (2012). *Photoelectrochemical hydrogen production* (Vol. 90). Springer. doi:10.1007/978-1-4614-1380-6

Venkatasatish, R., & Dhanamjayulu, C. (2022). Reinforcement learning based energy management systems and hydrogen refuelling stations for fuel cell electric vehicles: An overview. *International Journal of Hydrogen Energy, 47*(64), 27646–27670. doi:10.1016/j.ijhydene.2022.06.088

Verfondern, K. (2007). *Nuclear energy for hydrogen production*. Forschungszentrum, Zentralbibliothek.

Verma, S., Mishra, S., Gaur, A., Chowdhury, S., Mohapatra, S., Dwivedi, G., & Verma, P. (2021). A Comprehensive Review On Energy Storage In Hybrid Electric Vehicle. [English Edition]. *Journal Of Traffic And Transportation Engineering*, *8*(5), 621–637. doi:10.1016/j.jtte.2021.09.001

Veziroğlu, T. N. (2000). Quarter century of hydrogen movement 1974–2000. *International Journal of Hydrogen Energy*, *25*(12), 1143–1150. doi:10.1016/S0360-3199(00)00038-0

Veziroğlu, T. N. (2004). IJHE thrust toward meeting the needs of implementation. *International Journal of Hydrogen Energy*, *1*(29), 15. doi:10.1016/j.ijhydene.2003.09.003

Veziroğlu, T. N., & Şahin, S. (2008). 21st Century's Energy: Hydrogen Energy System. *Energy Conversion And Management*, *49*(7), 1820–1831. doi:10.1016/j.enconman.2007.08.015

Vincent, I., Lee, E. C., & Kim, H. M. (2020). Solutions To The Water Flooding Problem For Unitized Regenerative Fuel Cells: Status And Perspectives. *Rsc Advances*, *10*(29), 16844–16860. doi:10.1039/D0RA00434K PMID:35521448

Von Colbe, J. B., Ares, J.-R., Barale, J., Baricco, M., Buckley, C., Capurso, G., Gallandat, N., Grant, D. M., Guzik, M. N., & Jacob, I. (2019). Application of hydrides in hydrogen storage and compression: Achievements, outlook and perspectives. *International Journal of Hydrogen Energy*, *44*(15), 7780–7808. doi:10.1016/j.ijhydene.2019.01.104

Wang, H., Kong, H., Pu, Z., Li, Y., & Hu, X. (2020). Feasibility of high efficient solar hydrogen generation system integrating photovoltaic cell/photon-enhanced thermionic emission and high-temperature electrolysis cell. *Energy Conversion and Management*, *210*, 112699. doi:10.1016/j.enconman.2020.112699

Wang, H., Lin, H. J., Cai, W. T., Ouyang, L. Z., & Zhu, M. (2016). Tuning kinetics and thermodynamics of hydrogen storage in light metal element based systems – A review of recent progress. *Journal of Alloys and Compounds*, *658*, 280–300. doi:10.1016/j.jallcom.2015.10.090

Wang, H., Tong, Z., Zhou, G., Zhang, C., Zhou, H., Wang, Y., & Zheng, W. (2022). Research and demonstration on hydrogen compatibility of pipelines: A review of current status and challenges. *International Journal of Hydrogen Energy*, *47*(66), 28585–28604. doi:10.1016/j.ijhydene.2022.06.158

Wang, L., Zheng, C. X., Wei, S., Wang, B. C., & Wei, Z. X. (2015). Thermo-mechanical investigation of composite high-pressure hydrogen storage cylinder during fast filling. *International Journal of Hydrogen Energy*, *40*(21), 6853–6859. doi:10.1016/j.ijhydene.2015.04.018

Wang, S., & Jiang, S. P. (2017). Prospects of fuel cell technologies. *National Science Review*, *4*(2), 163–166. doi:10.1093/nsr/nww099

Wang, Y., Chen, K. S., Mishler, J., Cho, S. C., & Adroher, X. C. (2011). A Review Of Polymer Electrolyte Membrane Fuel Cells: Technology, Applications, And Needs On Fundamental Research. *Applied Energy*, *88*(4), 981–1007. doi:10.1016/j.apenergy.2010.09.030

Wang, Y., Yuan, H., Martinez, A., Hong, P., Xu, H., & Bockmiller, F. R. (2021). Polymer Electrolyte Membrane Fuel Cell And Hydrogen Station Networks For Automobiles: Status, Technology, And Perspectives. *Advances In Applied Energy*, *2*, 100011. doi:10.1016/j.adapen.2021.100011

Wang, Z., Roberts, R. R., Naterer, G. F., & Gabriel, K. S. (2012). Comparison of thermochemical, electrolytic, photoelectrolytic and photochemical solar-to-hydrogen production technologies. *International Journal of Hydrogen Energy*, *37*(21), 16287–16301. doi:10.1016/j.ijhydene.2012.03.057

Weeda, M., & Segers, R. (2020). *The Dutch hydrogen balance, and the current and future representation of hydrogen in the energy statistics*. Research Gate.

West, M., Al-Douri, A., Hartmann, K., Buttner, W., & Groth, K. M. (2022). Critical review and analysis of hydrogen safety data collection tools. *International Journal of Hydrogen Energy*, *47*(40), 17845–17858. doi:10.1016/j.ijhydene.2022.03.244

Wieczorek, A. K., Krystian, M., & Zehetbauer, M. (2006). SPD processed alloys as efficient vacancy-hydrogen systems. [). Trans Tech Publications Ltd.]. *Diffusion and Defect Data, Solid State Data. Part B, Solid State Phenomena*, *114*, 177–182. doi:10.4028/www.scientific.net/SSP.114.177

Wilberforce, T., El-Hassan, Z., Khatib, F. N., Al Makky, A., Baroutaji, A., Carton, J. G., & Olabi, A. G. (2017). Developments Of Electric Cars And Fuel Cell Hydrogen Electric Cars. *International Journal Of Hydrogen Energy*, *42*(40), 25695–25734. doi:10.1016/j.ijhydene.2017.07.054

Winkler, H. (2005). Renewable energy policy in South Africa: Policy options for renewable electricity. *Energy Policy*, *33*(1), 27–38. doi:10.1016/S0301-4215(03)00195-2

Włodarczyk, R. (2022). Corrosion Analysis Of Graphite Sinter As Bipolar Plates In The Low-Temperature Pem Fuel Cell Simulated Environments. *Journal Of Solid State Electrochemistry*, *26*(1), 39–47. doi:10.100710008-021-05015-8

Wolf, J. (2002). Liquid-hydrogen technology for vehicles. *MRS Bulletin*, *27*(9), 684–687. doi:10.1557/mrs2002.222

Wolniak, R., & Skotnicka-Zasadzień, B. (2022). Development of Photovoltaic Energy in EU Countries as an Alternative to Fossil Fuels. *Energies*, *15*(2), 662. doi:10.3390/en15020662

Wu, D., Peng, C., Yin, C., & Tang, H. (2020). Review Of System Integration And Control Of Proton Exchange Membrane Fuel Cells. *Electrochemical Energy Reviews*, *3*(3), 466–505. doi:10.100741918-020-00068-1

Wulf, C., Reuß, M., Grube, T., Zapp, P., Robinius, M., Hake, J. F., & Stolten, D. (2018). Life Cycle Assessment of hydrogen transport and distribution options. *Journal of Cleaner Production*, *199*, 431–443. doi:10.1016/j.jclepro.2018.07.180

Xiaoluan, Z., Farajian, H., Xifeng, W., & Ohshima, K. (2022). Scheduling Of Renewable Energy And Plug-In Hybrid Electric Vehicles Based Microgrid Using Hybrid Crow—Pattern Search Method. *Journal Of Energy Storage*, *47*, 103605. doi:10.1016/j.est.2021.103605

Xuan, J., Leung, M. K. H., Leung, D. Y. C., & Ni, M. (2009). A review of biomass-derived fuel processors for fuel cell systems. *Renewable & Sustainable Energy Reviews*, *13*(6-7), 1301–1313. doi:10.1016/j.rser.2008.09.027

Xu, J., Yu, J., Xu, J., Sun, C., He, W., Huang, J., & Li, G. (2020). High-Value Utilization Of Waste Tires: A Review With Focus On Modified Carbon Black From Pyrolysis. *The Science Of The Total Environment*, *742*, 140235. doi:10.1016/j.scitotenv.2020.140235 PMID:32629243

Xu, X., Zhou, Q., & Yu, D. (2022). The future of hydrogen energy: Bio-hydrogen production technology. *International Journal of Hydrogen Energy*, *47*(79), 33677–33698. doi:10.1016/j.ijhydene.2022.07.261

Xu, Z., Qiu, D., Yi, P., Peng, L., & Lai, X. (2020). Towards mass applications: A review on the challenges and developments in metallic bipolar plates for PEMFC. *Progress in Natural Science*, *30*(6), 815–824. doi:10.1016/j.pnsc.2020.10.015

Yadav, D., & Banerjee, R. (2018). Economic assessment of hydrogen production from solar driven high-temperature steam electrolysis process. *Journal of Cleaner Production*, *183*, 1131–1155. doi:10.1016/j.jclepro.2018.01.074

Yamada, E., Watanabe, S., Hayashi, A. K., & Tsuboi, N. (2009). Numerical analysis on auto-ignition of a high pressure hydrogen jet spouting from a tube. *Proceedings of the Combustion Institute*, *32*(2), 2363–2369. doi:10.1016/j.proci.2008.07.001

Yang, F. Y., Wang, T. Z., Deng, X. T., Dang, J., Huang, Z. Y., Hu, S., Li, Y., & Ouyang, M. G. (2021). Review on hydrogen safety issues: Incident statistics, hydrogen diffusion, and detonation process. *International Journal of Hydrogen Energy*, *46*(61), 31467–31488. doi:10.1016/j.ijhydene.2021.07.005

Yang, L., Yu, H., Jiang, L., Zhu, L., Jian, X., & Wang, Z. (2010). Improved Anticorrosion Properties And Electrical Conductivity Of 316l Stainless Steel As Bipolar Plate For Proton Exchange Membrane Fuel Cell By Lower Temperature Chromizing Treatment. *Journal Of Power Sources*, *195*(9), 2810–2814. doi:10.1016/j.jpowsour.2009.11.018

Yang, S. J., Kim, T., Im, J. H., Kim, Y. S., Lee, K., Jung, H., & Park, C. R. (2012). MOF-derived hierarchically porous carbon with exceptional porosity and hydrogen storage capacity. *Chemistry of Materials*, *24*(3), 464–470. doi:10.1021/cm202554j

Yang, Y., Hu, X., Pei, H., & Peng, Z. (2016). Comparison Of Power-Split And Parallel Hybrid Powertrain Architectures With A Single Electric Machine: Dynamic Programming Approach. *Applied Energy*, *168*, 683–690. doi:10.1016/j.apenergy.2016.02.023

Ying, Z., Wang, Y., Zheng, X., Geng, Z., Dou, B., & Cui, G. (2020). Experimental study and development of an improved sulfur–iodine cycle integrated with HI electrolysis for hydrogen production. *International Journal of Hydrogen Energy*, *45*(24), 13176–13188. doi:10.1016/j.ijhydene.2020.03.037

Yoo, E., Gao, L., Komatsu, T., Yagai, N., Arai, K., Yamazaki, T., Matsuishi, K., Matsumoto, T., & Nakamura, J. (2004). Atomic hydrogen storage in carbon nanotubes promoted by metal catalysts. *The Journal of Physical Chemistry B, 108*(49), 18903–18907. doi:10.1021/jp047056q

Yue, M. L., Lambert, H., Pahon, E., Roche, R., Jemei, S., & Hissel, D. (2021). Hydrogen energy systems: A critical review of technologies, applications, trends and challenges. *Renewable & Sustainable Energy Reviews, 146*. doi:ARTN 11118010.1016/j.rser.2021.111180

Yuzer, B., Selcuk, H., Chehade, G., Demir, M. E., & Dincer, I. (2020). Evaluation of hydrogen production via electrolysis with ion exchange membranes. *Energy, 190*, 116420. doi:10.1016/j.energy.2019.116420

Zeng, X., Li, M., Abd El-Hady, D., Alshitari, W., Al-Bogami, A. S., Lu, J., & Amine, K. (2019). Commercialization Of Lithium Battery Technologies For Electric Vehicles. *Advanced Energy Materials, 9*(27), 1900161. doi:10.1002/aenm.201900161

Zhang, C. Z., Cao, X. J., Bujlo, P., Chen, B., Zhang, X., Sheng, X. F., & Liang, C. (2022). Review on the safety analysis and protection strategies of fast filling hydrogen storage system for fuel cell vehicle application. *Journal of Energy Storage, 45*. doi:ARTN 10345110.1016/j.est.2021.103451

Zhang, B., & Wu, Y. (2017). Recent advances in improving performances of the lightweight complex hydrides Li-Mg-N-H system. *Progress in Natural Science, 27*(1), 21–33. doi:10.1016/j.pnsc.2017.01.005

Zhang, J. (2008). *PEM fuel cell electrocatalysts and catalyst layers: fundamentals and applications*. Springer Science & Business Media. doi:10.1007/978-1-84800-936-3

Zhao, F., Mu, Z., Hao, H., Liu, Z., He, X., Victor Przesmitzki, S., & Ahmad Amer, A. (2020). Hydrogen fuel cell vehicle development in China: An industry chain perspective. *Energy Technology (Weinheim), 8*(11), 2000179. doi:10.1002/ente.202000179

Zhao, Y., Kim, Y. H., Dillon, A. C., Heben, M. J., & Zhang, S. B. (2005). Hydrogen storage in novel organometallic buckyballs. *Physical Review Letters, 94*(15), 155504. doi:10.1103/PhysRevLett.94.155504 PMID:15904160

Zheng, J., Liu, X., Xu, P., Liu, P., Zhao, Y., & Yang, J. (2011). Development of high pressure gaseous hydrogen storage technologies. *International Journal of Hydrogen Energy, 37*(1), 1048–1057. doi:10.1016/j.ijhydene.2011.02.125

Zhiznin, S. Z., Shvets, N. N., Timokhov, V. M., & Gusev, A. L. (2023). Economics of hydrogen energy of green transition in the world and Russia. Part I. *International Journal of Hydrogen Energy, 48*(57), 21544–21567. doi:10.1016/j.ijhydene.2023.03.069

Zhiznin, S. Z., Vassilev, S., & Gusev, A. L. (2019). Economics of secondary renewable energy sources with hydrogen generation. *International Journal of Hydrogen Energy, 44*(23), 11385–11393. doi:10.1016/j.ijhydene.2019.03.072

Zhou, L. (2005). Progress and problems in hydrogen storage methods. *Renewable & Sustainable Energy Reviews, 9*(4), 395–408. doi:10.1016/j.rser.2004.05.005

Zhou, M., Chen, G., Dong, Z., Xie, B., Gu, S., & Shi, P. (2020). Estimation of surface albedo from meteorological observations across China. *Agricultural and Forest Meteorology*, *281*, 107848. doi:10.1016/j.agrformet.2019.107848

Zivar, D., Kumar, S., & Foroozesh, J. (2021). Underground hydrogen storage: A comprehensive review. *International Journal of Hydrogen Energy*, *46*(45), 23436–23462. doi:10.1016/j.ijhydene.2020.08.138

Züttel, A., Remhof, A., Borgschulte, A., & Friedrichs, O. (2010). Hydrogen: the future energy carrier. *Philosophical Transactions of the Royal Society A: Mathematical, Physical and Engineering Sciences, 368*(1923), 3329-3342.

Züttel, A. (2003). Materials for hydrogen storage. *Materials Today*, *6*(9), 24–33. doi:10.1016/S1369-7021(03)00922-2

Züttel, A. (2004). Hydrogen storage methods. *Naturwissenschaften*, *91*(4), 157–172. doi:10.100700114-004-0516-x PMID:15085273

Züttel, A., Sudan, P., Mauron, P., Kiyobayashi, T., Emmenegger, C., & Schlapbach, L. (2002). Hydrogen storage in carbon nanostructures. *International Journal of Hydrogen Energy*, *27*(2), 203–212. doi:10.1016/S0360-3199(01)00108-2

Zuttel, A., Wenger, P., Rentsch, S., Mauron, P., & Emmenegger, C. (2003). LiBH4 a new hydrogen storage material. *Journal of Power Sources*, *118*(1-2), 1–7. doi:10.1016/S0378-7753(03)00054-5

About the Contributors

Raluca-Andreea Felseghi is Scientific Researcher. The main research direction was the concern related to the implementation of hydrogen fuel cell technologies in practical applications. She is currently involved in teaching activities at the Technical University of Cluj-Napoca, Romania, Department of Mechanical Engineering - Faculty of Automotive, Mechatronics and Mechanical Engineering. At "Ştefan cel Mare" University of Suceava, Romania, she was a member of the research team in a project that aimed to integrate hydrogen energy into the electricity supply systems of electric vehicle charging stations (SMiLE-EV). Her main fields of interest are energy efficiency of buildings, stationary application of renewable energies, fuel cell hydrogen technology, hybrid power generation systems, impact of alternative energies on the environment. She is president of Scientific, Innovation and Technical Education Commission and member of the Romanian Association of Building Services Engineering - Transylvania Branch, a member of Romanian Association for Hydrogen Energy and a member of Women in Green Hydrogen community ().

* * *

Rakesh A. Afre completed his M.Sc (Physics with specialization in Energy Studies) from North Maharashtra University, Jalgaon, Maharashtra and PhD (Nanotechnology) from Nagoya Institute of Technology (NIT), Nagoya, Japan under guidance of Prof. Tetsuo Soga. After completion of M.Sc he joined as a JRF at IIT Mumbai to work on DST sponsored project entitled "Carbon nanomaterials for solar cell" under guidance of Prof. Maheshwar Sharon, Dept. of Chemistry. Later he moved to Japan to pursue his PhD degree under Japan Government Scholarship (Monbukagakusho). His doctoral studies were focused on "Synthesis of carbon nanotubes by renewable natural precursors" In 2007; he has been awarded with PhD degree and continued his further research on organic photovoltaic (OPV) devices at NIT. Then he moved to KRICT, South Korea for his next assignment to continue research on DSSCs and small molecule OPVs. Furthermore, he moved to Australia to continue his research on fabrication of transparent electrodes using nanomaterials for OPVs and then to Ethiopia as a Professor. He came back to India and joined Walchand Center for

Research in Nanotechnology and Bionanotechnology (wcRnb), WCAS, Solapur as a HOD of Nanotechnology department under eminent directors Dr. Madhuri Sharon and Prof. Maheshwar Sharon. Presently he is working as Professor and Deputy Director (Innovations) at Assam down town University, Guwahati, Assam. Having 15+ years of experience in research and development and published more than 25 high quality research papers and 2 international patents.

Ayşe Bayrakçeken Yurtcan is a professor in the Department of Chemical Engineering at Atatürk University, Türkiye. Her research focuses on electrocatalysts for fuel cells, carbon-based nanomaterials, nanomaterials for energy, and material characterization.

Amit Kumar Chaurasia obtained his PhD degree from IIT Roorkee, India (2015-2021). Currently, he is serving as Assistant Professor in the Amity Institute of Biotechnology, Amity University, Noida. Previously, he worked as Assistant Professor in Dept. of Chemical Engineering MVJ College of Engineering, Bangalore, India. Previously, he worked as Research Associate with AlchemiCarbons Noida, India, as teaching assistant at NPTEL for three years, VASCSC Ahmedabad as science communicator and Thyrocare Technologies Navi Mumbai as scientific officer. His doctoral research work based on the "Biohydrogen Production Using Electrodeposited Cathodes in Microbial Electrolysis Cells''. He has 16+ peer-reviewed international publications in high impact factors journals (IF: 5 to9.9, h-index 60-278) and presented his research work in 06 International conferences (Algeria, France and Thailand)/04 national conferences presentation/One poster presentation and participated in~10 seminar/workshop (GAIN Course, COP IX, TEQIP Course and short-term course). He has also filled one Indian patent in 2022. He is the managing editor and editorial members in the reputed international journals.

Ankita Jaiswal (she/her), is a Senior Research Fellow at the Department of Physics, Indian Institute of Technology (Indian School of Mines), Dhanbad, India. She has received her Bachelors's and Masters's degrees from Dept. of Physics, Banaras Hindu University. Her primary research interests include modelling functional materials like nanoclusters and fullerenes for applications in the field of hydrogen storage and optoelectronic devices.

Aarti Maurya is pursuing undergraduates' course in Biotechnology engineering and aspiring research scholar. Interested to work on industrial products synthesis and characterization looking to working in a growth-oriented organization which offers me challenging opportunities to enhance my profession and technical growth within organizational limit.

Traian Candin Mihaltan is an engineer in the field of Civil Engineering and Installations, a graduate of the Technical University of Cluj-Napoca, Romania. His main competencies and skills in the Civil Engineering and Building Services domain, in particular: assessment, calculation, and sizing of energy demand in stationary applications; energy analysis techniques and implementation practices of buildings with nearly Zero Energy Consumption (nZEB); energy management in the industry; renewable energy resources and associated energy conversion technologies; the conception, sizing, design, and operation of installations that integrate renewable energy sources; elaboration of energy efficiency programs, sources of financing of energy efficiency programs; contractual relations in the energy sector.

Athule Ngqalakwezi is an R&D engineer at Mintek. She hold a Masters degree from the University of the Witwatersrand in Chemical engineering. She has contributed to the body of knowledge through various publications and presenting at local and international conferences. She is also a science communicator and was recently named a top 3 positive role model on the South African10th Accenture gendermain streaming awards.

Ayşenur Öztürk Aydın holds a bachelor's degree from Hacettepe University (Ankara, Turkey), Chemical Engineering and Atatürk University (Erzurum, Turkey), Chemistry Departments. She also received her master's and doctorate degrees from the department of Chemical Engineering at Ataturk University. She has worked as a research assistant in the same department since 2014. She continues her studies on the development of different materials for the gas diffusion and catalyst layers of PEM fuel cells.

Sharad Dattatray Patil received a Ph.D. degree in mechanical engineering from IIT Kharagpur, India in 2014. He has been an Associate Professor in the Department of Mechanical Engineering at Rajarambapu Institute of Technology, India. His research interests include I C Engine, Heat Transfer, CFD, Thermal modeling, Metal Hydrides, Battery thermal management, and Nanofluids. He has filed a patent on high traction manual tiller. He has been serving as a reviewer in International and National Journals.

Maria Simona Raboaca is working as a Researcher at to National Research and Development Institute for Cryogenics and Isotopic Technologies ICSI Rm. Valcea, Hydrogen and Fuel Cell Department. Her Ph.D. is "Theoretical and Practical Contribution regarding sustaining with hybrid energy a Passive House" in the Faculty of Civil Engineering in Technical University of Cluj-Napoca, Romania. She was a project manager at ICSI for to project "Smart conductive charging station, fixed and mobile, for electric propulsion transportation (SMiLE-EV)" which proposes the deployment of fixed and mobile EV & PHEV charging stations to meet the mobility needs of tomorrow's society and to prepare active/potential industrial partners for knowledge/technology transfer at the component or system level in preparing to launch new products. She has been contributing to the field of Renewable Energy, Green Buildings, Passive House Concept, Hydrogen Energy for Stationary and Mobile applications. She is the author and co-author of relevant technical articles on the "clean automotive" topic, published in scientific conference proceedings and WoS journals.

Sridhar Sahu (he/his), earned his Ph.D. in theoretical and computational physics from the Indian Institute of Technology, Mumbai, after completing his master's degree in physics at Ravenshaw University in Odisha. He worked as a Visiting Scientist at the Max Planck Institute of Complex Systems in Dresden, Germany, and as a Postdoctoral researcher at Nanyang Technological University in Singapore before beginning his position as an Assistant Professor in the Department of Physics at the Indian Institute of Technology (I.S.M.) Dhanbad. In the same department, he is currently an Associate Professor. Dr. Sahu's research interests span different areas in computational material science: hydrogen storage, organic semiconductors, organic photovoltaics, and atomic and molecular clusters.

Nallin Sharma is highly motivated towards developing top notch functional devices using nanomaterials and materials engineering. Currently working as Institute of Eminence fellow at Indian Institute of Sciences, Bangalore (India). Awarded Ph.D for research in 2D Stanene based environment, energy and healthcare applications; year 2021, able guidance of prof. Hui Fen-Wu in College of Science, National Sun Yat-sen University (Taiwan). Received M.Tech Engineering Physics from GGS Indraprastha University, New Delhi (India). Gained research experience under Prof. Maheshwar Sharon and Dr. Madhuri Sharon from 2014-2017 for Carbon, Graphene based nanomaterials and device fabrication. Interested in synthesis of superior two-dimensional nanomaterials with tunable optical and electronic properties.

Rahul Uday Urunkar is pursuing a Ph.D. degree in Mechanical Engineering from Shivaji University, Kolhapur. He has been an Assistant Professor in the Department of Mechanical Engineering at Sanjeevan Engineering and Technology Institute, Panhala, India. His research interests include Heat Transfer, CFD, Thermal modeling, Metal Hydrides, and Nanofluids. He has been serving as a reviewer in International and National Journals.

Index

G

H

L

M

N

P

R

www.ingramcontent.com/pod-product-compliance
Lightning Source LLC
LaVergne TN
LVHW081338300926
849161LV00008B/2004